MODELLING AND SIMULATION IN THE SOCIAL SCIENCES
FROM THE PHILOSOPHY OF SCIENCE POINT OF VIEW

THEORY AND DECISION LIBRARY

General Editors: W. Leinfellner (*Vienna*) and G. Eberlein (*Munich*)

Series A: Philosophy and Methodology of the Social Sciences

Series B: Mathematical and Statistical Methods

Series C: Game Theory, Mathematical Programming and Operations Research

Series D: System Theory, Knowledge Engineering and Problem Solving

SERIES A: PHILOSOPHY AND METHODOLOGY OF THE SOCIAL SCIENCES

VOLUME 23

Series Editors: W. Leinfellner (Technical University of Vienna), G. Eberlein (Technical University of Munich); *Editorial Board:* R. Boudon (Paris), M. Bunge (Montreal), J. S. Coleman (Chicago), J. Elster (Oslo), J. Götschl (Graz), L. Kern (Munich), I. Levi (New York), R. Mattessich (Vancouver), J. Nida-Rümelin (Göttingen), A. Rapoport (Toronto), A. Sen (Cambridge, U.S.A.), R. Tuomela (Helsinki), A. Tversky (Stanford).

Scope: This series deals with the foundations, the general methodology and the criteria, goals and purpose of the social sciences. The emphasis in the Series A will be on well-argued, thoroughly analytical rather than advanced mathematical treatments. In this context, particular attention will be paid to game and decision theory and general philosophical topics from mathematics, psychology and economics, such as game theory, voting and welfare theory, with applications to political science, sociology, law and ethics.

The titles published in this series are listed at the end of this volume.

MODELLING AND SIMULATION IN THE SOCIAL SCIENCES FROM THE PHILOSOPHY OF SCIENCE POINT OF VIEW

edited by

RAINER HEGSELMANN

Universität Bremen, Germany

ULRICH MUELLER

Philipps-Universität, Marburg, Germany

and

KLAUS G. TROITZSCH

Koblenz-Landau Universität, Koblenz, Germany

KLUWER ACADEMIC PUBLISHERS

DORDRECHT / BOSTON / LONDON

A C.I.P. Catalogue record for this book is available from the Library of Congress

ISBN 0-7923-4125-2

Published by Kluwer Academic Publishers,
P.O. Box 17, 3300 AA Dordrecht, The Netherlands.

Kluwer Academic Publishers incorporates
the publishing programmes of
D. Reidel, Martinus Nijhoff, Dr W. Junk and MTP Press.

Sold and distributed in the U.S.A. and Canada
by Kluwer Academic Publishers,
101 Philip Drive, Norwell, MA 02061, U.S.A.

In all other countries, sold and distributed
by Kluwer Academic Publishers Group,
P.O. Box 322, 3300 AH Dordrecht, The Netherlands.

Printed on acid-free paper

Printed in the Netherlands

Table of Contents

Preface

More and more can model building in the social sciences rely on well elaborated formal theories. At the same time inexpensive large computational capacities are available nowadays. Both make computer based model building and simulation possible in social science, whose central aim is especially an understanding of social dynamics. Such social dynamics refer to public opinion formation, partner choice, strategy decisions in social dilemma situations and much more. In the context of such modelling approaches novel problems in philosophy of science arise which must be analyzed and among which are at least the following:

Idealizing models and real world: As a rule, the systematic analysis of properties of models is only possible if few parameters are involved. This in turn makes radical simplification necessary. Nevertheless, such models are instructive in a certain sense: They do not deliver any explanations of a single real world event, but they do lead to an understanding how certain types of events are possible in principle. But what kind of "understanding" is this? How are the perhaps quite exact explanations of a certain event *within* a strongly idealized model related to the *"analogous"* real world events? What can we learn at all from an analysis of radically simplified models and from experimenting in model worlds? What about *axiomatization* of complex theories in social science, and what about the concept of *law*, especially *statistical (stochastical) law* in the social sciences?

Novel accomplishments of scientific systematization in social science:
Explanation, prediction, and retrodiction are seen as the central accomplishments of scientific systematization in traditional philosophy of science. As a rule, all these accomplishments are related to *singular events*. Modelling of social dynamics is, however, not about singular events, but about *more general properties* of the modelled system. For example, we could be interested in a general survey of the reactions of a model on whole classes of parameter constellations, or to identify more precisely, in which parts of parameter space the model behaves chaotically, and in which parts it does not. For a representation of the insights gained under such respects, *visualization*, as known from chaos research, could be the only way to proceed sensibly. But how can

those accomplishments of scientific systematization, which have been achieved in model analyses, be described from the philosophy of science point of view? What is the status of visualizations? Does it really exceed mere didactics of representation?

Heuristics of model construction: Recently, philosophy of science is more and more interested in the problems of *rational heuristics*. Such heuristic problems are found in social science modelling to an extreme extent. Maybe more can be said about this problem today than that on the one hand, model assumptions must be simple, and on the other hand, the model must remain theoretically interesting. Can modelling techniques be described which are especially suited for certain, sharply outlined classes of problems? For which kinds of problems, for example, are low dimensioned cellular automata appropriate modelling tools? Which experiences have we made so far with the available modelling software? What about its further development?

In the context sketched above, it seemed sensible and fruitful for both sides, to bring together philosophers of science and social scientists working in the field of simulation and formal modelling. First version of all papers were discussed among all authors and afterwards rewritten, reviewed, and finally revised to be included into this volume.

It is a pleasure to express our thanks to Ali Achiri, computer science student at Koblenz University, who fulfilled the sometimes tedious task of preparing the authors' files (which, as usual, came in many different electronic formats) for the final camera-ready printout, as well as to all authors who at the same time acted as reviewers and made sure that all critical remarks of the enlightening discussions we had during the editing process were incorporated into the final versions of the contributions.

Besides, we want to thank the University of Bremen who funded the editing process in its first phase, and the University of Koblenz–Landau who payed for Ali Achiri's valuable work in the second phase, and last but not least the editors of the Theory and Decision Library who were interested to include this volume into their series at a very early stage (and thus motivated all authors to do their best possible job in writing and revising their contributions).

Bremen, Marburg, and Koblenz, October 1995

Rainer Hegselmann   Ulrich Mueller   Klaus G. Troitzsch

HOLISM, INDIVIDUALISM AND EMERGENT PROPERTIES

An Approach from the Perspective of Simulation

G. NIGEL GILBERT
University of Surrey at Guildford
Social and Computer Sciences Research Group
Guildford, United Kingdom

Circumstances make men just as much as men make circumstances. (Marx and Engels 1947, p. 29)

This paper begins by showing that sociologists have also struggled with one of the basic conceptual and philosophical problems encountered in simulating societies: the problem of understanding 'emergence' and, especially, the relationship between the micro and macro properties of complex systems. Secondly, I shall indicate ways in which some computer simulations may have oversimplified important characteristics of specifically human societies, because the actors (agents) in these societies are capable of, and do routinely reason about the emergent properties of their own societies. This adds a degree of reflexivity to action which is not present (for the most part) in societies made up of simpler agents, and in particular is not a feature of most current computer simulations.

The issues I shall address stem from the fact that both societies and the complex computational systems used to simulate societies and social phenomena are composed of many interacting agents (also known as people, actors or members). These systems can therefore be described either in terms of the properties and behaviour of the agents alone, or in terms of the system as a whole. The former mode of description focuses on the 'micro' level, that is, the features of individual agents and their local environments (which they can perceive directly), while the latter focuses on the 'macro' level, that is, the global patterns or regularities formed by the behaviour of the agents and their interactions. The problem is how to characterise the relationship between these two types of description, in particular when the macro properties can be said to 'emerge' from the micro level behaviour.

1

R. Hegselmann et al. (eds.),
Modelling and Simulation in the Social Sciences from the Philosophy of Science Point of View, 1–12.
© 1996 *Kluwer Academic Publishers. Printed in the Netherlands.*

1. Sociological Approaches to Emergent Properties

1.1. METHODOLOGICAL HOLISM

The relationship between micro and macro properties is one which has exercised sociologists since the foundation of the discipline. For example, Durkheim, one of its founding fathers, emphasised the external nature of social institutions (a macro property of a society) and argued that they imposed themselves on individuals at the micro level. After having defined a category of facts that he called 'social facts', he writes:

> Another proposition has been ... vigorously disputed ...: it is the one that states that social phenomena are external to individuals ... The states of the collective consciousness are of a different nature from the states of the individual consciousness: they are representations of another kind. The mentality of groups is not the mentality of individuals; it has its own laws. ... To understand the way in which a society conceives of itself and the world that surrounds it, we must consider the nature of society, not the nature of the individuals. (Durkheim 1895, pp. 6–65)

Thus for Durkheim, there are social representations which can be examined independently of the individuals that make up the society. These social facts are

> a category of facts with very special characteristics: they consist of ways of acting, thinking and feeling that are external to the individual and are endowed with a coercive power by virtue of which they exercise control over him. (Durkheim 1895, p. 69)

This view of the relation between individual and society was later developed by Parsons (1952) into a 'normative conception of order' which emphasised the role of internalised norms in ensuring the integration of society through shared values and obligations. Subsequent theorists have extended this tradition, which has become known as 'methodological holism' (O'Neill 1973). It asserts that the social behaviour of individuals should be explained in terms of the positions or functions of these individuals within a social system and the laws which govern it.

1.2. METHODOLOGICAL INDIVIDUALISM

In contrast, 'methodological individualists' see macro phenomena accounted for by the micro level properties and behaviour of individuals. The individualists' position demands that all the concepts in social theory are analysable in terms of the interests, activities etc. of individual members of society:

If social events like inflation, political revolution, 'the disappearance of the middle classes', etc., are brought about by people, then they must be explained in terms of people; in terms of the situations people confront and the ambitions, fears and ideas which activate them. In short, large scale social phenomena must be accounted for by the situations, dispositions and beliefs of individuals. (Watkins 1955, p. 58)

A rather sterile debate between these two camps continued for much of the 1970s and 1980s. With the benefit of hindsight, it is now possible to argue that while there was some truth in both, neither was a particularly helpful or revealing way of conceiving the relationship between macro and micro behaviour. It is the case, however, that most, if not all, current simulations of human societies essentially adopt one or other of these positions, often without making this explicit.

1.3. STRUCTURATION THEORY

Perhaps the most influential recent sociological approach to the relationship between macro and micro is the theory of structuration. The theory argues that there is a duality between society and knowledgeable human agents. Agents are seen as reproducing in action the structuring properties of society, thereby allowing social life to be reproduced over time-space (Giddens 1984). By 'structuring properties', Giddens means things like institutional practices, rules and resources. These 'properties' are the means by which social structure or 'society' is produced and reproduced. The word 'property' is perhaps confusing in this context, for a structuring property is not an inherent characteristic of a society, but is rather a social construction, produced through the actions of the members of the society. Human action has these structuring properties 'embedded' in it, so as people act, they contribute to the reproduction of society. Also, ('reflexively'), human action is both constrained and enabled by social structure, for this is the medium through which action is performed. So, structure is at the same time both the outcome of knowledgeable human conduct and the medium that influences how conduct occurs (Giddens 1984, p. 25).

1.4. AN EXAMPLE: HUMAN INSTITUTIONS

Institutions are among the most pervasive emergent features of human society. An institution is an established order comprising rule-bound and standardised behaviour patterns. Examples include the family, tribes and other collectivities, organisations, legal systems and so on. Sociological approaches to institutions can be considered in terms of the three positions described at the beginning of this paper: methodological holism, individualism and structuration theory.

In brief, the holists' position sees individuals being socialised into membership of an institution (for example, the family, or the nation) by parents and teachers. The socialisation process consists of the transmittal of the rules and norms of behaviour of the institution to new members. These rules and norms then guide members in institutional action. In contrast, the individualists see institutions as arising from negotiation among the members, each pursuing their interests. An approach from structuration theory might focus on the way in which individual action and interaction in everyday life generates institutions through rationalisations of action in which the institution itself becomes acknowledged by individuals as part of the process of accounting for their actions. As Bittner, although not himself an advocate of structuration theory, put it with reference to a particular kind of institution, the organisation:

> Organisational designs are schemes of interpretation that users can invoke in as yet unknown ways whenever it suits their purposes (Bittner 1974, p. 77).

In other words, institutions, as macro level features, are recognised by members and are used by them to warrant their own actions, thus reproducing those same features.

2. Complex Adaptive Systems

We can now consider whether structuration theory can be linked to or has anything to offer computational views of emergence in simulated societies. The question of emergence has probably been considered most systematically by those interested in complex systems (Stein 1989; Jen 1990; Stein and Nadel 1991). Such systems have the following general characteristics (Forrest 1990):

1. The system consists of a large number of interacting agents, operating within an environment. Agents act on and are influenced by their local environment.
2. There is no global control over the system. All agents are only able to influence other agents locally.
3. Each agent is driven by simple mechanisms, typically condition–action rules, where the conditions are sensitive to the local environment. Usually, all agents share the same set of rules, although because they may be in different local environments, the actions they take will differ.

A typical example of such a complex system is an array of cellular automata. Each cell can take on one of a number of internal states. State changes occur according to the results of the firing of simple condition–action rules dependent only on the states of neighbouring cells.

Some complex systems are also adaptive. In such systems, agents either learn or are subject to a process of mutation and competitive selection, or both. In learning systems, agents are able to modify their rules according to their previous 'success' in reacting to the environment. In systems involving competitive selection, agents' rules are altered, either at random or using some specific learning algorithm, and then those which are more successful are copied while the less successful are deleted from the system. An important characteristic is that agents adapt within an environment in which other similar agents are also adapting, so that changes in one agent may have consequences for the environment and thus the success of other agents. The process of adaptation in which many agents simultaneously try to adapt to one another has been called 'co-evolution' (Kauffman 1988).

2.1. EMERGENCE IN COMPLEX SYSTEMS

Because complex systems, whether adaptive or not, consist of many agents, their behaviour can be described either in terms of the actions of the individual agents or at the level of the system as a whole. In some system states the macro description may be very simple (e.g., if the agents are either not interacting or interacting in repetitive cycles, the macro description might be that 'nothing is happening') or exceedingly complex (e.g., if the agents are in complete disequilibrium). In some circumstances, however, it may be possible to discover a concise description of the macro state of the system. It is in these latter circumstances that it becomes possible to talk about the 'emergence' of regularities at the macro level.

An example relevant to social simulation can be found in the work of Nowak and Latané (e.g., 1993) which aimed to model the effect of social influence drawing on Latané's theory of social impact. This states that the impact of a group of people on an individual's opinion is a multiplicative function of the persuasiveness of the members of the group, their social distance from the individual and the absolute number of the members. For example, in a simulated world in which there are only two mutually exclusive opinions, the opinion adopted by an individual is determined by the relative impacts on the individual of those propounding the two opinions. At any moment in time, an agent's opinion is determined by the multiplicative rule that implements the theory of social impact, and is dependent only on the opinions of other agents. The simulation fits the definition of a complex system.

The outcome of their simulations is that coherent clusters of opinion emerge and remain in dynamic equilibrium, over a wide range of assumptions for the parameters of the model. The population does not move wholly to adopt one opinion and minority views are not completely expunged.

Nowak and Latané show that the clusters of individuals with the same opinion can be visualised as patches on a two-dimensional grid in which the nodes represent the members of the population. These clusters can be said to have emerged and to form regularities.

In systems in which there is emergent behaviour, it is convenient to think of the emergent properties of the system as influencing the actions of the agents. Thus, not only do the agents' actions at the micro level, when aggregated and observed at the macro level, constitute the emergent behaviour, but also the macro emergent behaviour can be said to influence the micro actions of the agents, in a form of feedback. For example, in Nowak and Latané's simulation, one can speak of an opinion cluster influencing the opinions of individuals outside the cluster. But it must be remembered that in the standard complex system, there is no explicit mechanism for such feedback to take place: agents are affected only by their local neighbours. The idea of feedback, or 'feeddown', from macro to micro level is merely a convenient shorthand.

2.2. LEVELS OF EMERGENCE

So far, for simplicity, it has been assumed that there are clearly distinct 'levels': the micro and the macro. In fact the situation is more complex than this. For example, it may be the case that individual identity is best regarded as an emergent phenomenon, where the micro-level 'agents' are sub-cognitive, such as neurons. And societies are perhaps best considered as emergent phenomena arising from the interaction of social institutions. Thus it is better to consider a hierarchy of levels of emergence, rather than a straightforward division between micro and macro.

2.3. A DEFINITION OF EMERGENCE?

Some criterion is required which will distinguish emergent behaviour from behaviour which is predictable from the simple aggregation of individual characteristics of the agents. Formulating a clear definition of emergence is a problem that has exercised philosophers of science for some time. Stöckler (1991), in a review of the history of the concept, notes that its first use was in 1874 and that an analysis is provided by Nagel (1961) who argues that a property of an object is emergent if it is impossible to deduce this property from even a complete knowledge of the components and their relationships. A similar definition of emergence arises from the work on complex adaptive systems: that it should not be possible to derive analytically emergent behaviour solely from consideration of the properties of agents. In other words, emergent behaviour is that which cannot be predicted from knowledge of the properties of the agents except as a result of simulation. This kind of definition of course provides an obvious explanation for the popularity of simulation as an investigative technique in the field of complex adaptive systems.

This criterion is, however, not sufficient to end the worries of those who are concerned about whether they have really achieved emergence. For it is always possible that at some future time analytical methods will be developed that can be used to derive macro properties from micro ones. In a few cases, this seems to have already happened. For example, Forrest and Miller (1990) define a mapping between classifier systems (i.e., systems consisting of nodes that use the genetic algorithm to synthesise patterns of connections that collectively co-evolve to perform some function) and Boolean networks whose properties are, relative to classifier systems, much better understood. This mapping opens the way to predicting the emergent properties of classifier systems where previously predictions were impossible. The example shows that if we define emergence in terms of an inability to find an analytical solution, any particular emergent property stands the risk of being demoted from the status of emergence some time in the future. This suggests that emergence may be neither a stable nor an especially interesting property of complex systems: what is interesting are the systems' macro properties and the relationship of those macro properties to the micro ones.

3. Complex Adaptive Systems and Human Societies

The above account of complex adaptive systems is generic, in the sense that it is intended to apply equally to many disparate domains. For example, there has been work applying these ideas to the evolution of antibodies, stock market crashes, ecological dynamics, computer networks and so on (Waldrop 1992). Can these same accounts be applied unchanged to human societies, or is there something special about human societies which would indicate that such models are completely inappropriate, or must be amended or developed in order to provide insights into the structure and functioning of human society?

Debating whether there is a boundary between humans and other animals and, if so, its nature has become a matter of some controversy, with the argument that great apes deserve the same moral status as humans (Vines 1993). The question of whether human societies differ in any fundamental way from non-human societies threatens to become mired in the same moral debate. Nevertheless, it is clear that, while an exact boundary line may be difficult to draw and some characteristics of human societies are also to be found to some degree in some animal societies, human societies are significantly different from any animal society. Does this also mean that present theories of complex adaptive systems, while perhaps useful for understanding non-human societies, are lacking in some significant way as models of human societies?

I believe that the answer to this question is that such theories do at present fail to model a crucial feature of human societies and that the preceding discussion, especially the notion of structuration, can help to identify what this is and how it may be corrected.

A fundamental characteristic of humans, one that is so important that it makes the societies they form radically different from other complex systems, is that people are routinely capable of detecting, reasoning about and acting on the macro-level properties (the emergent features) of the societies of which they form part. The overall lesser intelligence of animals prevents this occurring in non-human societies. The consequences of this capability to 'orient' to macro-level properties are far-reaching.

Our discourse is thoroughly permeated by references to the macro-level properties of our society. For instance, we are not only able to speak a language using a common lexicon in which words have shared meaning, but we are also able to describe and comment on our lexicon; that is, as members of a human society, we can observe and reason about this macro-level property of our society.

Let us now consider another example, one where a simulation has been used to examine a theory relating to human society and where the researchers have tried to provide a relatively sophisticated approach to modelling human sociality. I shall compare its implicit characterisation of the relationship between macro and micro with sociological accounts that theorise society in terms of structuration.

4. The EOS Simulation

We can now examine an example of a simulation to see whether it fits neatly into any of these theoretical perspectives on the relationship between macro and micro. I have been associated with the EOS (Emergence of Organised Society) project since its inception, although Jim Doran and Mike Palmer are the people who have done all the work (Doran et al. 1993). The objective of the project is to investigate the growth of complexity of social institutions during the Upper Palaeolithic period in Southwestern France when there was a transition from a relatively simple egalitarian hunter-gatherer society to a more complex one, with centralised decision-making, specialists and role differentiation, territoriality and ethnicity (Cohen 1985; Mellars 1985; Doran 92).

The project is intended to examine a particular theory about the causes of this transition, one which takes seriously the effect of human cognitive abilities and limitations. Consequently, the simulation that is at the heart of the EOS project is based on a computational model which incorporates a number of agents having capabilities that enable them to reason about and act on their (computational) environment. The agents, implemented using a production system architecture, are able to develop and store beliefs about themselves and other agents in a 'social model'. Doran *et al.* (1993) describe the social model thus:

> The social model is particularly important. It is designed to record an agent's beliefs about the existence of groups of agents and the identities of their leaders. Associated with beliefs about leaders is other information, for example, the size of its immediate following. Some followers may themselves be leaders. An agent may well appear in its own social model as a leader or follower in particular groups. Also held within the social model are beliefs about territories. There is no expectation that the information held in an agent's social model will be either complete or accurate.

The social model is important because it has an effect on how an agent behaves in relation to other agents. For example, an agent's membership of a semi-permanent group depends on the agent becoming aware of its membership of that group and then treating fellow members differently from non-members.

Agents start without any knowledge of groups or other agents, but with a need to obtain a continuing supply of 'resources', some of which can only be secured with the cooperation of other agents. If an agent receives a request to cooperate with the plans of another agent, it will do so, or it will attempt to recruit agents to a plan of its own. Agents that successfully recruit other agents become group leaders and the other agents become followers. Such temporary groups continue to work together, with group membership recorded in the members' social models, unless other processes (such as the agent's movement out of the proximity of other members) intervene.

A number of observations can be made about this simulation. First, in terms of the definition offered previously, it is a complex system, although not an adaptive one. Secondly, group membership is an emergent property of the behaviour of the agents. Thirdly, agents pursue their own interests to obtain resources, forming 'groups' as and when necessary through negotiation with other agents. These groups have no existence other than their representation in agents' social models.

In particular, the agents, even though their membership in a group is recorded in their social model, have no awareness of the group as an entity in its own right with which they can reason. Agents not in a group have no way to recognise that the group exists, or who its members are. In short, the simulation permits agents to reason about micro properties that they perceive in their local environment, and from this reasoning and the consequent action, macro levels properties emerge. But because the agents have no means of perceiving these macro properties, the model as a whole fails to match the capabilities of human societies.

Structuration theory begins to suggest what additional functionality could be built into the agents. It proposes that actions are sedimented into 'structuring properties' such as institutional practices, rules and typifications. Thus it would be necessary for the EOS agents to recognise that they are members of groups and to realise what this implies about members' actions; and do so explicitly and in a way that results in a representation that they can subsequently reason with. Their own actions, as group members, would reinforce ('reproduce') these explicit representations of the group and its practices.

In the simulation, a 'group' emerges from agents' actions, but it remains an implicit property of the agents' social models. As would be expected from a simulation that tacitly adopts the individualistic approach, the existence of the 'group' can be observed by the researcher either by looking at the patterns of interactions at the macro level or by 'looking inside the heads' of the agents, but it is not visible to the agents themselves. An alternative way of implementing the simulation, rightly rejected by the research team, would have been to build the notion of a group into the system at a macro level so that the agents' actions were constrained by their group membership. In this case, which corresponds roughly to the holists' position, the group would not have been an emergent property of the simulation, but a constraining 'social fact'.

A third possibility is to design the simulation so that agents themselves are able to detect the emergence of groups and react ('orient') to this structure, so that they become no less privileged in this respect than the observer. This latter course is the approach that might be advocated by structuration theorists.

5. Conclusion

This paper originated from a feeling that, despite the sophistication of some of the present day work, the models being used to simulate human societies are lacking in a crucial respect. However, pinning down clearly, and in a computationally precise way, the nature of the difference between non-human and human societies is not straightforward.

The suggestion of this paper is that sociological debates about the nature of human society can contribute to our understanding of these issues, even if a certain amount of effort is required to 'translate' these debates into forms that are accessible to computational modelling. We have seen that one theory — structuration — provides some useful suggestions about how the relationship between macro and micro level properties might be conceptualised and even some broad indications about how a simulation might be implemented. It must be noted, however, that structuration theory is certainly not the only position on these issues to enjoy respect among sociologists and other present day ideas may be equally or more fruitful.

A further suggestion is that one capacity of humans which is probably not widely shared by other animals is an ability to perceive, monitor and reason with the macro properties of the society in which they live. It seems likely that this ability is related, and is perhaps even a consequence of humans' ability to use language. Much of the lexicon of modern adults is, for instance, composed of concepts about macro properties.

A final suggestion, linking the rest of my argument, is that human societies are dynamic: that is, they are (re-)produced by people as they proceed through their everyday life. Structures, institutions and so on are not reflections of some external reality but are human constructions. In particular, the 'macro properties' that are recognised by members of human societies are not fixed by the nature of the agents (as they would be in other complex systems), but are an accomplishment, the outcome of human action.

6. Acknowledgements

This paper was originally presented at the conference on 'Modelling and simulation in the Social Sciences from the point of view of the philosophy of the social sciences', Bremen, Germany February 17–19, 1994. I am grateful to the EOS project team, especially Mike Palmer and Jim Doran, and the participants at SimSoc '92 and '93 for opportunities to discuss these issues. The conclusions are, however, entirely my own.

References

Bittner, E. (1974) The Concept of Organisation. R. Turner (eds.), *Ethnomethodology.* Harmondsworth: Penguin.

Cohen, M.N. (1985) Prehistoric hunter-gatherers: the meaning of social complexity. T. Douglas-Price and J. A. Brown (ed.), *Prehistoric hunter-gatherers: the emergence of cultural complexity*, New York: Academic, pp. 99–119.

Doran, J.E., Palmer, M., Gilbert, N., and Mellars, P. (1993) The EOS Project: modelling Upper Palaeolithic change. G. N. Gilbert and J. Doran (eds.), *Simulating societies: the computer simulation of social phenomena*, London: UCL Press.

Doran, J.E. (1992) A computational investigation of three models of specialization, exchange and social complexity. J.-C. Gardin and C. S. Peebles (eds.), *Representations in archaeology*, Indiana University Press, pp. 315–328.

Durkheim, E. (1895) *The Rules of Sociological Method.* Readings from Emile Durkheim, Chichester: Ellis Horwood.

Forrest, S. (1990) Emergent computation. *Physica D* 42, pp. 1–11.

Forrest, S., and Miller, J.H. (1990) Emergent Behaviour in Classifier Systems. S. Forrest (ed.), *Emergent computation*, Cambridge, MA: MIT Press.

Giddens, A. (1984) *The Constitution of Society*. Cambridge: Polity Press.

Jen, E. (ed.) (1990) *1989 Lectures in Complex Systems*. SantaFe Institute Studies in the Sciences of Complexity. Redwood City, CA: Addison-Wesley.

Kauffman, S. (1988) The evolution of economic webs. P. W. Anderson and K. J. Arrow (eds.), *The economy as an evolving complex system*. Reading, MA: Addison-Wesley, pp. 125–145.

Marx, K. and Engels, F. (1947) *The German Ideology*. New York: International Publishers.

Mellars, P. (1985) The Ecological Basis of Social Complexity in the Upper Palaeolithic of Southwestern France. T. Douglas-Price and J. A. Brown (eds.), *Prehistoric hunter-gatherers: the emergence of cultural complexity*, New York: Academic, pp. 271–297.

Nagel, E. (1961) *The structure of science*. New York: Harcourt, Brace and World.

Nowak, A. and Latané, B. (1993) Simulating the emergence of social order from individual behaviour. N. Gilbert and J. Doran (eds.), *Simulating Societies: the computer simulation of social phenomena*, London: UCL Press.

O'Neill, J. (ed.) (1973) *Modes of individualism and collectivism*. London: Heinemann.

Parsons, T. (1952) *The Social System*. New York: Free Press, Glencoe.

Stein, D.L. (ed.) (1989) *Lectures in Complex Systems*. Santa Fe Institute Studies in the Sciences of Complexity. Redwood City, CA: Addison-Wesley.

Stein, D.L. and Nadel, L. (eds.) (1991) *1990 Lectures in Complex Systems*. Santa Fe Institute Studies in the Sciences of Complexity. Redwood City, CA: Addison-Wesley.

Stöckler, M. (1991) A short history of emergence and reductionism. E. Agazzi (ed.), *The problem of reductionism in science*, Amsterdam: Kluwer.

Vines, G. (1993) *Planet of the free apes?* New Scientist 5 June 1993 138 (1876), pp. 39–42.

Waldrop, M.M. (1992) *Complexity: the emerging science at the edge of chaos*. New York: Simon and Schuster.

Watkins, W.J.N. (1955) Methodological individualism: a reply. *Philosophy of science*, p. 22.

SIMULATION AND RATIONAL PRACTICE

HARTMUT KLIEMT
FB Philosophie, Religionswissenschaften,
Gesellschaftswissenschaften
Gerhard Mercator Universität GHS Duisburg
Duisburg, Germany

1. Purpose of the Paper

In this paper I shall raise a very basic question about simulation and rational practice: If life is the best teacher under what circumstances and for what kinds of practical lessons is simulation at least a reasonably good one?

I propose that in dealing with such issues we should make a basic distinction between "thick" and "thin simulations". This distinction is illustrated in the paper. It is argued that completely different things can be learnt from thick and thin simulations respectively.

2. On the Potential Usefulness of "Thick Simulations"

2.1. BASIC FEATURES OF THE EXAMPLE

For a while I worked as a student assistant for a marketing specialist. He hired me because he hoped that I would apply my alleged skills as a student of operations research to solving the problem of "how many clerks on a floor?" This problem emerges frequently in the management of department stores. One should find out an optimal number of clerks maximizing long run expected profits. We started with studying a particular department store (actually with one floor of this store).

There were empirical data about customers, statistics on when and in what numbers they would typically arrive etc. To make a long story brief one can state that the real world system was quite well known and quite well understood in its standard workings.

Since there were different departments and several floors in the department store under study customers could themselves adapt to a certain

R. Hegselmann et al. (eds.),
Modelling and Simulation in the Social Sciences from the Philosophy of Science Point of View, 13–27.
© 1996 *Kluwer Academic Publishers. Printed in the Netherlands.*

extent to the lengths of queues at every point in time. They could for instance arrange their shopping in different order. At the same time clerks could to a certain — though very limited — extent be shuffled between service counters or even departments.

Waiting time before being served was regarded as a crucial variable. It was assumed that customers might be willing to wait for some moderate amount of time but not for long. Otherwise they would walk away and not return for the same item. It was also assumed that frustrations arising from futile trips to the store or long waiting times would render the store less popular and thus have some long run impact on pay-off.

Though there was a well-specified objective function to be maximized there was no way to formulate a classical optimization model like for instance within a linear or non-linear programming approach. The complexities of the situation, the stochastic fluctuations of numbers of incoming customers, the formation of queues could not adequately be modeled with traditional methods. Therefore we turned to simulation experiments as a substitute for classical optimization. At that time GPSS (general purpose simulation system), basically a somewhat buggy set of FORTRAN subroutines, was available on the UNIVAC 1108 to which we had access. GPSS seemed just fine for modeling queuing problems like "how many clerks on a floor?" So I decided on using this macro language rather than starting from scratch.

Very simply speaking I constructed a kind of mapping between service counters and subroutines (or macros) – for each counter one subroutine or at least a separate call of the routine. One could virtually see the store when reading and interpreting the code. Random number generators would create customers in ways coherent with the statistical data.

In a first step I tried to reproduce simply the conditions and patterns described in the data set. When looking at runs of the simulation analogues of the real world phenomena could be observed in the simulation. The same patterns of queues extending too much and idle clerks hanging around for a while emerged. At the same time the pattern of changes in the objective function corresponded to the pattern in the data etc.

After accomplishing the task of somehow reproducing the random events of the world in an appropriate random way and bearing in mind the one to one mapping between subroutines and counters one could trust that the model captured the essential characteristics of the underlying process. Given this evidence it made good sense to run experiments by changing the "capacity" parameters of the subroutines – modeling swifter service by more clerks or the opposite by reducing their number etc. and testing this against the objective function.

Well, about that time, I quitted working on the project because I took my final exams in management science and afterwards turned to philosophy. But even the few things I have learnt from my work on the problem of "how many clerks" should there be "on a floor" may suggest some down "to floor" philosophical conclusions about this type of simulation models.

2.2. CHARACTERISTICS OF THICK SIMULATIONS

As a reminder, let us first recall a classical definition of a system of objects like the one given, for instance, in Kleene (1952, p. 8): "By a system S of objects we mean a (non-empty) set D (or possibly several such sets) of objects among which are established certain relationships." Thus, in the most simple case a system is an ordered pair (D, r), where "r" denotes some relation on D. The system is abstract if "the objects of the system are known only through the relationships of the system... (w)hat is established in this case is the structure of the system, and what the objects are, in any respects other than how they fit into the structure, is left unspecified". Moreover, "any further specification of what the objects are gives a *representation* (or *model*) of the abstract system, i.e. a system of objects which satisfy the relationships of the abstract system and have some further status as well."

In the case of "how many clerks on a floor?" we can, with some justification, argue that the simulation model (using the term model here in the non-technical sense of a kind of rudimentary theory or description in some representational language) as well as reality are *both* models (now in the technical sense) of the same abstract system. That is, we can take recourse to the set theoretic concept of a model in characterizing how a simulation model manages to simulate the underlying reality. (Henceforth I shall assume that it is obvious from the context which concept of a "model" I have in mind when using that term.)

A "simulation model" works basically in the same way as a "ship model" in one of those specialized big tanks in which researches traditionally and still nowadays try to find out optimal shapes for ships. That the one kind of experiments is done on a computer and the other kind in real water seems to be of minor importance. And, of course a good computer model may work as a substitute for the water tank as well after formulating an appropriate "simulation model".

In both cases systematic variation of parameter values is viable. One can find out the effects and try to reach better and better results according to some objective function measuring performance along the relevant dimensions. One could nowadays even try to use evolutionary methods in searching the parameter space for improvements.

As far as we have empirical data on how well the simulated reality corresponds to the underlying "real reality" we can quite well assess the reliability of conclusions drawn from the simulated experiments. Due to the fact that the "simulation model" and the underlying reality are models of the same abstract system and to that extent exhibit the same structure the simulated experiments are actually more or less like standard experiments. The main advantage of such simulations is the fairly standard one that one can experiment without bearing the full costs of experimentation.

It seems obvious to me that simulations of the type discussed so far are legitimate instruments of research almost to the same extent as standard experimenting techniques. There is quite a bit of independent knowledge of the elements of the real world systems that are modeled. Besides and independently of the simulation we have access to other theories and relatively thick descriptions of the real world system. We basically understand what is going on in the simulated system. There is also a natural mapping between elements of real word systems and elements of the simulation (systems on the computer). Last but not least, for certain parameter constellations the simulated behavior can be tested against real world behavior (partly but not exclusively by retrodiction etc.).

We have good reasons to trust the predictions of such simulations because they implicitly provide explanations of what is going on in the systems under scrutiny. For convenience let us refer to such simulations as "thick simulations".

The potential usefulness of "thick simulations" for planning possible effects of intervention into the simulated process can be hardly denied. If in fact the simulations *are* thick they will in general show how alternative interventions will affect certain aims, ends or values represented approximately in an objective function.

2.3. HOW TO GENERALIZE EXPERIENCES FROM "THICK SIMULATIONS"

Evidently a typical thick simulation offers only very limited and very specific knowledge. For instance in the problem of how many clerks should be allocated to service on a floor we learn something about the specific processes prevailing in a specific department store. A thick simulation does not per se tell us anything about other department stores and the processes prevailing there. The simulation of the department store X may implicitly contain a kind of theory of *that* store. But this theory, of course, is not a general one. Even though we might rightly call it a theory it is not even a mid-range theory. It is the theory of X and not of Y and all the other stores.

Still, from thick simulations that necessarily are based on specific descriptions of specific conditions prevailing in a specific system something general may be learnt too.

For instance a company specializing on counseling department stores could keep records about how well its suggestions worked in practice. In pilot studies it might even have access to one store of a large chain and try out the suggestions developed in a simulation model under real world conditions. This company might then know almost as well as those engineers who have been experimenting on ship models in tanks for a long time that the results of their simulations carry over quite reliably to the real world.

In some cases it may be possible to formulate explicitly what has been learnt on the meta-level. The counseling company for instance may perhaps be able to lay down the "rules of the art" of forming thick models in a kind of advice book containing the "do" and "do nots" deemed conducive to the success of the practices of the company.

There are still other ways of how knowledge can be transferred to other systems than the simulated one. Thick models of the type "how many clerks on a floor" might be used to teach people how to find out efficient ways to allocate clerks even without formulating a simulation model themselves. For instance we might let students – or, for that matter, store managers – run experiments on a thick simulation model of one department store. Evidently there is almost no hope that these students later on in their careers will encounter the very same department store again. Still they might learn something for their future practical work. Experimenting with the computer model they gain insights about what in principle might work and what not.

Thus, even though it might be hard to give an explicit theoretical account of how to run department stores there may be ways of training on a simulated job. Even though this training does not teach trainees explicit optimization methods or rules, those who are so trained may afterwards have acquired some kind of implicit knowledge that makes them perform better.

Referring to the notorious example (supposedly Fritz Machlup's) of the person who learns how to ride a bicycle without learning how to solve differential equations we may say that running simulated stores can turn us into better store managers without learning a theory of how to manage stores optimally. After practicing as managers of a simulated environment we will be better equipped to run stores. We may become better practitioners who know how to do things rightly without acquiring an explicit theory of running stores.

The philosophical manager might sum this up in the classical line "omne agens agendo perficitur" – whatever acts perfects its capabilities in acting. Acting in the theater of virtual reality may perfect our abilities to act in the theater of life.

Like in athletic training it may be a good policy to got to some extremes too. "Unrealistic" modifications of the statistics about queuing in the realistic model may then be used. Trainees might learn how to adapt optimally the allocation of clerks in the specific department store modeled in the simulation even under extreme parameter constellations, i.e. a stampede of housewives who "cut it loose" after the most recent add on children's underwear. Becoming experienced in solving these tasks trainees may well be able to act more skillfully in managing different systems under standard and non-standard conditions.

2.4. THICK SIMULATIONS AND PRACTICAL SCIENCE

If one recalls how strongly a good education in any of the so-called practical sciences like management science — or, for that matter medical science (cf. on the latter Kliemt 1986) — depends on case studies, on solving specific cases etc. it is obvious that thick simulations should have very useful applications in these fields.

The assumption underlying such ways of teaching of course is that there is a kind of carry-over of experiences from one realm to another one. This expectation about teaching and learning is plausible if we take into account the importance of acquiring implicit knowledge. It may well be the case that — like medical doctors use to tell you — every case is unique. There is no general theory that could take into account all practically relevant parameter constellations. However, as doctors insist not without all plausibility the experienced practitioner will know how to deal prudently with these diverse constellations.

In the same way there may be no generally applicable theory of optimal allocation of clerks on a floor. Still, experimenting with simulations may provide some experience in fields in which we otherwise could not gain experience — at least not at acceptable costs — and where no general theory predicting alternative courses of events under alternative parameter constellations is available either.

Such uses of simulation techniques in practical sciences that try to train individuals in how to intervene wisely into complex systems seem to me completely acceptable and their more widespread usage desirable. In particular, in times of dramatically decreasing prices for computer equipment it seems foolish not to make use of thick simulations as a substitute for more traditional techniques. I feel that in those practical sciences where general theories are still lacking but we nevertheless intend to intervene in the course of nature *reading* about practical cases is evidently inferior to practical experience with a thick simulation. It is desirable that there be some kind of rudimentary theory cum experience. — And, did not Hume use "experimental" and "experiential" synonymously?

But, perhaps so-called practical science in the last resort is not at all science in the proper sense of that term. It prudently uses scientific results in scientifically informed practices but it is in itself not a science. That it is taught in institutions related to the teaching of science proper is merely a contingent fact which does not render practical training in complicated practices scientific. From this point of view the American way of somewhat separating business school, law school and medical school from university in general may be a good idea.

Summing up the preceding discussion we may state: From a methodological point of view thick simulations seem to be on quite firm ground. Their functions as instruments of prediction, training and the like are quite obvious. Thick simulations are impervious to fundamental objections to the extent that they as well as reality itself are models of an abstract system rich enough to contain all or most relationships deemed relevant for the practical purposes at hand. At least I cannot see any principal methodological reasons why we should not put them to work whenever this holds good.

If *we in fact manage to provide thick simulations of relevant realms* they might be very useful. But given the state of social science and our poor knowledge of the workings of realistically complex social systems the assumption that a thick simulation is available may simply amount to self-deception. Let us therefore leave practical science and turn to general social science and the potential usefulness of simulations as research strategies or practices in that realm.

3. Possible Functions of Thin Simulations in Social Science

Often no thick simulations are available in the realm of general social science because the underlying social systems are not sufficiently well understood. Most of the time our understanding of complex social systems only faintly echoes our detailed knowledge of the workings of a run of the mill department store.

We just have some vague ideas about which kinds of mechanisms might conceivably lead to certain phenomena but there are several potential explanations for the phenomena under scrutiny. In such situations we may still be inclined to apply simulation techniques. However, we use them now for different purposes. Their application is not a way of experimenting with a process well understood but rather is itself instrumental to acquiring some understanding of a process which otherwise is not understandable at all or only very vaguely so because there is no or not sufficient independent knowledge of the process.

To simulations that contrary to thick simulations cannot be controlled

by independent knowledge and an independent understanding of the real
world system to which the simulation at least purportedly refers I shall
refer as "thin simulations" henceforth. Let me try to explore some of the
possible or conceivable uses of thin simulations within the realm of social
science – deliberately including some seemingly exotic ones.

3.1. DISCIPLINING THEORY FORMULATION

As in the case of so-called practical science one may have some second
thoughts about the "science" in much of social science. Everybody who
has endured some courses in traditional sociology, who has struggled with
pompous but ultimately vacuous terminologies and still somehow man-
aged to keep his common sense intact must once in a while have felt the
urge "to run some nice simulations to become clean again" (I pinched this
phrase from David Levy's response to so-called "hermeneutic economics"
— though he regrettably favors running regressions).

Whatever has a sobering effect on those working in the field should be
greeted as an accomplishment. Thus, if people are somehow forced to cast
their speculations about causal relations and interdependencies in social
reality into simulation models this is definitely advantageous.

This is not to say, however, that there could not be any problems and any
theses that might be hard to capture within a simulation approach. Still, as
far as they go, computer simulations have the merit of being formulated in
a precise way in a programming language and thus their content is at least
intersubjectively accessible. From this point of view they may be a kind
of substitute for the formal language of mathematics which we otherwise
might prefer as a device of precise and reliable communication.

Though it is no minor aspect of simulation studies that they are intersubjectively accessible and in general lead to what might be called "disciplined speculation" — as opposed to undisciplined — this is not all that
may be said in favor of using simulations in realms where they are not thick
but rather very thin and speculative.

3.2. YELLOW PAPER SIMULATIONS

Americans have the nice custom of using yellow paper for first approaches
to a problem. First and second drafts are written on yellow paper and this
is where the "real scientific process" goes on. The final outcome on white
paper may be more polished but it is not on white paper where we develop
our ideas. We scribble nice little graphs in two dimensions on yellow paper.
We sketch some equations for simplified cases in the yellow — and often in
the "blue" so to say.

Simulations may sometimes serve approximately the same purposes. We use them to get some feeling for what is going on. Like on yellow paper we may first formulate a quite austere model. Toying with that simple model we may hit upon some general regularities. We then might even be in a position to formulate a general model in analytical terms. Sometimes it may even be the case that our toy simulation may guide us towards formulating a general formal theorem that we can strictly and generally prove.

For obvious reasons I suggest to refer to such simulations as yellow paper simulations. Whenever we do not have some intuitive access to how complex interactions between several elements of a complex social reality might be related to each other yellow paper simulations can potentially serve important heuristical purposes in theory formation. It is clear too that yellow paper simulations as such are not of scientific interest. They are an intermediate not an end product. It is not them but the theories that are put to the test either on white paper or in the real world.

This does not imply that yellow paper simulations are without interest. Quite to the contrary. Like there are many important products in our world that might not exist without intermediate products there may be important theories that will not exist without adequate heuristics preceding their formulation.

Still, besides serving disciplining and heuristical functions thin simulations may be put to more radical uses too. Let me refer to them then as "isolating simulations".

3.3. ISOLATING MODELING AND ISOLATING SIMULATION

One interesting practice of some of the best modeling in social science is based on radical simplifications. To give just three examples let us first briefly recapitulate Geoffrey Brennan's and James Buchanan's Leviathan model of state behavior. In their book on *the power to tax* (1980) they imagine a Leviathan government which aims at maximizing tax revenue. In particular it is not primarily interested in the common good of the governed or other aims. They then try to find out the implications of such an austere model by driving it to its extremes.

A likewise unrealistic assumption was already proposed by David Hume as a foundation of all general political theory in his marvelous essay on "how a science of politics is possible". Let us consider this as a second example.

Hume first remarks that no general theory of politics would be viable at all if the good or bad performance of political institutions would be completely dependent on the personal characteristics of those who exercise power. He then goes on to suggest that in matters political "everybody should be supposed to be a knave" (1985, p. 42).

Again Hume knew quite well that in fact not everybody is a knave or if according to some normative standard everybody were a knave then for empirical purposes we would still need some way to discriminate between the relatively more knavish and those who are relatively less so. But nevertheless Hume held that it was a sound norm of political theorizing to proceed on the counterfactual assumption that everybody is a knave.

To give a final example of the same genre let me briefly draw attention to the likewise counterfactual assumption about inter-firm competition on a market. In his seminal 1950 paper Armen Alchian analyzes the process of weeding out less successful firms as an evolutionary process in which totally myopic players interact with each other. None of the players intentionally optimizes. Nevertheless, since only certain strategies will survive selection, the outcome is as if all had been optimizing all the time.

I feel that simplifications like the three examples discussed before are valuable precisely because they distort reality in a certain way. The counterfactual character of the underlying assumptions should be boldly acknowledged. It seems that for certain theoretical purposes distorting reality is a virtue rather than a vice. But what might those purposes be?

In our more ironic moods we might be willing to say that the Leviathan model of the tax collecting state is not counterfactual at all. It rather faithfully approximates what governments are all about. But even involuntary connoisseurs of German politics must agree that this view *is* a distortion.

Therefore it comes as no surprise that Brennan and Buchanan were blamed for distorting social reality. But Brennan and Buchanan themselves would be the first to agree that they did and intended to do just that. As stated before Hume was well aware of the counterfactual character of his suggestion for those who engage in normative institutional analysis in politics. Armen Alchian would otherwise be among the first to insist that human actors are not totally myopic, he would object to the traditional sociologists that humans are not following fixed behavioral programs determined by internalized values. He would insist that in fact they anticipate future causal effects of their choices and thus act in teleological (strategical) rather than completely myopic ways. Still, he would go on to defend his counterfactual view as useful for certain purposes.

I intuitively share such views but I have some difficulty in characterizing the purposes explicitly and precisely. Nevertheless, as far as simulation in social science is concerned it is worthwhile to make one obvious point here: If the models characterized before are legitimate tools of social science then thin computer simulations which do nothing but isolate certain factors and study their implications by taking them to the extreme are legitimate tools too. Since the models described before do form paradigm examples of what is regarded as good social theory it follows that thin simulations could in principle be regarded as paradigm cases of good social science too. That Axelrod's simulations were so well received may serve as an indication that they in fact are if they are marketed well.

Since according to the preceding considerations thin simulations are comparable to well entrenched social science modeling there seems to be no need for those who work in the field to act humbly. They can point out that their ways of modeling are not in any obvious way inferior to common practices of some of the best of social science.

Though I feel that an argument pointing to a comparable common scientific practice can be quite strong as a methodological defense many people of course would say that this is not a great accomplishment at all. That a is like be b is not much of a justification if there is no good argument in favor of a either. If that were so one could likewise defend the social science "adventures in word music" which I rejected in my introductory remarks about vacuous terminology. One could insist here, though, that the three models mentioned before were at least built on counterfactual assumptions in a precise way and that simulations that try to isolate counterfactually certain factors and study their interaction do the same. Still, it would be better to give a kind of independent defense for the practices of more radical modeling based on counterfactual assumptions.

3.4. THIN SIMULATIONS AND THE INSTRUMENTALIST MIND SET

Let us be quite explicit about the fact that the distortions that I have in mind are not of the harmless type to which we may refer as idealization. They are more extreme than assuming frictionless motion. The latter is meant as an approximation. We command some empirical knowledge about how it approximates reality and how much it distorts reality. We have experience of a continuum of cases, experiments in the vacuum etc. The underlying forces like gravity are still present but they now are meant to work without interference.

But, for instance, Alchian cannot plausibly claim that his model is an approximation of that kind. His model abstracts away strategic behavior altogether. No effort is made to bring in strategic behavior as an explanatory factor and to show how it affects outcomes in a continuum of cases approaching the idealized situation as a limiting case. And, as mentioned before, Armen Alchian himself would otherwise reject explanations of social phenomena for the simple reason that they are not taking into account the factor he is eliminating in his thought experiment.

Though it is hard to say *precisely* how the cases of the assumption of frictionless motion and completely myopic behavior differ they are definitely located at opposite extremes of a spectrum of possible theoretical abstractions. When using theoretical idealizations like frictionless motion we know ways to estimate how well they approximate real phenomena while models like the three mentioned before are most of the time constructed as a purely conceptual exercise. They are used to explore what is conceivable in principle under some very stringent hypothetical assumptions.

As compared with state of the art physics or even the physics of the Newtonian era we are simply lagging behind in social science. If we are using abstractions we more often than not are simply speculating about merely conceivable processes. But a lot is conceivable in principle. We can imagine possible worlds of all sorts. Are we any better than magicians and story tellers if we indulge in the discussion of such images of the merely conceivable? Is such speculation about models really superior to those speculative practices of antiquity and medieval times which never came to the testing of hypotheses? Is speculation even if it is disciplined speculation as introduced above really enough to make a science?

As we know there may be some, in particular some economists, who would claim that the only purposes of science including social science are explanation and prediction. However, since a proper explanation in the standard sense of the term must be based on factual rather than counterfactual premises the models presently under scrutiny cannot count as standard explanations — not even approximate ones. This raises the question of how we can defend these models and their central place in social science.

In view of this some scientists tend to go one step further. They argue that the counterfactual character of the premises does not matter. If the predictions of the models are correct they may also be used as explanations. Some economists and, for that matter some physicians including some witch doctors, seem to argue in favor of purely instrumentalist views. They would say however thin a model or a simulation is if it is useful as an instrument of prediction or even weaker, if a practice incorporating such a thin model or simulation delivers some goodies it is justified.

Without going through the ritual of discussing instrumentalism let me simply remark that I am totally unconvinced. But then the question of how to defend thin simulations — and for that matter thin models —is open still. Besides the respectable reason that these models are more precise than their competitors and besides the less respectable reason that I and many of my colleagues simply like these models better than verbal speculation are there more respectable reasons? As I said I do regard this as an open question but here is one very tentative answer which I submit as a somewhat exotic final remark.

3.5. CRITIQUE OF COMMON SENSE ASSUMPTIONS

May be, as Hume speculated in his *dialogues concerning natural religion*, that our world is the outcome of the efforts of some young gods in their apprenticeship who badly failed in producing something reasonable. May be they are simulating something they are interested in. In any case we do not have any experience of that kind. We cannot step out so to say and look at the system as a whole. Likewise as far as some fundamental issues of social science are concerned we cannot step out of the system and run experiments on society as a whole. We all do have, however, certain views and opinions about what makes the social and political world go round. These views and opinions shape what we regard as serving our own interests as well as those of the community at large. Like religious creeds they make us more or less inclined to accept certain policies as reasonable and others as unreasonable.

The latter in itself is in no way irrelevant. What becomes real and accepted in day to day politics is dependent on what we regard as plausible and feasible. Because, as David Hume (1985, p. 32) had it "it is on opinion only that government is founded". This actually and even more surprisingly echoes a remark of Hobbes who said in his Behemoth (1682/1990, p. 16)

> ... the power of the mighty hath no foundation but in the opinion and belief of the people

Of course, if somebody was ever aware that interest reigns the world it is Hobbes and Hume. What they are trying to say then must be made consistent with the rest of their theorizing. As far as this is concerned the answer seems quite simple: interest determines the "deep structure" of social interaction while within the "surface structure" opinions are of importance. But, what happens on the surface influences the deeper interests too. Thus it is in no way unimportant for the purposes of predicting, explaining and understanding. Quite to the contrary the interplay between deep and surface structure is of crucial importance.

Moreover, it must be noted that humans frame their opinions according to their interests. They act upon what Hume called their "opinion of interest" (cf. 1985, p. 33). Sometimes such an opinion is shaped by direct experiences and observations. But as far as the most fundamental questions of social organization are concerned more often than not experience becomes very indirect and thin indeed. Here our common sense itself operates on thin models which determine what we conceive as viable.

Even though many issues of practical relevance may be beyond the powers of our minds and imagination, as a matter of fact we will not cease to speculate about these matters. To provide images and to have some access to constrained and precise forms of speculation may be of the utmost value here. Lacking a better alternative we must try to substitute controlled experiments on reality itself by controlled speculation. Simulations may be viewed then as a kind of tool for this. They may be used in confronting simple common sense pseudo-theories with their implications in a likewise simplified but more precise and intersubjectively accessible manner and to illustrate the implications of competing simplified conceptions of the fundamental mechanics of social interaction. As far as practical purposes are concerned and as long as there are no superior alternatives it is futile to criticize simplifying models of science for not meeting certain high scientific standards.

In matters practical we have to go for the *comparatively* better. What we can achieve may be bad enough still but that it is not good is no good criticism as long as there is no better alternative. Taking resort to thin simulations may often be the best we can do. Of course, there is no reason to trust them blindly nor to be seduced by their high-tech image. As everything else in social science they should be taken with a grain of salt.

Certainly I do not expect the public at large to be open to such things as computer simulations. However, opinion is shaped in a more indirect way too. There are opinion leaders. Some of them will be educated at social science departments. Given the average state of such departments at least in Germany one may regret that but it still is a fact. Now given your own interests and own opinions what would you prefer students to be taught Luhmann and Habermas or Axelrod? What, do you feel will lead to better results from the point of view of rational practice and therefore be a better practice itself?

4. Acknowledgement

I should like to express my gratitude for helpful comments to the discussion group at the "brown bag luncheon" at the Center for the Study of Public Choice, George Mason University, Fairfax, VA. Roger Congleton corrected my fundamentally mistaken view of engineering while Wolfgang Balzer from the University of Munich sent me extremely helpful written comments that should have considerably improved the paper if I as a receiver behaved as well as the sender of the message. Of course the conventional disclaimer applies.

References

Alchian, A.A. (1950) *Uncertainty, Evolution, and Economic Theory*, Journal of Political Economy, 58, p. 211 ff.

Brennan, G.H., and Buchanan, J.M. (1980) *The Power to Tax*. Cambridge: Cambridge University Press.

Hobbes, Th. (1682/1990) *Behemoth*. Chicago: Chicago University Press.

Hume, D. (1985) *Essays Moral, Political, And Literary*, ed. by E. F. Miller, Indianapolis: Liberty Press.

Kleene, St.C. (1952/71) *Introduction to Metamathematics*. Bibliotheca Mathematica 1, Amsterdam and London: North-Holland Publishing Company.

Kliemt, H. (1986) *Grundzüge der Wissenschaftstheorie. Eine Einführung für Mediziner und Pharmazeuten*. Stuttgart: Gustav Fischer.

EPISTEMIC CULTURES IN THE SOCIAL SCIENCES

The Modelling Dilemma Dissolved

KARL H. MÜLLER
Institute of Advanced Studies
Wien, Austria

In the academic year 1939/1940, Joseph A. Schumpeter organized a *non*-dying Harvard Seminar[1] on the topic of rationality in the social sciences, including economics. After one year of intense lectures from different scientific fields, Talcott Parsons was to edit a collection of articles originating from this seminar. Schumpeter himself contributed a draft version where he introduced, by systematic ordering, a table of four elementary problem areas relevant for the topic of rationality. Accordingly, Schumpeter distinguished between two areas of observation (observer/observed), and, moreover, between an internal and external perspective, where the *internal* or *subjective* point of view referred to inner states, intentions or preferences of individuals, be they on the observer or on the observed side, and the *external* or *objective* side could be qualified as *ascriptive*. Thus, Schumpeter arrived at the *identification* of four main problem areas on rationality. However, Schumpeter did not succeed in providing *satisfying* accounts and solutions for each of the four rationality fields. Moreover, the planned publication underwent the process of a *dying* Harvard or Non-Harvard collection

---

[1]The phrase of the *dying seminar* in Harvard refers to the following narrative by Thomas C. Schelling (1978) —

Somebody organizes a group of twenty-five who are eager to meet regularly to pursue a subject of common interest. It meets at some hour at which people expect to be free. The first meeting has a good turnout, three quarters or more, a few having some conflict. By the third or fourth meeting the attendance is not mich more than half and pretty soon only a handful attend. Eventually the enterprise lapses, by consent among the few at a meeting or by the organizers' giving up and arranging no more (Schelling 1978, p. 91f.) —

where an unequal distribution of critical threshold values for attendance leads to a continuous fading away of participants.

R. Hegselmann et al. (eds.),
Modelling and Simulation in the Social Sciences from the Philosophy of Science Point of View, 29–63.
© *1996 Kluwer Academic Publishers. Printed in the Netherlands.*

volume.[2] Consequently, the rest *had* to be silence ...[3]

This small historical episode has been chosen as an introduction for two reasons. First, it offers a *systematic* summary of essential problem areas with respect to the utilization of rationality-assumptions in the social sciences.[4] And second, the episode makes abundantly clear that a highly problematic configuration, subsequently introduced and defined as *modeling dilemma*, has persisted for decades, not only within economics, but within the social sciences in general. Moreover, the utilization of rationality assumptions must be considered only as part of a *wider* methodological conundrum, since a *large* number of *heterogeneous* simplifying components, ranging from statistics to the needs and peculiarities of the algorithms used, enters into the *actual* model-building processes ...

At this stage, it seems *highly* appropriate, to introduce the central focus of the present article, namely the concept of the *modeling dilemma*, in closer detail. Put in a conventional methodological perspective by separating between *empirical* and *normative* domains of discourse, the following dilemma arises, seemingly by necessity, within the social sciences and its model-building operations[5], past and present:

---

[2] In a mode of analogy, the phrase of the *dying collection volume*, Harvard and otherwise, could relate to the following complex configuration —

> *Somebody organizes a group of twenty-five who are eager to write a paper on a subject of common interest. He sets a dead-line at which people are expected to have completed their contribution. The first dead-line has a good turnout, two thirds or more, a few having some conflict. By the third or fourth dead-line the additional manuscript-turnout becomes negligible and pretty soon no one sends a manuscript any more. Moreover, a small number of authors withdraws their contributions. At this stage, the enterprise eventually lapses, since an interesting dynamic development is set in motion ...* —

where unequal distributions of critical threshold values for *deadly* deadlines on the one hand and for necessary necessities to get articles published lead to all sorts of dynamic trajectories, *including* the fading away of manuscripts ...

[3] For a brief summary of the Schumpeter seminar, see Swedberg (1994), p. 175f.

[4] In the following article, the term *social science* refers to all types of scientific disciplines which, in one way or the other, are concentrating on human actions, interactions and their respective results. Thus, economics as well as sociology, political science or psychology form *core* elements for a comprehensive set of social science disciplines. Moreover, the examples in this article are deliberately chosen from a wide array of disciplinary fields, ranging from economic theory or econometrics to sociology and political science. Thus, any example from a disciplinary segment$_i$ which is supposed to exemplify a specific piece of information, should be taken as *pars pro toto*, since highly similar configurations could be identified within disciplinary areas$_j$, too.

[5] The expression *model-building* is confined to those social science frameworks only that are characterized by a comparatively high degree of formalization and, moreover, are utilized within an *explanatory* context (See, e.g. Fararo 1989; Troitzsch 1990). Thus, the set of models under consideration ranges from classical macro- or microeconomic equilibrium models (Weintraub 1977), to the growing stock rational choice-versions both in political science or in sociology (Elster 1985, 1986), or to the rapidly expanding class of complex models across the social sciences (Anderson *et al.* 1988; Casdagli and Eubank

On the one side of the horn, any *empirical* interpretation of the normal model-applications in economics, sociology, political science or demography is confronted with the immediate objection that *essential* model-building operations violate even the most tolerant test-conditions. Consider the following excerpts from a fairly recent book on macroeconomics which reflects the current state of the art of macro-economic modeling (Frisch and Wörgötter 1993) and which, due to its *very* characteristic features, has been quoted extensively.

> Consider a small open economy inhabited by a large number of identical individuals. The lifetime utility of the representative individual is given by ... (Calvo and Vegh 1993, p. 10)
>
> In order to account for the international trade in capital goods, we assume that both domestically-produced and imported goods can be converted into an investment good according to a constant returns-to-scale technology ... (Gavin 1993, p. 31)
>
> There are three regions: Germany, France (which together make up Europe) and the United States ... here representing the rest of the world ... Exchange rate developments are perfectly anticipated apart from the effects of initial shocks (Hallett *et al.* 1993, p. 49)
>
> Domestic producers are assumed to maximize profit ... by optimally choosing the variable inputs L_t and N_t ... We assume that domestic residents allocate their financial wealth between domestic money, domestic bonds and foreign bonds (Hof 1993, p. 73f.)
>
> Production is carried out by many identical competitive firms. For notational simplicity, the number of firms is equal to the size of the population (also equal to the size of the labor force) (Hoon and Phelps 1993, p. 97)
>
> We will assume a small open economy, and hypothesize a state of perfect capital mobility, that is, perfect substitutability between domestic and foreign interest-bearing assets (Claassen 1993, 137)
>
> Investment depends on a number of non-quantifiable factors such as political stability and the industrial relations climate. Assuming these are favorable, we may write investment as a function of expected prices and their variance ... (Worrell 1993, p. 161)
>
> The model contains a rudimentary construction sector supplying a durable good producing housing services ... For simplicity, no rental market for housing is included, that is to say, all dwellings are owner-occupied (Nielsen and Sorensen 1993, p. 205)
>
> We assume that there are two classes of speculators. One class is called 'chartists' (noise traders), the other 'fundamentalists' ... The

1992; Casti 1992, 1994; Haag *et al.* 1992a; Jen 1990; Müller and Haag 1994; Stein 1989; Weidlich and Haag 1983) ...

> 'chartists' use the past of the exchange rates to detect patterns which
> they extrapolate into the future. The 'fundamentalists' compute the
> equilibrium value of the exchange rate (Grauwe and Dewachter 1993,
> p. 355)

> Agents are assumed to have a qualitative (intuitive) understanding
> of the models. This is formalized by assuming that they use their own
> estimates (guesses) of the parameters in place of the true values. The
> assumption is that these parameter estimates have the same algebra-
> ic signs as those of at least one of the RE reduced forms. (Goldberg
> and Frydman 1993, p. 384)

> It is supposed that there are two players on the market, and 'market
> expectations' are determined as a weighted sum of rational expecta-
> tions and chartist expectations. (Vijayraghavan 1993, p. 401)

Thus, it seems quite obvious that any interpretation for these stage setting
modeling assumptions as empirically *well*-founded or *valid* runs into the
serious risk of having to accept *any* account, be it astrology, be it in the
creationist spirit or be it, more generally speaking, of the type characterized
by Martin Gardner as *bogus science* (Gardner 1981), as genuinely *empirical*.

Turning now to the *other* side of the dilemma by characterizing the
prevalent modes of operation as *normative*, one is immediately confronted with a different and similarly devastating inconsistency. Why? Take,
for example, the three region assumption in the article by Hallett *et al.*
Here, one cannot find *any* normative commitment that the world *should*
consist of three major nation-states only. In a similar manner, the Nielsen
and Sorensen-paper does not state that it would be *rational* to exclude a
rental market for housing. Consequently, any normative interpretation of
the modeling practice imposes an unjustifiable and, in most instances, *highly* implausible account of the role and function of essential model-building
assumptions[6]. Moreover, any normative assessment clearly violates the actual practices with respect to the output of economic, political or sociological model-building which, once again, is not phrased in terms of goals,
reachability and choices of appropriate means but which, in most instances,
is couched in a conventional *explanatory framework* by focussing on specific
interaction patterns or on a set of specially interlinked causes.[7]

---

[6]It might be argued, however, that rationality assumptions qualify as *natural* candidates for a *normative* interpretation. But in most instances, models do not consist of
rationality assumptions only, but of a variety of *additional* components which are strictly independent from the rationality elements and which, taken as a whole, *re-iterate* the
dilemmatic choice between empirical and normative domains ... .

[7]In a personal conversation, Rainer Hegselmann has suggested to use the term *modeling paradox*, highlighting the peculiar situation that model builders in the social sciences
try to understand the *complexities* of the surrounding socio-economic universe by employing, in the beginning, *very* simplifying and empirically ill-founded devices. Thus, the

Consequently, Schumpeter's fourfold rationality domains turn out to be a comparatively small sub-set of a wider class of modeling components, rational[8], whose crucial deficiency and difficulty lies in their unresolved status, placing them apparently in the nowhereland *between* empirical and normative social science applications.

1. Preliminary Considerations

In order to transform Schumpeter's draft version and, moreover, the many escape-strategies from the modeling dilemma developed so far[9] into a *satisfying* escape from the modeling dilemma, a requirement of *historical reachability* will be introduced. The subsequent dissolution-sketch of the modeling dilemma will utilize only those cognitive elements that have been available in Schumpeter's days, too. In other words, the dissolution will be accomplished *via* a reconfiguration and rearrangements of cognitive components well known and ready at hand fifty years ago.[10]

The starting point for the successful dissolution of the modeling-dilemma in the social sciences lies, first, in the utilization of a very well-known and, by now, classical distinction, originally made famous by Rudolf Carnap already in the 1940's, namely the differentiation between *pure* forms of semantic analyses and their *descriptive* counterparts. (See esp. Carnap 1975) By analogy, a similar differentiation will be introduced here with respect to *pragmatics* by differentiating between two main roads of analysis, namely between*pure* pragmatics on the one hand and *descriptive* pragmatics on the other hand.[11] Thus, the main ingredients for a *pragmatic* analysis

---

modeling paradox can be summarized as an alchemical GIGO-Principle. (Garbage in — Gold out)

[8]It must be emphasized, once again, that *many* of the modeling parts which have been quoted in the introductory chapter like the equalization of the number of firms with the size of the population (Hoon and Phelps 1993, p. 97) *cannot* be attributed to the field of *rational behavior* and the domains of *homo oeconomicus.*

[9]For a summary, see the expanded version of the present article in Müller 1995c, p. 4ff.)

[10]In certain contexts within the history of science, especially in the case of dilemmata, it becomes worthwhile to stress the point that, by an *adequate* process of permutations and reconfigurations, or, to borrow a phrase from Douglas R. Hofstadter, by a suitable *variation on a thema* (Hofstadter 1985), a successful problem-solution would have been ready at hand already a long time ago.

[11]Thus, the two main roads for *pragmatic* analyses can be formulated, following Carnap's original demarcations, in the subsequent variational manner:

By descriptive pragmatics we mean the description and analysis of the pragmatic features either of some particular given scientific language games, e.g. during the period of the French Enlightenment, or of all historically given scientific language games in general .. Thus, descriptive pragmatics describes facts; it is an empirical science. On the other hand, we may set up a system of pragmatic rules, whether in close connection with a historically given scientific language games or freely invented;

of science lie in the area of *scientific language games* and *rule-systems* which characterize the essential moves, operations or practices of such language games.[12]

One of the most interesting and heuristically fruitful tools for *both* types of pragmatic analyses of scientific language games consists, as has been indicated already by another member of the Vienna Circle, namely by Otto Neurath, in a *morphological* analysis[13] (Neurath 1981). Neurath has elaborated on the morphological method in various articles around the time Schumpeter had finished his *Vergangenheit und Zukunft der Sozialwissenschaften* (1915). Proceeding along the *analytical* branch of morphological analyses in the field of pragmatics, i.e. scientific rule systems, one could, in principle, define a large *number* of different cross-tables. Take, for example, the classical dimension, dating back to David Hume, between the empirical and the normative realms (Hume 1989; Streminger 1994), take, as a special instance, the subsequent differentiation between two areas of decision theory as a reference point —

> Normative decision theory ... is deductive. It postulates certain criteria of optimality or rationality or equity and derives strategies or methods of allocation or methods of aggregating preferences that are supposed to satisfy these criteria. A descriptive theory starts with observations of how actors choose in given classes of decision situations and attempts to describe their behaviour as systematically as possible. (Rapoport 1989, p. 5f.) —

and take, finally, a distinction put forward by Mario Bunge (Bunge 1977, 1979, 1983a, 1983b) who differentiated strictly between two basic types of systems approaches, namely concrete and conceptual ones —

> A system, then, is a complex object, the components of which are interrelated rather than loose. If the components are conceptual, so is the

we call this a pragmatic system. The construction and analysis of pragmatic systems is called pure pragmatics ... Pure pragmatics consists of rule constructions ... and their consequences; therefore, in contradistinction to descriptive pragmatics, it is entirely analytic and without factual content. (A variation to Carnap 1975, p. 11f.)

---

[12]Since the general focus lies in a pragmatic analysis, it will become clear, in due course, that the classical controversies of the Carnap-Quine debate (Quine 1961; Creath 1990) simply *cannot* arise. Why? Because the pure or analytical dimension will receive, due to the *pragmatic* perspective, a thoroughly *Quinean* interpretation and justification ...

[13]For a morphological analysis, be it along the empirical or along the analytical path, the necessary research steps can be summarized in the following manner: *First*, a small number of different and heterogeneous dimensions must be identified and arranged as cross-tables, offering a $m \times n$ *array of distinct combinations*. *Second*, these cross-tables can then be used in different ways, ranging from the empirically observable and measurable distribution of these different combinations up to the detection of white spots which have not been realized so far ... (See also Dubach 1977)

system; if they are concrete or material, then they constitute a concrete or material system. A theory is a conceptual system, a school a concrete system of the social kind. These are the only kingdoms we recognize: conceptual and concrete. (Bunge 1979, p. 4)

Such a two-fold separation can lead to an interesting cross-table in which important groups of scientific practices or language games can be distributed across the resulting four areas. More concretely, these four areas imply a

TABLE 1. A Morphological Space for Language Games in the Social Sciences

		Dimension$_2$	
		empirical	*normative*
	concrete	Area I	Area II
Dimension$_1$			
	conceptual	Area III	Area IV

division of research practices within the social sciences according to which large parts of evaluation research, planning or optimizations occupy Area II, in which most components of linguistics and language-based research are situated in Area III, in which formal areas like mathematics or statistics lie within Area IV — and in which the main proportion of the scientific output resides within the first area.[14]

With respect to the modeling dilemma itself, the four problem areas of the Schumpeter sketch on rationality can be generalized and arranged in the following manner:

The problems of treating rationality assumptions and similar simplification components within the model building-operations in the social sciences can thus be separated into four areas. The modeling dilemma itself, while situated in all four domains can be treated on two distinct levels, namely on a general and on a *specific* niveau. According to the *specific* levels, modeling components, like rationality assumptions and the like, are utilized *within* one of the four particular dilemmata areas of Table 2, whereas in the *general* version the peculiarities of specific utilization contexts do not play a significant role. For reasons of deductivity, the modeling dilemma will be

---

[14]It might be asked to which field the morphological explorations of the present article belong, since it is neither empirical in nature nor concerned with the conceptual structures of social science research nor, for that matter, normative in character ... It will become clear, within the subsequent chapters, that *another* dimension must be added to the two-dimensional matrix of Table 1 so that a *satisfying* location of the morphological approach can be undertaken. By chance, this additional dimension will also effectively dissolve the modeling dilemma with which the present article had started.

TABLE 2.   A Morphological Space for the Modeling Dilemma

		Dimension$_2$	
		observer	*observer*
Dimension$_1$	*internal*	Dilemma I	Dilemma II
	external	Dilemma III	Dilemma IV

dissolved subsequently in its general form since this *overall* solution will, simultaneously, pave the way for the more specific four variants, too.

2. The Dissolution–Sketch for the Modeling Dilemma

Having arrived at a general framework for the identification of language games within the scientific system and their relevant rule systems, it should become feasible, finally, to produce a new type, and, it must be added, a more *successful* type of a *general* dissolution of the modeling dilemma. Moreover, the following remarks will shed new light on the principal heuristics for different ways of socio-economic world-making and for the separate roles of model-building in the social sciences where these different approaches can and must be qualified as *equally* necessary and indispensable.

The starting point consists in the introduction of a new dimension, which comes originally from modal logic and which has started in its modern form with C.I. Lewis already around 1920. Accordingly, a differentiation into two reference areas will be undertaken, namely into *possible world* domains and the realm of the *actual* world.[15] (See also Hughes and Cresswell 1985, p. 75ff. or, for social science modeling, Gilbert 1981, p. 1ff.) Thus, a very simple two-dimensional morphological space assumes the following form:

Even at this point, no cognitive *cash value* (Wilfried Sellars) will be recognizable since the partitioning into Field I- and Field III-games must, in all probability, lead to a complete marginalization of the model-building efforts within the social sciences, too. By characterizing the bulk of modeling activities as a scientific language game within *possible* worlds, the modeling

[15] It must be stressed from the beginning that the distinction between *actual world* and *possible worlds* is *not* to be confused with a *realistic* commitment. On the contrary, *actual world* and *possible worlds* are differentiated in a purely *pragmatic* manner by focussing on differences with respect to *rule systems* and *evaluation criteria* for both types of operations. To be more precise, the rule systems for *actual* world practices focus on observation rules, measurement-rules, testing-rules and the like which are largely *absent* in the case of *possible* worlds-practices.

TABLE 3. A Modal-based Morphological Space for Four
Types of Language Games in the Social Sciences

		Dimension$_2$	
		empirical	*normative*
Dimension$_1$	*possible world*	Field I	Field II
	actual world	Field III	Field IV

activities lie, apparently, within the same camp as any type of *bogus* science
which, after all, can be qualified as science-*fiction*, too. But at this stage,
an explicit reference to the underlying basic pragmatic concepts, namely
that of rules, rule systems and evaluation criteria, becomes essential. In the
subsequent paragraphs, it will be demonstrated in an *a priori* fashion that,
on the one hand, rule systems differ radically for research within Field I
and within Field III and that, moreover, specific sets of *non*-trivial evaluation criteria can be found which differentiate clearly between interesting,
fruitful research in each of the two respective fields from their unattractive,
trivial or *bogus* counterparts.

2.1. MAIN DIFFERENCES IN THE RULE SYSTEMS

Consequently, it will and must become the task of the subsequent part
to highlight some of the main differences with respect to the rule systems
for Field I-activities on the one hand and for Field III-operations on the
other hand. The easiest way to identify major differences consists in the
elaboration of two paradigmatic modeling examples, one from the side of
actual world-modeling, the other from the possible worlds-area. In this
manner of *exemplar based learning* (Charniak and McDermott 1985), two
social models will be presented on the domain of education, one long term
model of the Austrian education system focussing on the distribution of
pupils across various school-types[16], the other one on the French education
system and the transition process from the end of High School either to the
universities or to the *Instituts universitaires de Technologie* (IUT).[17]

Using a master-equation framework (Haken 1981, 1983; Weidlich and
Haag 1988), the distribution of pupils across various school-types in
Austria from 1970 to 1990 has been modeled by building up three factor

---

[16]For the first model and its subsequent revisions, see especially Müller and Lassnigg
(1992b); Haag and Müller (1992b); Müller and Haag (1994); Müller (1995b).
[17]For the French model, see esp. Boudon (1979), p. 122ff.

sets determining group behavior in a large scale social system. The following explanatory schema has been used —

Changes in school type$_i$: $f\{$attractivities, barriers, global mobility$\}$

It would require too much space within this article to present a formally accurate account of the overall modeling framework[18], so it must be sufficient, at this stage, to characterize it in a qualitative manner. The set of barriers, restricting the *movement* of pupils across or between school-types, consisted of

$\{$Legal restrictions[19], delays in the schooling career[20],
gender separation in school types[21], dualism[22] $\}$

whereas the attractivity set has been composed of the following elements:

$\{$Synergy parameter (agglomeration)[23], expected duration time[24],
capacities[25], sectoral distribution of the employment system[26]$\}$

Finally, the outcome of the model-estimation consisted in a reference-scenario for the Austrian system of education in which, by introducing assumptions like *average* migration flows and the like, the distribution of pupils exhibited the development pattern of Table 4:

The Boudon-model starts, first, with actual data from France (Boudon 1979, p. 112) on the distribution of students in various university segments

[18] For more details, see the references quoted in Footnote 16.

[19] This factor has been specified in form of a matrix of legally allowed transition between school types, with value 1 for legally admitted transition and 0 otherwise.

[20] A second matrix on so called-*Schullaufbahnverluste* has been constructed in the following manner: Given the reference years for different school forms like five years for Upper Vocational Schools and the like, the time delay associated with each of these transitions has been calculated. Consequently, the probability for a change of the school type $j \to i$ is assumed to depend also on those expected delays.

[21] According to this factor, transitions are assumed to slow down significantly in the case of highly segregated schools, the one being predominantly male, the other mainly female.

[22] *Dualism* refers to the split between the university linked school forms and, with the exception of the Primary School, the remaining segments and has been formalized as a 1/0 matrix, too.

[23] Here, a logistically shaped diffusion process is postulated, assuming that in the case of education systems too, agglomeration effects play an essential role.

[24] To put it very briefly, those schools which offer, on the average, a long potential time span within the education system are considered to be more attractive than other forms, where, again on the average, a comparatively short time span within the education system is to be expected.

[25] This factor is related to the supply side in the education system and assumes that capacity problems and bottlenecks within a certain school type contribute significantly to the attractivity of the school type.

[26] The operationalization of this factor used the distribution of past and present shares of the primary, the secondary and the tertiary sector as an appropriate explanatory variable.

TABLE 4. Changes in the Stock of Pupils in the Austrian school-system 1990–2005

REFERENCE PATTERN					
SCHOOL COMPOSITION IN THE YEAR 1990 (IN %)			SCHOOL COMPOSITION IN THE YEAR 2005 (IN %)		
PS:	33.31	IV_1: 2.03	PS: 32.91 (-0.4)	IV_1: 1.65	(-0.38)
SP:	.72	IV_2: .91	SP: .45 (-0.27)	IV_2: .44	(-0.47)
GS:	21.94	IV_3: .92	GS: 16.93 (-5.01)	IV_3: .76	(-0.16)
AS_1:	8.61	IV_4: .40	$AS_1$1: 13.61 (+5.00)	IV_4: .25	(-0.15)
SS:	.90	UV_1: 4.29	SS: .75 (-0.15)	UV_1: 6.81	(+2.52)
DV_1:	3.34	UV_2: .29	DV_1: 2.05 (-1.29)	UV_2: .17	(-0.12)
DV_2:	.29	UV_3: 3.26	DV_2: .12 (-0.17)	UV_3: 7.98	(+4.72)
DV_3:	7.85	UV_4: .84	DV_3: 5.84 (-2.01)	UV_4: 1.64	(+0.80)
DV_4:	2.22	AS_2: 5.76	DV_4: 1.38 (-0.84)	AS_2: 4.49	(-1.27)
PTS:	2.11		PTS: 1.76 (-0.35)		

SIMULATIONS			
ENTRANCE-VARIATION (SMALL INCREASES IN MIGRATIONS) (%-CHANGES TO REFERENCE SCENARIO)		ENTRANCE-VARIATION (SMALL INCREASES IN BIRTH RATES) (%-CHANGES TO REFERENCE SCENARIO)	
PS: +1.75	IV_1: +2.81	PS: +7.16	IV_1: +2.37
SP: +1.81	IV_2: +3.11	SP: +0.00	IV_2: +2.36
GS: +4.76	IV_3: +2.37	GS: +3.49	IV_3: +2.94
AS_1: +0.00	IV_4: +2.98	AS_1: +4.85	IV_4: +2.95
SS: +3.57	UV_1: +0.58	SS: +0.67	UV_1: +1.53
DV_1: +1.97	UV_2: +1.44	DV_1: +1.78	UV_2: +1.75
DV_2: +2.63	UV_3: +1.04	DV_2: +2.18	UV_3: +0.87
DV_3: +2.22	UV_4: +0.46	DV_3: +1.21	UV_4: +2.72
DV_4: +1.91	AS_2: +0.25	DV_4: +1.82	AS_2: +2.73
PTS: +4.46		PTS: +3.23	

The abbreviations stand for: PS (Primary School), SP (Special Primary School), GS (General Secondary School), AS_1 (Academic Secondary School I, SS (Special Secondary School), DV_{1-4} (Dual Vocational School), PTS (Polytechnical School), IV_{1-4} (Intermediate Vocational School), UV_{1-4} (Upper Vocational School), AS_2 (Academic Secondary School II). Furthermore. the index number 1 designates *multiple* school forms whereas the indices 2 to 4 stand for *singular* schools (2: agriculturally related, 3: industrially linked, 4: service-oriented) and the IUT which would indicate a Field-I approach, too. The decisive step towards a genuine Field III-enterprise is made, then, with the introduction of relevant modeling-assumptions and with the subsequent elaboration of an explanatory$_p$ framework. First, assumptions are postulated with respect to the potential benefits of universities for a cohort of twenty *identical* students in a typical possible worlds-manner: 6 students are assumed to earn a benefit of 2 money units, 8 students a benefit of one money unit and six will reap no benefits at all, whereas the benefits for the IUT are postulated to be one money unit for the entire group of 20. Due to this distribution of potential benefits, the explanatory scheme highlights a significant aggregation problem, since the decision-configuration of a single person out of the cohort of 20 takes the following form:

TABLE 5. Utility Matrix for a Single Student in the Cohort of 20

Strategies for the student$_i$	Number of students who, aside from student$_i$, opt for the university										
	0	1	2	...	7	...	10	....	15	...	19
IUT	1	1	1	...	1	...	1	...	1	...	1
University	2	2	2	...	1.86	...	1.55	...	1.25	...	1

The values in the utility matrix are to be understood as expected utilities. Thus, the value 1,86 is the result of a configuration in which a student has the chance of 6/7 to earn 2 money units and 1/7 to obtain only one money unit. Consequently, $(6/7) \times 2 + (1/7) \times 1 = 1,86$ money units. The other values in the utility matrix are computed in the same manner. (See also Boudon 1979, p. 134)

It is interesting to note that Boudon makes the limitations of his explanatory scheme[27] as well as the Field-III interpretation of his results[28] *abundantly* clear. He concedes that the modeling framework is a typical possible world-approach only and, more importantly, he confines the results to

---

[27]Boudon, or to be more precise: Boudon's translator, wrote the following remarks *highly* characteristic for a *possible* worlds-approach:

Dieses Modell stellt selbstverständlich eine Idealisierung dar: Es trifft selbstverständlich nicht zu, daß jeder Student sich für genau so gut wie alle anderen Studenten hält; es trifft nicht zu, daß die Studenten vollständig über die Spielregeln informiert sind. (Boudon 1979, p. 135)

[28]With respect to the interpretation of the results, Boudon remains consistently within the confines of a possible worlds-language game:

Möglicherweise ist die Ursache für das Scheitern der Kurzstudienlehrgänge eher in Paradoxien der Aggregation individueller Entscheidungen zu suchen als in kulturalistischen Erklärungen (Entwertung technischer Bildung usw.). (Boudon 1979, p. 135)

It should not be considered as a mere coincidence however, that Boudon and so many other writers have followed *consistently* along the basic distinctions which have been introduced in Table 2 and along the different rule systems and criteria which are presented in Table 6.

the mere *possibility* that problems of aggregation rather than socio-cultural factors lie at the heart of the low acceptance of IUT in France. The decisive point at this stage lies in the fact that Boudon's modeling$_P$ account should be and must be considered as *successful* and *important* simply because it fulfills criteria like *the reduction of complexity*, i.e. the reduction of a highly complex socio-economic transition process to one control-variable only or the *counter-intuitive* nature of the results since traditional theories of school transitions (Arrow 1973; Bowman 1981; Psacharopoulos 1987) have focussed on many different areas *except* for an *aggregation* problem of individual choices.

With these two examples it should become easier to follow the distinctions introduced via Table 6 where the main *differences* in the scientific practices within Field I-modeling work and Field III-activities are summarized.

Phrased in a *very* analogical manner, the rule systems for Field I-work and for Field III-studies differ like the *standard* recipes for conducting empirical social research (Bortz 1984) from Paul K. Feyerabend's *Anything goes*-rule (1978, 1985).

2.2. MAIN DIFFERENCES IN THE EVALUATION CRITERIA

But the rule systems for Field I-research and Field III-practices do not constitute the sole *principium divisionis*, the evaluation criteria[29] for these two types of scientific language games differ radically too. Introducing, in the spirit of historical reachability, the famous Hempel-Oppenheim articles on explanation (Hempel 1942; Hempel and Oppenheim 1948), the main evaluation criterion for Field I-investigations can be stated in a straightforward manner for it lies in the successful proliferation of explanations, predictions and retrodictions where "successful" should refer to the *simultaneous* fulfillment of the four explanatory requirements put forward by Hempel and Oppenheim, especially the *truth condition*.[30] Field III-evaluations, however, focus on explanations, too, but here a significant elimination must take place since the truth condition$_A$ has to be abandoned. Additionally, an important set of material criteria has to be introduced in which

---

[29] *Evaluation criteria* as a pragmatic concept refers to the set of all those attributes by which the *actual* moves and practices within the context of a language game and its rule system can be evaluated. Thus, turning to Wittgenstein's famous characterization of *games* and *family resemblances* (Wittgenstein 1971a, PI p. 66f.), one may, in similar fashion, introduce the term of evaluation criteria in an analogical manner, stressing the *variability* of evaluation criteria across different types of games.

[30] It goes (almost) without saying that, due to the overall pragmatic context, the truth condition has to be interpreted in a pragmatic manner too, i.e. with reference to *rule-systems*. Consequently, truth means the satisfaction or the following of *specific* rule systems.

TABLE 6. Rule Systems for Two Types of Modeling Practices

FIELD I: MODELING OPERATIONS FOR ACTUAL WORLD		FIELD III: MODELING OPERATIONS FOR POSSIBLE WORLDS	
$Set_{1,A}$:	*Strong* Data Requirements with respect to — — Observability — Testability — Measurements — Tests	$Set_{1,P}$:	No Data Requirements *Freedom to choose*
$Set_{2,A}$:	*Strong* Requirements for *Non*-Theoretic Modeling Components$_A$ with respect to — — Observability — Testability — Measurements — Tests	$Set_{2,P}$:	No Requirements for *Non*-Theoretic Modeling Components$_P$- *Freedom to choose*
$Set_{3,A}$:	Requirements for Theoretical Modeling Components$_A$ Consistent Links to the Body of Non-Theoretical Elements$_A$ and Data$_A$	$Set_{3,P}$:	Requirements for Theoretical Modeling Components$_P$ Consistent Links to the Body of Non-Theoretical Elements$_P$ and Data$_P$

As a point of illustration, the data work necessary for the Field I-education model was in the range of twelve months and required an intensive cooperation between statistical offices and the IAS-team. For the Boudon-model however, none of the data-collection and data-adaptation procedures was required. Here, an inspired afternoon is sufficient to find an appropriate distribution of utility values and the like.

evaluation rules like policy relevance, simulations, generalizability, medium or high degrees of formalization and the like occupy a *predominant* role. Moreover, Field III-approaches yield useful and indispensable exploratory services which, almost by necessity, cannot be rendered by Field I-models. Why? Simply because Field I-applications, due to the very strong requirements on measurements and observations, require a well *established* research domain (*normal science*[31]) whereas Field III-work, due to its empirically

[31] The distinction between normal and exploratory science does not follow exactly the distinctions by Thomas S. Kuhn in which a basic differentiation has been introduced between *normal* and *extra-ordinary* science. In the present context, *normal* science refers to long established fields of research whereas *exploratory* science is linked to newly emerging

unrestricted character, can be performed in newly developed or poorly understood areas as well.

Thus, model accounts, ranking high on the criteria sets$_P$, will be qualified as *successful, innovative,* or *imaginative irrespective* of their extremely
low correspondence to *actual* data, observations or measurements. A rational choice account on the *optimal* course of a psychic depression or on an
optimal allocation of manic-depressive phases during one's life cycle will
and must become, if successful along the evaluation set$_P$, a potentially *interesting* focus of discussion. Once again, a counter-intuitive example might
be helpful to demonstrate the universal applicability of the *rational* mode
of attribution for *evidently* inappropriate configurations. The example itself
is devoted to the topic of *altruism* and is couched, moreover, in a *typical*
Field III-manner:

First, a possible world situation is built up, by postulating two persons, father and son, by focussing on intergenerational transfers and,
moreover, by assuming a single consumption commodity, —

> Let C denote the sole consumption good, corn, the total amount of
> which we fix arbitrarily. Suppose all this corn is initially under the
> father's control. The level of corn consumed by an individual affects
> his pleasure. We refer to this direct pleasure as 'felicity' and describe
> it by functions .... (Stark 1995, pp. 15f.)

Second, altruism is introduced via a simple scalar β_i —

> the weight that one places on the utility of the other relative to one's
> own felicity. (Stark 1995, p. 16)

Third, the optimal consumption level both for the consumption of the
father as well as that of the son is calculated, yielding a solution where
—

> the father's optimal allocation is such that he wishes to consume a
> larger proportion of corn than his son wishes him to consume (Stark
> 1995, p. 18)

Finally, several implications are drawn from this analytic solution, by
pointing out to the beneficiary role of altruism for producing mutually
agreeable transfers, by showing that altruism, while reducing conflict,
does not eliminate it, or by demonstrating that a rise in altruism *may*
result in a worse outcome for both parties involved. (Stark 1995, p. 19ff.)

More generally, *any* phenomenon of the social worlds, from marriages
and family life (Becker 1981) to seemingly remote areas like health and sick-
ness, feelings, including, *pace* Elster (1992), those of shame, can *legitimately*
become an object of a possible worlds-investigation within the context of

---

and, in most instances, *weakly* understood topics.

TABLE 7. Evaluation Criteria for Two Distinct Scientific Modeling Practices

FIELD I: EVALUATION OF MODEL SOLUTIONS FOR THE ACTUAL WORLD	FIELD III: EVALUATION OF MODEL SOLUTIONS FOR POSSIBLE WORLDS
$Set_{1,A}$: FORMAL CRITERIA$_A$ (Prediction, Retrodiction)	$Set_{1,P}$: FORMAL CRITERIA$_P$ (Forward- and Retrosimulation$_P$)
Main Explanatory Requirements: — Explanans$_A$ — Explanandum$_A$ — Consistency — Truth$_A$	Main Explanatory Requirements: — Explanans$_P$ — Explanandum$_P$ — Consistency
$Set_{2,A}$: MATERIAL CRITERIA (Improved Understanding of the Socio-economic Universes) CONTEXT OF NORMAL SCIENCE Simulations$_A{}^a$ Innovation$_A$ Policy Relevance$_A{}^c$ Generalizability$_A{}^e$ Surprising$_A$ Results	$Set_{2,P}$: MATERIAL CRITERIA (Improved Understanding of the Socio-economic Universes) CONTEXT OF NORMAL SCIENCE Simulations$_P{}^b$ Innovation$_P$ Policy Relevance$_P{}^d$ Generalizability$_P{}^f$ Counter-intuitive$_P$ Results (Paradoxes, Critical Thresholds, etc.)
Mastering of Complexity Above Average Degree of Formalization^g	Reduction of Complexity *High* Degree of Formalization^h Testability in Principle ("Anschlußfähigkeit")
Congruency$_A{}^i$	Congruency$_P{}^j$
CONTEXT OF EXPLORATION	CONTEXT OF EXPLORATION Qualitative Understanding of Poorly Understood Socio-Economic Domains Explorations with Respect to Theory Choice

Footnotes on next page.

TABLE 7.   Evaluation Criteria for Two Distinct Scientific Modeling Practices
(Footnotes)

---

[a] It should be emphasized that the variations, introduced in a Field I-simulation, must follow the normal Field I-criteria.

[b] Likewise, simulations$_P$ are typically un-restricted and can be put forward with the fullest freedom of choice.

[c] Within Field I, a *direct* connection can be established between model$_A$-work, simulations$_A$ and the corresponding policy advices$_A$.

[d] For Field III, policy advice can be given too, which must be couched, however, in an indirect manner, i.e. as possible worlds-advice on potential problem areas, etc. For a typical advice of this type, see the discussion of Robert Axelrod's book in the following chapter.

[e] Generalizability$_A$ means the extension from one area of application$_A$ to other empirical$_A$ domains.

[f] Contrary to the Field I-meaning, generalizability$_P$ consists mainly in the removal of specific restrictions by their non-restricted counterparts (e.g., from 2 person games to n person games, from perfect information to imperfect information, etc.)

[g] For Field III, the linking of modeling work with recent *advances* in mathematics or statistics must be seen as an essential and *very* important evaluation criterion.

[h] Within Field I however, it is regarded, normally, as a special *success* if *medium* types of formalization can be utilized for empirical$_A$ data.

[i] Congruency$_A$ is a typical residual category, comprising *additional* evaluation standards for Field I. In order to present an operational definition, congruency$_A$ can be equated with those elements which one may find in David F. Hendry (1987), but *not* in Table refkhmtab7.

[j] Similarly, congruency$_P$ can be considered as the set of those evaluation criteria for Field III-investigations which are explicitly mentioned in Hendry (1987), but not in Table 7.

---

the *rational* stance. Likewise, everyday calculations, management behaviour or the activities of scientists can be subject to a *normative stance* (Elster 1990), an *emotional stance* (Elster 1992), or an *ethnographic stance* [32].

On the other hand, complex Field I-models focussing, for example, on *group* behavior and a *habitual* stance, can be extended to areas *outside* the existing set of paradigmatic applications$_A$: from migration (Weidlich and Haag 1988), sectoral employment (Müller and Haag 1994) into more remote areas like politics (Erdmann 1986; Hofinger and Grützmann 1994), innovation and diffusion dynamics (Weidlich and Haag 1983; Mensch *et al.* 1991; Zhang 1991), cognitive dynamics (Müller 1992, 1993). Moreover, successful Field I-models can and must be, if possible, combined to comparatively larger scale models, leading thus to a steady increase for predictions$_A$ and simulations$_A$.

---

[32] On concepts like *intentional stance, rational stance, emotional stance*, etc. see especially chapter 4 of the present article.

3. Epistemic Cultures in the Social Sciences

Having reached, by now, a *successful* dissolution of the modeling dilemma by focussing on two highly differentiated sets of rules and evaluations, the next step will carry the preceding result one step further by generalizing it even further. For this purpose, the introduction of a new concept will become necessary, namely that of an *epistemic culture*, which has been proposed first by Karin Knorr-Cetina in order to describe *clusters* of essential scientific activities and practices (Knorr-Cetina 1992b).[33] An epistemic culture, then, consists of a set of basic scientific research operations which must be considered as highly typical with respect to the fabrication of a particular knowledge domain and, more general, to the orchestration and organization of particular areas of investigation.

> The notion (of *epistemic culture*, K.H.M.) foregrounds not only the difference between the notion of a laboratory and the concept of experiment traditionally defined, it also foregrounds the disunity of the sciences in regard to the meaning of the empirical, the enactment of object relations, the construction and fashioning of the social within science.(Knorr-Cetina 1992b, p. 3)

More specifically, epistemic cultures typify special and unique relations to their cognitive as well as to their outside environment. In this sense, Knorr-Cetina distinguishes, within the context of the *natural* sciences, two dominant epistemic cultures, one located in the area of high-energy physics and described in terms of a closed and self-contained type of knowledge production, the other situated in areas like molecular biology, where open processes of *trial and error* dominate the research scenes.Once again, the principal ways to arrive at a meaningful and empirically grounded notion of epistemic cultures can proceed along two different lines:

> The first route, undertaken for example by Knorr-Cetina, makes use of an intensive, methodologically sophisticated inspection[34] of the *actual* practices of day-to-day activities in the context of laboratories —
>
> > The laboratory allowed ... to consider the technical activities of science within the wider context of equipment and symbolic practices within which they are embedded ... In other words, the study of laboratories has brought to the fore the full spectrum of activities involved in the production of knowledge. (Knorr-Cetina 1992b, p. 3)

[33]It must be noted that the introduction of the recent concept of epistemic culture does not violate the principle of historical reachability since, once again, Otto Neurath had already in the 1930's and early 1940's proposed a *Gelehrtenbehavioristik* which comes already close to the notion of epistemic cultures (Neurath 1981).

[34]In Knorr's case, ethnomethodology, phenomenology of the Merleau-Ponty style as well as a strong reference to Michel Foucault's work form the theoretical background for the interpretative work on *science observed*.

Via a rich theoretical background, Knorr-Cetina achieves a successful *ordering* and *clustering* of the mass of empirical protocols, and arrives, in the end, at an *empirically* well-founded separation between main types or clusters of scientific research activities.

The second way however, does proceed in a *morphological* and *a priori* manner, separating between principal components of scientific operations and arriving, then, at a variety of *possible* configurations and recombinations. These *possible* types of scientific operations possess the status of *potentially* fruitful and enlightening conjectures which need further empirical collaborations to determine their *actual* distribution or their historical development path.

Since the subsequent remarks will follow along a possible worlds-strategy, the evaluation criteria for the usefulness of the identification of epistemic cultures in the social sciences are clear. The distinctions must clearly exhibit a heuristic *surplus-value*, i.e. they must rank high with respect to evaluation criteria like *reduction of complexity, counter-intuitive insights, innovative content, "Anschlußfähigkeit", qualitative understanding, exploratory theory-assessments* and the like ...

Again, like in the preceding chapters on two types of modeling, an *exemplar*-based approach will be chosen in which two sets of products, highly typical for two different *epistemic* social science cultures, will be introduced and discussed.

The first example comes from a classic on classics, namely from John Madge's book on the origins of scientific sociology (Madge 1962). In this volume, Madge gives a detailed account of path-breaking studies from Emile Durkheim's analysis of suicide and anomy (Durkheim 1983) to Leon Festinger's and Harold H. Kelley's investigations on attitude changes through social contacts (Festinger and Kelley 1951). Between these two poles, one finds, within Madge's volume, a large number of sociological projects, ranging from William I. Thomas'and F. Znaniecki's books on *The Polish Peasant in Europe and America* (Thomas and Znaniecki (1918-1920)), from the Chicago School and, more specifically, from H.W. Zorbough's *The Gold Coast and the Slum* (1929) to the studies on the rise and the roots of fascism by the Frankfurt Institute (Ackermann *et al.* 1976) and to Robert F. Bales' account on *Interaction Process Analysis* (Bales 1951) In a final chapter on *The Lessons*, Madge turns to the question of similarities and characteristic traits for major works in sociology and arrives, after reviewing the methods and techniques used in a comparative manner, at the following conclusion.

> Against this background the characteristic novelty of the works introduced in this book becomes apparent. Each item of research is unremittingly empirical and, like the products of the social-survey

movement, almost all the studies are immediately concerned with the alleviation of current social problems. At the same time, almost without exception, each study makes a concurrent contribution to verifiable knowledge. (Madge 1962; p. 537)

The social sciences, by employing documents, interviews and observations from and within their socio-economic environments, have apparently developed, over the last hundred years, a distinctive epistemic culture which is, following Madge's observations, both *empirically$_A$* and *policy$_A$* oriented.[35]

The second example has more recent origins and is the widely cited and acclaimed book by Robert Axelrod on the evolution of cooperation (Axelrod 1984). The research operations necessary for this type of study differ *very* significantly from the first series of classical sociological investigations. First, the *direct* connex with social problems and their reduction is not given any more although Axelrod discusses at length an *actual world*-example, namely *the live-and-let-live system in trench warfare in World War I*(Axelrod 1984, p. 73ff.). Second, the data base is not established via questionnaires, observations or interviews from the manifold of life-worlds but within the social laboratory itself.[36] More precisely, a typical possible worlds-configuration is set up for iterated games of the PD (prisoner's dilemma) which offers, for each player, two strategies, namely cooperation (C) and defection (D) and which exhibits a payoff-distribution of 5 (T for temptation to defect), 3 (R for reward for cooperation), 1 (P for punishment for mutual defection) and 0 (S for sucker's payoff). Third, the prevalent mode of investigation lies in a *computer* simulation, i.e. in simulations which take place *within* the context of the laboratory itself. In the case of Axelrod, the design of a computer tournament has been chosen in which basically all strategies admitted were allowed to play against each other. Fourth, and *very* importantly, the *results* of the tournament, viz. a rank-ordering of strategies, were subject to a second order analysis with respect to common attributes which could be identified for successful tournament strategies, a validation move which might be qualified as *typical* for possible worlds-modeling. Fifth, a set of rules has been for-

---

[35] On the importance of the policy-side and on the close linkages between state apparatus and social sciences, see esp. Wagner (1990).

[36] One might object immediately that Axelrod's computer tournament required a substantial *postal* input from outside, namely all the strategy suggestions for the Prisoner's Dilemma. But this specific detail is irrelevant to the present questions of data sources since a morphological analysis within the laboratory itself would have yielded a similar combination of strategies. One might argue, however, that Axelrod's computer tournament has to say very much on the sociology of science as well because it reveals the preferences of scientists engaged in game theory, evolutionary biology and the like ...

mulated which one should follow in socio-economic configurations of the PD-type with pay-offs sufficiently *similar* to a 5,3,1,0-distribution —

> Don't be envious (110ff.), Don't be the first to defect (113ff.), Reciprocate both cooperation and defection (118ff.), Don't be too clever (120ff.)

Sixth, and finally, the fourth part of Axelrod's book is devoted, under the heading of *How to Promote Cooperation* (Axelrod 1984, p. 124) — to *reformers* and offers advice *consistent* with the design and the result of his study. Thus, Axelrod has definitely an extremely important *political* advice to offer —

> Enlarge the shadow of the future (126ff.), Change the payoffs (133f.), Teach people to care about each other (134ff.), Teach reciprocity (136ff.), Improve recognition abilities (139ff.) —,

although the results and the scientific basis for these devices have been confined, throughout the study, to the laboratory alone

From these two highly significant examples it becomes relatively easy to find an inductive generalization to the notion of epistemic cultures. In social science areas like sociology or in political science[37], the traditional and predominant mode of knowledge production has been clustered within an epistemic culture which can be described by attributes like a focus on the actual world (empirical$_A$) and openness (data generation for processes outside the social science laboratory), whereas the Axelrod book is one of the most prominent examples of an entirely different epistemic culture with opposite attributes: with a focus on possible worlds (empirical$_P$), and closure (data generation within the social science laboratory itself). Via a two case-inductivism, one is led, therefore, to the following table which, once again, highlights significant differences between two epistemic cultures within the social sciences.[38]

Seen from a history of science-perspective, it would become an extremely challenging, albeit rewarding research task to describe the evolution of

---

[37] It must be pointed out that economics, since the introduction of the utility synthesis during the 1870's, may be considered as the leading discipline *away* from the confines and principles of the traditional epistemic culture. Moreover, it would be an *extremely* enlightening research objective to frame the debates, starting around the so-called *Methodenstreit* until to the present time, within the conceptual apparatus of *dominant* epistemic cultures, *assimilation* attempts to it and a new epistemic culture in the making ...

[38] It should be added, once again, that the distinctions between the two epistemic cultures just introduced should be considered by no means as exhaustive. On the contrary, especially within the social sciences, it should become very useful to distinguish between a variety of epistemic cultures, some of them being confined to special territories, some of them to a *specific* cluster of disciplines, etc. With respect to *modeling* operations however, the separation between just *two* epistemic cultures should be considered both as necessary *and* sufficient.

TABLE 8. Two Epistemic Cultures in the Social Sciences I — Principal Components

| | | Dimension$_2$ | |
		open	*closed*
	actual world	Epistemic Culture I	[Intermediaries I]
Dimension$_1$			
	possible worlds	[Intermediaries II]	Epistemic Culture II

theoretical economics or econometrics as a consecutive path from the upper left area, via the intermediary station II, to the lower right side of Table 8. Moreover, from the preceding table, a conjecture can be put forward that only over the last decades, starting with the diffusion of new information processing technologies, a separation into at least two major epistemic cultures has been established which can be summarized via Table 9.

Thus, the hypothetical identification of *at least* two *epistemic* cultures *within* the social sciences must be considered as the most *general* dissolution of the modeling dilemma. Due to this *overall* separation, one is invited, therefore, to distinguish clearly between *two* types of modeling and simulation within the contemporary social sciences, one labelled modeling$_A$ and adhering to the traditional, *open*-oriented epistemic culture$_A$, the other one characterized as *modeling$_P$* and being the core element of a *new* type of epistemic culture$_P$, namely that of a *closed* laboratory setting$_P$. Table 10 summarizes the separation of two modeling approaches in which many essential ingredients like explanations, simulations, predictions, retrodiction, control, etc. should be used with appropriate subindices — A or P — in order to facilitate the identification of the scientific language games pursued.

A final argument can be put forward which should demonstrate, once again, the *heuristic* value of the basic differentiations introduced here. Within the social sciences, the next decades will experience a tremendous surge in modeling approaches, which, to varying degrees, belong to the set of *complex* models. (See, aside from the literature quoted in Footnote 6, also Campbell *et al.* 1992; Crilly *et al.* 1991; Kaye 1993.) This modeling revolution which has been well under way for the last decade already will make it almost imperative to separate clearly between model$_P$-approaches and their model$_A$-counterparts for one finds many model families which can be utilized within a modeling$_A$-environment *as well* as in a modeling$_P$ context. Moreover, the emergence of *virtual* laboratories —

A virtual laboratory can be divided into two components: the application programs, data files and textual descriptions that describe the experiments; and the system support that provides the framework on

TABLE 9. Two Epistemic Cultures in the Social Sciences II — Main Characteristics

EPISTEMIC CULTURE I:	Open Systems for the Actual World	EPISTEMIC CULTURE II	Closed Systems for Possible Worlds:
Main Focus with Respect to			
DATA	Observable , Non-Experimental Processes Data with Strong Quality restrictions (Reliability, Validity)		Laboratory Data, Experimental Data, Artificial Data
THEORY AND MODELING	Grounded Theory; Middle-range Theories Grand Foundations ^{a}		Highly Formalized Theories
GENRES ^{b}	*Thick* Descriptions: ^{c} Explanation sketches$_A$ Model-Explanations$_A$ Model-Simulations$_A$		*Thin* Formalizations: Model-Explanations$_P$ Model-Simulations$_P$ Explanation sketches$_P$

^{a} Especially within sociology, a remarkable feature lies in an *excessive* pre-occupation of reconfiguring the *entire* discipline *anew*. From Talcott Parsons *Theory of Social Action* (1961) onward, one finds, over the last decades, a wide range of *very* comprehensive *foundation* attempts, culminating in voluminous works by Habermas (1981); Luhmann (1984); Münch (1988, 1993), etc.

^{b} Concluding, in an analogical spirit, the horizontal tour on games, rules, evaluation criteria and *genres*, one must, finally, stress the variability of the material requirements and the outfits for different types of games.

^{c} The concept of a *thick description* refers to the phenomenon that within the traditional epistemic culture a *very* large number of books has been produced which cover hundreds and hundreds of pages and which, by and large, qualify as *descriptive* frameworks. (See e.g. the extremely stimulating discussion on the impossibility of transforming a thick and voluminous description, namely S:P. Huntington's book on modernization (Huntington 1968), into a thin and consistent model in the article by Krause in this volume).

which these domain-dependent experiments are built (Prusinkiewicz and Lindenmayer 1990, p. 194) —

will exert a considerable impact on the rapid development of the epistemic culture$_P$ and, consequently, on the course of the co-evolution *between* the new epistemic culture and its long established counterpart.[39]

[39] Again, a final remark becomes appropriate to point to the fact of a *multitude* of epistemic cultures within the contemporary social sciences, especially at the *regional* or the *gender* level. With respect to *modeling* activities however, the separation between *two* dominant cultures should be, so the argument, *both necessary – and* sufficient.

TABLE 10. Two Modeling Approaches in the Social Sciences

		Dimension$_2$	
		open	*closed*
	actual world	Epistemic Culture$_A$	
Dimension$_1$		{Modeling$_A$}	
	possible worlds		Epistemic Culture$_P$
			{Modeling$_P$}

With these distinctions it should be easier to accept that modeling in the social sciences adheres to *different* epistemic cultures and that, moreover, modeling can be performed in two highly differentiated manners: either as a Field I-practice which sticks to the main principles of the traditional epistemic culture or as a Field III-operation which follows a *different* set of rules and, even more importantly, of evaluation criteria.[40] With the present article it is hoped for that the basic distinctions between modeling$_P$ and modeling$_A$, between modeling rules$_A$ and modeling rules$_P$, and, finally, between evaluation criteria$_A$ and evaluation criteria$_P$ can be considered both as necessary and sufficient for a *satisfying* dissolution of the modeling dilemma and, consequently, for an *adequate* understanding and interpretation of the role and function of rationality assumptions or other simplification devices within the social sciences.

4. Schumpeter Revisited

So far, the article has concentrated on the *overall* solution for the modeling dilemma and, more specifically, to the four Schumpeterian problem areas of rationality only. The *general* dissolution, however, has the distinctive advantage of being *applicable, mutatis mutandis*, to the specific four areas-set, introduced in Table 2, too.

In order to stick to the format of a single article and not to the reference frame of a booklet on social science methodology, only a *bare* methodology

---

[40]It would be an extremely interesting task to apply the notion of *epistemic regimes* (Bjørn Wittrock) to the cognitive as well as to the socio-economic transformations of the period from 1970 to the present time in order to arrive at a comprehensive general framework in which the emergence of a *new* epistemic culture, based on *closed* laboratory research and *possible* worlds-accounts, could be related to massive changes in the technological settings of laboratories across the social sciences as well as to a growing disenchantment between the predominant discourse-coalitions of the late 1960's and early 1970's. (For the process of consecutive epistemic regimes from 1800 to 1970, see esp. Wittrock 1993) Likewise, the distinctions between two different modes of knowledge production, namely mode I and mode II (Gibbons *et al.* 1994; Nowotny 1995), might serve as an interesting *overall* reference frame, too.

sketch can be presented to arrive at similar *satisfying* answers for the *specific* four problem fields of the original Schumpeter article. The most important move, which, however, would have been a highly unlikely one for the *realist* Schumpeter (Schumpeter 1989), consists in a *radically constructivist* turn by pointing to the unavoidable and necessary role of the observer.[41] Following more recent advances initiated by Heinz von Foerster (1985), Ernst von Glasersfeld (1986), Humberto R. Maturana (1985), Jean Piaget (1973, 1983, 1985, 1992) or Francisco J. Varela (1989), the role of the observer must be transformed from an unavoidable nuisance backstage to that of a central main stage-actor. According to this turn of *Bringing the observer back in* (Watzlawick and Krieg 1991), any account of the socio-economic worlds, by necessity, is bound to be observer-*dependent*. In this spirit, rational decision theory (see, e.g. Bacharach and Hurley 1994) becomes an *external* mode of attribution no less than the intentional stance (Daniel C. Dennett) which, following Dennett (1987), can be used as an *attribution* strategy for the whole animate and, at least *partly*, for the *inanimate* world, too:

> Do people actually use this strategy? Yes, all the time. There may some-day be other strategies for attributing belief and desire and for predict-ing behavior, but this is the only one we all know now. And when does it work? It works with people almost all the time ... The strategy al-so works on most other mammals most of the time. For instance, you can use it to design better traps to catch those mammals, by reason-ing about what the creature knows or believes about various things, what it prefers, what it wants to avoid. The strategy works on birds, and on fish, and on reptiles, and on insects and spiders, and even on such lowly and unenterprising creatures as clams ... It also works on some artifacts ... The strategy even works for plants ... It even works for such inanimate and apparently undesigned phenomena as lightning. An electrician once explained to me how he worked out how to pro-tect my underground water pump from lightning damage: lightning, he said, always wants to find the best way to ground, but sometimes it gets tricked into taking second-best paths. (Dennett 1987, p 21f.)

Thus, a *contemporary* partitioning[42] of the four areas for rationality problems in particular and modeling components in general can, then, be

---

[41] In Chapter 4, the condition of historical reachability is to be abandoned. This require-ment has been employed to demonstrate that all *essential* cognitive ingredients would had been available in the 1940's or 1950's already to solve the modeling dilemma. The sub-sequent remarks in the chapter on *Schumpeter Revisited* are directed to a *contemporary* solution of the rationality problems in four specific areas of investigation.

[42] It must be noted that the requirement of historical reachability, after having *dissolved* the modeling dilemma in a *satisfying* manner, is dropped for the subsequent modeling sketch which, therefore, will be concerned mainly with *contemporary* advances within the social sciences and related cognitive domains.

TABLE 11. A contemporary morphological space for the four rationality problems

| | | Dimension$_2$ | |
		observer/ observer (self-referential)	observer/ observed (referential)
	internal (neural states)	Area I	Area II
Dimension$_1$			
	external (modes of attribution for actions or practices)	Area III	Area IV

put forward in the subsequent fashion, where the Schumpeterian internal/external dimension is operationally defined in the following manner: Descriptions of *neural* states or *emergent* descriptions of such states must be located on the *internal* pole and modes of *behavior*-attribution — the *intentional, emotional, ethnographic* stances ... — on the *external* side. Likewise, relations between an observer and her or his environment can be qualified as *self*-referential whereas relations between a scientific observer and her or his fields *outside* one's own environment is to be categorized as *referential*.

Not surprisingly, a wide array of research topics and of different modeling approaches, depending on their Field I- or on their Field III-localization, can be used. Starting with the problem classes for which all of the examples have been chosen[43], the following specific additions and qualifications become necessary:

Area IV: Classically, Area IV belongs to the *core*-domain of micro-sociology, micro-economics, micro-political science and the like. The preceding discussion should have pointed out the *heuristic* value and the *usefulness* for distinguishing clearly between Field I-approaches and Field III-analyses especially in this domain. Moreover, models of the *homo oeconomicus* variety are most efficiently utilized within Field III-work where they fulfil the necessary rule-requirements and evaluation criteria in a surprisingly *successful* manner.[44] Only *very* rarely however,

---

[43] It should be added that the dissolution of the modeling dilemma is not confined to the examples from micro-economics or micro-sociology alone, but can be, in principle, reformulated for any level in the social science-complex: from its macro-macro-levels well to the level of pico-economics, femto-sociology and the like ...

[44] Thus, the verve of contemporary criticism against *homo oeconomicus* modeling, like the one from Etzioni (1994) or, to a lesser extent, from Friedberg (1995), must be seen as *valid* with respect to its comparative *disadvantages* in Field I and as *highly invalid* with

models, focussing on rational decision procedures or on game-theoretic frameworks, should and can be employed in Field I-investigations.

Area II: For these fields, a cognitive revolution on cognitivism or, to use a book-title by Michael Gazzaniga, on the *cognitive neuro-sciences* has occurred (Gazzaniga 1995) whereby the internal neuro-states of individuals become subject to a rapidly increasing variety either of Field I- or to Field III-explorations. Taking a separation from the domain of Artificial Life (Langton 1989; Langton *et al.* 1992; Langton 1994; Varela and Bourgine 1992) it becomes useful to separate the research-areas, aside from the traditional micro-level, into two *additional* domains. On the one hand, a *basic* or *femto*-area can be distinguished, where

> tasks like ... wandering, avoiding obstacles, wall following, looking for a certain object, delivering some object, cleaning the floor, following someone, etc. (Brooks 1992, p. 436)[45]

become the central focus of investigation. On the other hand, a *meso* or *pico*-domain (see, e.g. Ainslie 1992) can be identified whose main research interest lies in the problem of *task-integration.* Thus, the following three areas of investigations can be put forward for this relatively recent area of *neural*-based social sciences:

> *First, tasks* (Rodney A. Brooks), *drafts* (Daniel C. Dennett) or *agents* (Marvin Minsky), at the femto-, pico-, and micro-level, especially for senso-motoric processes like walking, seeing, grasping, hearing, and the like ...

> *Second, recursive couplings,* especially, but, *pace* Luhmann (1988, 1990), not exclusively communications, at the femto-, pico-, and micro-level ...

> Third, *disturbances,* again on all three levels of investigation, and their corresponding neural settings. (For more details, see Müller 1991)

In Area II, *rationality* will play a major role, especially in the form of principles of *maximization, minimization* or *optimality* which become essential for the explanatory frameworks for the interaction patterns of neural groups, *both* for Field I- and for Field III-studies. (See, esp. Edelman 1989, 1992, 1993)

respect to the *successes* within Field III.

[45] It must be stressed, however, that in the articles by Rodney Brooks one finds a separation into *micro*-domains, *macro*-areas and the *ecological* level which, following the terminology introduced here, corresponds to the *femto-, pico-* and *micro*-distinctions. The new terminology has been chosen for two reasons. *First,* Ainslie's book on *pico-economics* (1992) has become a well-known social science standard for problems of conflicts *within* persons. *Second,* the micro- macro-dualism is very much entrenched in the current social science literature and has acquired, by now, relatively clear boundaries.

The remaining two domains belong to the discipline of science of science since they focus on the actions and practices of scientists (Area III) or on the neuro-settings of scientists (Area I). More precisely, the following research topics can be identified which, once again, can be dealt within a modeling$_A$ or modeling$_P$ approach:

Area III: Once again, the same model-types which are at the disposal for Area IV, can be applied to the scientific realm, too. Moreover, extremely interesting moves toward *self*-referentiality could be accomplished since the modeling frameworks$_{A,P}$ can be used for purposes of *self*-explanations, too. Take, for example, a Field I-model which is couched in a master-equation scheme and which is specified to capture the cognitive dynamics within a scientific domain or discipline (Müller and Haag 1994), then, via a *consistent* process of self-specification and data-collection, an explanatory scheme for the most likely diffusion trajectories of this type of model can be built up. Like in the case of Area IV, *homo oeconomicus*-variations will play an essential role in Field III-explorations on the economics of research, on the detection of possible *critical* limits in diffusion processes, on paradoxical results with respect to innovation patterns and the like ....

Area I: Finally, establishing links between advances in the cognitive sciences with an in-depth analysis of the neural ensembles of scientists might turn out, in the future, as an extremely valuable research road. Again, typical Field III-approaches like the utilization of genetic algorithms (Holland *et al.* 1986; Holland 1992; Koza 1992) or, alternatively, PET-tomography and similar experimental routes along Field I should become a frequently used research tool for neuro-based investigations of *science in action* (Bruno Latour). Once again, rationality will occupy a central stage, especially since principles of *maximization, minimization* or *optimality* will become essential explanatory frameworks for the interaction patterns of neural groups, *both* for Field I- and for Field III-investigations. (See, e.g. Hanson and Olson 1990; Koch and Segev 1991, or Wise 1987)

Thus, the unfinished article by Joseph A. Schumpeter could be completed *both* in a general version, utilizing only cognitive components available in the 1940's already, and in a special manner, particularly adapted to the methodological and theoretical environment of the 1990's.

5. Concluding Outlooks

Seven years ago, two nobel prize winners in economics, Lawrence Klein and Maurice Allais, delivered, independently from each other, lectures on the future course of economics at the Institute for Advanced Studies. At first sight, the upshots of their lectures seemed utterly contradictory: Klein advocated more interdisciplinary cooperation between economists, psychologists, sociologists, historians and the like, more powerful computers, more data ... Allais, on the other hand, favored more powerful models with few, but highly significant components, more axiomatizations, more thinking ... The present article offers, among many other features, a convenient way to reconcile *both* perspectives by attributing Klein's vision as a rational *widening* strategy for scientific endeavors within Field I, whereas Allais' plea for *deepening* should be considered as a highly relevant agenda for Field III-operations. In the end, *both* roads to the socio-economic universes around us, be they on the micro — or on the macro-scale, are in almost desperate need for rapid reconfigurations, adaptations, and modernizations.

6. Acknowledgement

Thanks go to Jim Coleman, Jon Elster, Heinz von Foerster, David Hausmann, Rainer Hegselmann, Karin Knorr-Cetina and Jonathan Turner as well as to Raimund Alt and Christoph Hofinger (both at the IAS), who, at various stages, have kept the basic ideas in this article alive.

References

Ackermann, N.W., et al. (1976) *Der autoritäre Charakter. Studien über Autorität und Vorurteil*, 2 Bde. Amsterdam: de Munter.

Ainslie, G. (1992) *Picoeconomics. The Strategic Interaction of Successive Motivational States within the Person.* Cambridge: Cambridge University Press.

Anderson, P.W., Arrow, K.J., and Pines D. (1988) (eds.) *The Economy as an Evolving Complex System.* Redwood City: Addison-Wesley.

Arrow, K.J. (1973) Higher Education as a Filter, In *Journal of Public Economy* 2, pp. 193–216.

Axelrod, R. (1984) *The Evolution of Cooperation.* New York: Basic Books.

Bacharach, M., and Hurley, S. (1994) (eds.) *Foundations of Decision Theory. Issues and Advances.* Oxford: Blackwell Publishers.

Bales, R.F. (1951) *Interaction Process Analysis.* Reading: Addison-Wesley.

Becker, G. (1981) *A Treatise on the Family.* Cambridge: Harvard University Press.

Berger, J., Zelditch, M., and Anderson, B. (1989)(eds.) *Sociological Theories in Progress. New Formulations.* Newbury Park: Sage Publications.

Bewley, T.F. (1987) (ed.) *Advances in Econometrics.* Fifth World Congress. Cambridge: Cambridge University Press.

Bortz, J. (1984) *Lehrbuch der empirischen Forschung für Sozialwissenschaftler.* Berlin: Springer.

Boudon, R. (1979) *Widersprüche sozialen Handelns.* Darmstadt: Luchterhand.

Boudon, R. (1980) *Die Logik des gesellschaftlichen Handelns. Eine Einführung in die soziologische Denk- und Arbeitsweise.* Neuwied: Luchterhand.

Bowman, M.J. (1981) *Educational Choice and Labor Markets in Japan.* Chicago: The University of Chicago Press.

Brooks, R.A. (1992) Artificial Life and Real Robots In Varela, F.J., and Bourgine, P. (1992) (eds.) *Toward a Practice of Autonomous systems. Proceedings of the First European Conference on Artificial Life.* Cambridge: The MIT Press, pp. 3–20.

Bryant, C.G.A., and Jary, D. (1991) (eds.) *Giddens· Theory of Structuration: A Critical Appraisal*. London: Routledge.

Bunge, M. (1977) *Treatise on Basic Philosophy. Ontology I: The Furniture of the World*. Dordrecht: Reidel.

Bunge, M. (1979) *Treatise on Basic Philosophy. Ontology II: A World of Systems*. Dordrecht: Reidel.

Bunge, M. (1983a) *Treatise on Basic Philosophy. Epistemology and Methodology I: Exploring the World*. Dordrecht: Reidel.

Bunge, M. (1983b) *Treatise on Basic Philosophy. Epistemology and Methodology II: Understanding the World*. Dordrecht: Reidel.

Calvo, G.A., and Vegh, C.A. (1993), Exchange-rate Based Stabilisation under Imperfect Credibility, In Frisch, H., and Wörgötter, A. (1993) (eds.) *Open-Economy Macroeconomics. Proceedings of a Conference Held in Vienna by the International Economic Association*. London: Pinter, pp. 3–28.

Campbell, D.K., Ecke, R.E., and Hyman J.M. (1992) (eds.), *Nonlinear Science: The Next Decade*. Cambridge: The MIT Press.

Carnap, R. (1975) *Introduction to Semantics & Formalization of Logic. Two Volumes in One*. Cambridge: Harvard University Press.

Casdagli, M., and Eubank, S. (1992) (eds.) *Nonlinear Modeling and Forecasting*. Redwood City: Addison-Wesley.

Casti, J.L. (1989) *Alternate Realities. Mathematical Models of Nature and Man*. New York: John Wiley & Sons.

Casti, J.L. (1992) *Reality Rules*, 2 , New York: John Wiley.

Casti, J.L. (1994) *Complexification. Explaining a Paradoxical World through the Science of Surprise*. New York: Harper Collins.

Charniak, E., and McDermott, D. (1985) *Introduction to Artificial Intelligence*. Reading: Addison-Wesley.

Claassen, E.M. (1993) Real Shocks and the Real Exchange Rate, In Frisch, H., and Wörgötter, A. (1993) (eds.) *Open-Economy Macroeconomics. Proceedings of a Conference Held in Vienna by the International Economic Association*. London: Pinter, pp. 137–152.

Coleman, J.S. (1990) *Foundations of Social Theory*. Cambridge: Harvard University Press.

Cook, K.S., and Levi, M. (1990) (eds.) *The Limits of Rationality*. Chicago: The University of Chicago Press.

Creath, R. (1990) (ed.) *Dear Carnap, Dear Van. The Quine-Carnap Correspondence and Related Work*. Berkeley: University of California Press.

Crilly, A.J., Earnshaw, R.A., and Jones, H. (1991) (eds.) *Fractals and Chaos*. New York: Springer.

Dennett, D. (1987) *The Intentional Stance*. Cambridge: The MIT Press.

Dennett D.C. (1991) *Consciousness Explained*. Cambridge: The MIT Press.

Dubach, P. (1977), Morphologie als kreative Methode in der Langfristplanung, In G. Bruckmann(1977) (ed.), *Langfristige Prognosen. Möglichkeiten und Methoden der Langfristprognostik komplexer Systeme*. Würzburg: Physica, pp. 112–125.

Durkheim, E. (1983) *Der Selbstmord*. Frankfurt: Suhrkamp.

Edelman, G.M. (1989) *The Remembered Present. A Biological Theory of Consciousness*. New York: Basic Books.

Edelman, G.M. (1992) *Bright Air, Brilliant Fire. On the Matter of the Mind*. New York: Basic Books.

Edelman, G.M. (1993) *Unser Gehirn — ein dynamisches System. Die Theorie des neuronalen Darwinismus und die biologischen Grundlagen der Wahrnehmung*. München: Piper.

Elster, J. (1985) *Sour Grapes. Studies in the Subversion of Rationality*. Cambridge: Cambridge University Press.

Elster, J. (1986) (ed.) *Rational Choice*. Oxford:BasilBlackwell.

Elster, J. (1989) *Nuts and Bolts for the Social Sciences.* Cambridge: Cambridge University Press.

Elster, J. (1990) When Rationality Fails, In Cook, K.S., and Levi, M. (1990) (eds.) *The Limits of Rationality.* Chicago: The University of Chicago Press, pp. 19–51.

Elster, J. (1992) Egonomics. The Study of Conflict within Persons. (mimeo)

Erdmann, G. (1986) Dynamische Wahlprognosen mittels eines synergetischen Verhaltensmodells, dargestellt am Beispiel der Niedersachsenwahl 1986. Zürich: Workingpaper Nr.86/66 of the Institute for Economic Research at the ETH Zürich.

Esser, H., and Troitzsch, K.G. (1991) (eds.) *Modellierung sozialer Prozesse. Neuere Ansätze und Überlegungen zur soziologischen Theoriebildung.* Bonn, Informationszentrum Sozialwissenschaften.

Etzioni, A. (1994) *Jenseits des Egoismus-Prinzips. Ein neues Bild von Wirtschaft, Politik und Gesellschaft.* Stuttgart: Schäffer-Poeschel.

Fararo, T.J. (1989) *The Meaning of General Theoretical Sociology. Tradition and Formalization.* Cambridge: Cambridge University Press.

Festinger, L., and Kelley, H.H. (1951) *Changing Attitudes through Social Contact.* Ann Arbor: University of Michigan Press.

Feyerabend, P.K. (1978) *Der wissenschaftstheoretische Realismus und die Autorität der Wissenschaften.* Braunschweig: Friedr.Vieweg & Sohn.

Feyerabend, P.K. (1985) *Philosophical Papers* 2, Cambridge: Cambridge University Press.

Foerster, H.v. (1985) *Sicht und Einsicht. Versuche zu einer operativen Erkenntnistheorie.* Braunschweig: Friedr.Vieweg & Sohn.

Friedberg, E. (1995) *Ordnung und Macht. Dynamiken organisierten Handelns.* Frankfurt: Campus.

Frisch, H., and Wörgötter, A. (1993) (eds.) *Open-Economy Macroeconomics. Proceedings of a Conference Held in Vienna by the International Economic Association.* London: Pinter.

Gardner, M. (1981) *Science: Good, Bad and Bogus.* New York: Avons Books.

Gavin, M. (1993) Devaluation, the Terms of Trade and Investment in a Keynesian Economy In Frisch, H., and Wörgötter, A. (1993) (eds.) *Open-Economy Macroeconomics. Proceedings of a Conference Held in Vienna by the International Economic Association.* London: Pinter, pp.29–45.

Gazzaniga, M.S. (1995) (ed.) *The Cognitive Neurosciences.* Cambridge: The MIT Press.

Gibbons, M., et al. (1994) *The New Production of Knowledge. The Dynamics of Science and Research in Contemporary Societies.* London: Sage Publications.

Gilbert, G.N. (1981) *Modelling Society. An Introduction to Loglinear Analysis for Social Researchers.* London: George Allen & Unwin.

Glasersfeld, E.v. (1987) *Wissen, Sprache und Wirklichkeit. Arbeiten zum radikalen Konstruktivismus.* Braunschweig: Friedr.Vieweg & Sohn.

Goldberg, M., and Frydman, R. (1993) Theories, Consistent Expectations and Exchange Rate Dynamics, In Frisch, H., and Wörgötter, A. (1993) (eds.) *Open-Economy Macroeconomics. Proceedings of a Conference Held in Vienna by the International Economic Association.* London: Pinter, pp. 377–399.

Grauwe, P. de, and Dewachter, H. (1993) A Chaotic Monetary Model of the Exchange Rate, In Frisch, H., and Wörgötter, A. (1993) (eds.) *Open-Economy Macroeconomics. Proceedings of a Conference Held in Vienna by the International Economic Association.* London: Pinter, pp. 353–376.

Green, D.P., and Shapiro, I. (1994) *Pathologies of Rational Choice Theory. A Critique of Applications in Political Science.* New Haven: Yale University Press.

Haag, G. (1989) *Dynamic Decision Theory: Applications to Urban and Regional Topics.* Dordrecht: Kluwer.

Haag, G., Mueller, U., Troitzsch, K.G. (1992a) (eds.) *Economic Evolution and Demographic Change. Formal Models in Social Sciences.* Berlin: Springer.

Haag, G., and Müller, K.H. (1992b) Employment and Education as Non-Linear Network-

Populations, Part I & II, In Haag, G., Mueller, U., Troitzsch, K.G. (1992) (eds.)
    Economic Evolution and Demographic Change. Formal Models in Social Sciences.
    Berlin: Springer, pp. 349–409.
Habermas, J. (1981) *Theorie kommunikativen Handelns*, 2 , Frankfurt:Suhrkamp.
Haken, H. (1981) *Synergetik. Eine Einführung.* Berlin: Springer.
Haken, H. (1983) *Advanced Synergetics. Instability Hierarchies of self-Organizing Systems
    and Devices.* Berlin: Springer.
Hallett, A.J.H., et al. (1993) Sheet Anchors, Fixed Price Anchors and Price Stability
    in a Monetary Union In Frisch, H., and Wörgötter, A. (1993) (eds.) *Open-Economy
    Macroeconomics. Proceedings of a Conference Held in Vienna by the International
    Economic Association.* London: Pinter, pp. 46–69.
Hanson, S.J., and Olson, C.R. (1990) (eds.) Connectionist Modeling and Brain Functions:
    The Developing Interface. Cambridge: The MIT Press.
Hempel, C.G. (1942) The Function of General Laws in History, In *The Journal of Phi-
    losophy* 39, pp. 35–48.
Hempel, C.G., and Oppenheim, P.(1948) Studies in the Logic of Explanation, In *Philos-
    ophy of Science* 15, pp. 135–175.
Hendry, D.F. (1987) Econometric Methodology: a Personal Perspective, In Bewley, T.F.
    (ed.) *Advances in Econometrics. Fifth World Congress*, Cambridge: Cambridge Uni-
    versity Press, pp. 29–48.
Hof, F.X. (1993) Foreign Supply Shocks, Wage Indexation and Optimal Monetary Policy,
    In Frisch, H., and Wörgötter, A. (1993) (eds.) *Open-Economy Macroeconomics. Pro-
    ceedings of a Conference Held in Vienna by the International Economic Association.*
    London: Pinter, pp. 70–91.
Hofinger, C., and Grützmann, K. (1994) Das Politik-Modell: Attraktivitäten als De-
    terminanten von Wählerbewegungen in Österreich 1970 - 1990, In *WISDOM* 3/4,
    pp. 79–89.
Hofstadter, D.R. (1985) *Metamagical Themas. Questing for the Essence of Mind and
    Pattern.* New York: Basic Books.
Holland, J., Holyoak, K.J., Nisbett, R.E., and Thagard, P.R. (1986) *Induction. Processes
    of Inference, learning, and Discovery.* Cambridge: The MIT Press.
Holland, J.H. (1992) *Adaptation in Natural and Artificial Systems. An Introductory Anal-
    ysis with Applications to Biology, Control, and Artificial Intelligence.* Cambridge: The
    MIT Press.
Hoon, H.T., and Phelps, E.S. (1993) The Impact of Fiscal and Productivity Shocks on
    the Natural Rate of Unemployment in a Two Country World, In Frisch, H., and
    Wörgötter, A. (1993) (eds.) *Open-Economy Macroeconomics. Proceedings of a Con-
    ference Held in Vienna by the International Economic Association.* London: Pinter,
    pp. 95–118.
Hughes, G.E., and Cresswell, M.J. (1985) *An Introduction to Modal Logic..* 5th ed,
    London-New York.
Hume, D. (1989) *Ein Traktat über die menschliche Natur.* 2 Bde., Hamburg: Meiner.
Huntington, S.P. (1968) *Political Order in Changing Societies.* New Haven: Yale Univer-
    sity Press.
Jen, E. (1990) (ed.) *1989 Lectures in Complex Systems.* Redwood City: Addison-Wesley.
Kaye, B. (1993) *Chaos & Complexity. Discovering the Surprising Patterns of Science and
    Technology.* Weinheim: VCH.
Knorr-Cetina, K. (1984) *Die Fabrikation von Erkenntnis. Zur Anthropologie der Natur-
    wissenschaft.* Frankfurt:Suhrkamp.
Knorr-Cetina, K. (1992a) The Couch, the Cathedral, and the Laboratory: On the Re-
    lationship between Experiment and Laboratory in Science, In Pickering, A. (1992a)
    (ed.) *Science as Practice and Culture.* Chicago: The University of Chicago Press,
    pp. 113–138.
Knorr-Cetina, K. (1992b) *Epistemic Cultures. How Scientists Make Sense.* (mimeo)

Koch, C., and Segev, I. (1991) (eds.) *Methods in Neural Modeling. From Synapses to Networks.* Cambridge:The MIT Press.

Koza, J.R. (1992) *Genetic Programming. On the Programming of Computers by Means of Natural selection.* Cambridge: The MIT Press.

Kreuzenkamp, H.A., and Magnus, J.R. (1995) (eds.) *The Significance of Testing in Econometrics.* Annals of Econometrics (Journal of Econometrics 67). Amsterdam:Elsevier.

Langton, C.G. (1989) (ed.) *Artificial Life.* Redwood: Addison-Wesley.

Langton, C.G., Taylor, C., Farmer, J.D., and Rasmussen, S. (1992) (eds.) *Artificial Life II.* Redwood: Addison-Wesley.

Langton, C.G. (1994)(ed.) *Artificial Life III.* Reading: Addison-Wesley.

Latour, B. (1987) *Science in Action. How to Follow Scientists and Engineers through Society.* Cambridge: Harvard University Press.

Lave, J. (1988) *Cognition in Practice. Mind, Mathematics and Culture in Everyday Life.* Cambridge: Cambridge University Press.

Luhmann, N. (1984) *Soziale Systeme. Grundrißeiner allgemeinen Theorie.* Frankfurt: Suhrkamp.

Luhmann, N. (1988) *Die Wirtschaft der Gesellschaft.* Frankfurt: Suhrkamp.

Luhmann, N. (1990) *Die Wissenschaft der Gesellschaft.* Frankfurt: Suhrkamp.

Lyotard, J.F. (1982) Das postmoderne Wissen. Ein Bericht, In *Theatro Machinarum* 3/4.

Madge, J. (1962), *The Origins of Scientific Sociology.* New York: The Free Press.

Maturana, H.R. (1985) *Erkennen: Die Organisation und Verkörperung von Wirklichkeit. Ausgewählte Arbeiten zur biologischen Epistemologie.* 2. Aufl., Braunschweig: Friedr.Vieweg & Sohn.

Mensch, G., Weidlich, W., and Haag, G. (1991) The Schumpeter Clock. A Micro-Macro-Model of Economic Change, Including Innovation, Strategic Investment, Dynamic Competition and Short and Long Swings in Industrial transformation — Applied to United States and German Data, In OECD (1991) (ed.), *Technology and Productivity. The Challenge for Economic Policy.* Paris:OECD.

Michalski, R.S., Carbonell, J.G., and Mitchell, T.M. (1986) (eds.) *Machine Learning. An Artificial Intelligence Approach,* 2, Los Altos: Morgan Kaufmann.

Müller, K.H. (1991) Elementare Gründe und Grundelemente für eine konstruktivistische Handlungstheorie, In Watzlawick, P., and Krieg, P. (1991) (eds.) *Das Auge des Betrachters. Festschrift für Heinz von Förster.* München: Piper, pp.191–245.

Müller, K.H. (1992a) *Expeditionen in die Wissenschaftsdynamik.* Vienna: IAS Reserach Papers.

Müller, K.H., and Lassnigg, L. (1992b) (eds.) *Langfristige Szenarienanalyse des österreichischen Bildungssystems.* Vienna: IAS Project Papers.

Müller, K.H. et al.(1993) *Wissenschaft als System. Von der Black Box zum Black Hole und retour?* Vienna: IAS Research Papers.

Müller, K.H., and Haag, G. (1994) (eds.) *Komplexe Modelle in den Sozialwissenschaften.* Special edition of WISDOM 3/4.

Müller, K.H. (1995a) *Sozialstrukturanalysen und komplexe Modelle. Vermittlungsdesigns.* Berlin: Sigma.

Müller, K.H. (1995b) *Neuschätzung des Bildungsmodells.* Vienna: IAS Project Papers.

Müller, K.H. (1995c) *Epistemic Cultures in the Social Sciences. The Modeling Dilemma — Dissolved.* Vienna: IAS.

Münch, R. (1988) *Theorie des Handelns. Zur Rekonstruktion der Beiträge von Talcott Parsons, Emile Durkheim und Max Weber.* Frankfurt: Suhrkamp.

Münch, R. (1993) *Die Kultur der Moderne* 2 Bde., Frankfurt:Suhrkamp.

Neurath, O. (1971) Foundations of the Social Sciences, In O. Neurath, R. Carnap, C. Morris (1971) (eds.), *Foundations of the Unity of Science. Toweard an International Encyclopedia of Unified Science,* 2, Chicago : The University of Chicago Press, pp. 1–51.

Neurath, O. (1981) *Gesammelte philosophische und methodologische Schriften,* 2 volumes, Wien: Hölder-Pichler-Tempski.

Nielsen, S.B., and Sorensen, P.B. (1993) Capital Taxation, Housing Investment and Wealth Accumulation in a Small Open Economy, In Frisch, H., and Wörgötter, A. (1993) (eds.) *Open-Economy Macroeconomics. Proceedings of a Conference Held in Vienna by the International Economic Association*. London: Pinter, pp. 203—223.

Nowotny, H. (1995) *The Dynamics of Innovation. The Multiplicity of the New*. Budapest: Collegium Budapest

Parsons, T. (1961) *The Structure of Social Action. A Study in Social Theory with Special Reference to a Group of Recent European Writers*. New York: Free Press.

Piaget, J. (1973) *Einführung in die genetische Erkenntnistheorie*. Frankfurt: Suhrkamp.

Piaget, J. (1983) *Biologie und Erkenntnis. Über die Beziehungen zwischen organischen Regulationen und kognitiven Prozessen*. Frankfurt: Fischer.

Piaget, J. (1985) *Weisheit und Illusionen der Philosophie*. Frankfurt:Suhrkamp.

Piaget, J. (1992) *Das Erwachen der Intelligenz beim Kinde. Mit einer Einführung von Hans Aebli*. München: dtv.

Pickering, A. (1992a) (ed.) *Science as Practice and Culture*. Chicago: The University of Chicago Press.

Pickering, A. (1992b) From Science as Knowledge to Science as Practice, In Pickering, A. (1992a) (ed.) *Science as Practice and Culture*. Chicago: The University of Chicago Press, pp. 1–26.

Prusinkiewicz, P., and Lindenmayer, A. (1990) *The Algorithmic Beauty of Plants*. New York:Springer.

Psacharopoulos, G. (1987) (ed.) *Economics of Education. Research and Studies*. Oxford:Pergamon Press.

Quine, W.V.O. (1961) *From a Logical Point of View. Logico-Philosophical Essays*. 2. Aufl., New York:Harper & Row.

Rapoport, A. (1989) *Decision Theory and Decision Behaviour. Normative and Descriptive Approaches*. Dordrecht:Kluwer.

Ritzer, G. (1990) (ed.) *Frontiers of Social Theory. The New Syntheses*. New York: Columbia University Press.

Schelling, T. (1978) *Micromotives and Macrobehavior*. New York: W.W. Norton & Company.

Schumpeter, J.A. (1915) *Vergangenheit und Zukunft der Sozialwissenschaften*. München:Duncker & Humblot.

Schumpeter, J.A. (1975) *Kapitalismus, Sozialismus und Demokratie. Einleitung von Edgar Salin*. 4. Aufl., München: Francke Verlag.

Schumpeter, J.A. (1989) *Essays on Entrepreneurs, Innovations, Business Cycles, and the Evolution of Capitalism, ed. by R.V. Clemence. With a New Introduction by R. Swedberg*. New Brunswick: Transaction Publishers.

Sims, C.A. (1987) Making Economics Credible, In Bewley, T.F. (1987) (ed.) *Advances in Econometrics. Fifth World Congress*. Cambridge: Cambridge University Press, pp. 49–60.

Stark, O. (1995) *Altruism and Beyond. An Economic Analysis of Transfers and Exchanges within Families and Groups*. Cambridge:Cambridge University Press.

Stein, D.L. (1989) (ed.) *Lectures in the Sciences of Complexity*. Redwood City: Addison-Wesley.

Streminger, G. (1994) *David Hume. Sein Leben und sein Werk*. Paderborn: Ferdinand Schöningh.

Swedberg, R. (1994) *Joseph A. Schumpeter. Eine Biographie*. Stuttgart: Klett Cotta.

Thomas, W.I., and Znaniecki, F. (1918-1920), *The Polish Peasant in Europe and America* 5, Boston: Gorham Press.

Troitzsch, K.G. (1990) *Modellbildung und Simulation in den Sozialwissenschaften*. Opladen: Westdeutscher Verlag.

Varela, F.J., and Bourgine, P. (1992) (eds.) *Toward a Practice of Autonomous systems. Proceedings of the First European Conference on Artificial Life*. Cambridge: The MIT

Press.

Vijayraghavan, V. (1993), Bandwagon Effects, or Rational Expectations in the Exchange Rate?, InFrisch, H., and Wörgötter, A. (1993) (eds.) *Open-Economy Macroeconomics. Proceedings of a Conference Held in Vienna by the International Economic Association.* London: Pinter, pp. 400–407.

Wagner, P. (1990) *Sozialwisenschaften und Staat. Frankreich, Italien, Deutschland 1870 - 1980.* Frankfurt: Campus.

Waters, M. (1994) *Modern Sociological Theory.* London: Sage.

Watzlawick, P., and Krieg, P. (1991) (eds.) *Das Auge des Betrachters. Festschrift für Heinz von Förster.* München: Piper.

Weidlich, W., and Haag, G. (1983) *Concepts and Models of a Quantitative Sociology. The Dynamics of Interacting Populations.* Berlin: Springer.

Weidlich, W., and Haag, G. (1988) (eds.) *Interregional Migration. Dynamic Theory and Comparative Analysis.* Berlin: Springer.

Weintraub, S. (1977)(ed.) *Modern Economic Thought.* University of Pennsylvania Press.

Wise, S.P. (1987) (ed.) *Higher Brain Functions. Recent Explorations of the Brain·s Emergent Properties.* New York: John Wiley & Sons.

Wittgenstein, L. (1971a) *Philosophische Untersuchungen.* Frankfurt:Suhrkamp.

Wittgenstein, L. (1971b) *Über Gewißheit.* Frankfurt:Suhrkamp.

Wittrock, B. (1993), The Modern University: the Three Transformations, In S. Rothblatt, B. Wittenrock (1993) (eds.), *The European and American University since 1800. Historical and Sociological Essays.* Cambridge: Cambridge University Press, pp. 303–362.

Worrell, D. (1993) Economic Adjustment and Growth in Small Developing Countries, In Frisch, H., and Wörgötter, A. (1993) (eds.) *Open-Economy Macroeconomics. Proceedings of a Conference Held in Vienna by the International Economic Association.* London: Pinter, pp. 153–163.

Zapf, W. (1994) *Modernisierung, Wohlfahrtsentwicklung und Transformation. Soziologische Aufsätze 1987–1994.* Berlin: Sigma.

Zhang, W.B. (1991) *Synergetic Economics. Time and Change in Non-Linear Economics.* Berlin: Springer.

IMPOSSIBLE MODELS

ULRICH KRAUSE
Department of Mathematics
University of Bremen
Bremen, Germany

1. The Real World and the World of Models

It is a widespread belief among all kinds of people, including scientists that reality, though difficult to capture, can nevertheless finally be approximated by making models better and better. The meaningfulness of the wordings "approximated" and "better" in this context, however, may be doubted. To be sure, there is a real world supplying us with a great variety of inputs. Also, we are more or less able to act within the real world, applying thereby guidelines stemming from such diverse sources as everyday experience and full fledged science. Living in the real world and making models of parts of it, however, does not mean that reality is approximated by the models used. The latter would require the notion of a distance between reality and models which makes little sense, I think. (For an opposite view see [Niiniluoto, 1987]. For drawing my attention to Niiniluoto's work as well as for other valuable remarks I would like to thank Stephan Hartmann.)

Instead of trying to compare the two different worlds of reality and models, it seems more sensible to compare models with each other. It is a little bit like the situation after a traffic accident: Something happened, no doubt, but it is over now; policemen try to find out what happened by looking for witnesses and by comparing their testimonies. Whether these evidences contradict or are in line with each other, if so, to what extent — those considerations play an important role in "capturing reality". Similarly, it is meaningful and worthwhile to explore the relationships *within* the complex world of models. Contradiction and consistency are particularly important aspects of models. Other interesting relationships among models are inclusion and exclusion, homomorphy and isomorphy. Furthermore, one may look for integrating non–contradicting models into a supermodel,

R. Hegselmann et al. (eds.),
Modelling and Simulation in the Social Sciences from the Philosophy of Science Point of View, 65–75.
© 1996 *Kluwer Academic Publishers. Printed in the Netherlands.*

or one wants to find the common kernel of different models, or one wants to judge competing models — and so on. These are only catchwords indicating a program which differs very much from all approaches trying to deal directly with an imaginary relationship between reality and models, as it is the case with the approximation approach.

In the present paper I only want to hint at such a program and I do not want to pursue it further here. Instead, I would like to address an interesting case of internal consistency or meaningfulness for a special model. As a minimal requirement for a good model one may consider its consistency on logical grounds. (Ironically, models used by working scientists do not always satisfy this seemingly innocent requirement.) Another minimal requirement is that a model should be compatible with conditions coming with the kind of measurement for the variables of the model. The latter is particularly relevant in the social sciences where the attribution of numerical values to variables is often not an obvious task. Therefore, a scale has to be specified which usually is an ordinal scale or an interval scale or a ratio scale; this means that meaningful relationships among variables should be invariant for strictly increasing transformations or strictly increasing affine transformations or similarity transformations of the scale. Those and other invariance conditions are the object of measurement theory, dimensional analysis and the theory of meaningfulness. (Cf. Krantz *et al.*, 1971, Narens, 1981). Sometimes it is a difficult task to make a model compatible with the appropriate invariance conditions; sometimes it turns out after rigorous analysis that a model satisfying the wanted conditions of invariance is impossible. A famous example is Arrow's Impossibility Theorem which, roughly speaking, states that a set of seemingly very natural properties for a social welfare function is not compatible with individual preferences as ordinal variables. Another case of an impossible model will be treated in detail in the present paper.

Section 2 takes up a controversy between the two social scientists S.P. Huntington and H.A. Simon on the one side and the two mathematicians N. Koblitz and S. Lang on the other side. This controversy, which in parts was quite heated, is about a simple model proposed by Huntington and refined thereafter by Simon to meet sharp criticism raised by Koblitz and Lang against Huntington. Though the criticism by Koblitz and Lang is partly justified, it does, however, not address the heart of the problem, namely: Is the model as refined by Simon possible at all? Section 3 of the paper analyzes this question in detail and concludes that even when further refined the wanted model is impossible. This conclusion rests on a Theorem characterizing invariant orderings on the two–dimensional space. A proof for this Theorem is presented in Section 4. The final section 5 is devoted to some speculative remarks concerning the critical capacity of mathematics. Mathematics is mainly esteemed as a universal slave in building up models; it is argued, however, that in the future mathematics will have its role also in the X–ray screening of models for meaningfulness.

2.   The Huntington/Simon–Koblitz/Lang Controversy

In an influential book the political scientist S.P. Huntington establishes the following relationships ([Huntington, 1968, p. 55]):

1. $\dfrac{\text{Social mobilization}}{\text{Economic development}} = \text{Social frustration}$

2. $\dfrac{\text{Social frustration}}{\text{Mobility opportunities}} = \text{Political participation}$

3. $\dfrac{\text{Political participation}}{\text{Political institutionalization}} = \text{Political instability}$

These equations were criticized by the mathematician N. Koblitz [Koblitz, 1981] who doubted that any of the terms involved could be measured by numerical values and who asked for the units of measurement. Koblitz also accused Huntingon's use of the equations as producing mystification, intimidation, an impression of precision and profundity.

This criticism was taken up by the mathematician S. Lang who used it in preventing Huntington from being elected as a member of the National Academy of Science in the U.S. [Lang, 1987]. (Lang had published before a documentation [Lang, 1981] about fallacies in the quantitative methodology of the sociologist S.M. Lipset.) Against the attack by Koblitz and Lang, Huntington's approach was defended by the economist H.A. Simon. According to Simon, Huntington "makes use of ordinal variables in a way that has excited the criticism and ridicule of some professional mathematicians although it appears to fall entirely within the framework of the analysis set forth here." [Simon, 1987, p. 5] On Huntington's equations and the variables involved, Simon comments as follows: "In the accompanying text, the measurement of these variables is discussed in such a way as to make it obvious that they are defined only up to monotonic transformations. Hence it certainly makes no sense, as the critics point out, to speak of ratios of these variables. But a sympathetic reading of Huntiongton's text reveals that, at most, he is guilty of the sin of using unorthodox notation. For in his reasoning about the social and political processes represented by these formalisms, he makes use only of the signs of the partial derivatives of the dependent on the independent variables, the derivatives being positive for variables in the numerators of his "fractions" and negative for variables in the denominator." [Simon, 1987, pp. 5–6]. I don't want to go into the heated debate between Koblitz and Simon which developed afterwards in the pages of the Mathematical Intelligencer (see Koblitz 1988a, Koblitz 1988b,

Koblitz 1988c, Simon 1988a, Simon 1988b, Simon 1988c). Although not without interest in itself, this debate does not go into the heart of the matter, which is, as I see it, whether the "ordinal interpretation of Huntington's relations" undertaken by Simon [Simon, 1987] is possible at all.

Briefly, this interpretation by Simon is as follows: Variables which are defined only up to positive strictly monotonic transformations Simon calls *ordinal variables*. In terms of the ordinal variables M, F, D for social mobilization, social frustration and economic development respectively, Simon rewrites Huntington's first equation $F = [M] \oslash [D]$. Here the notation $z = [x] \oslash [y]$ is introduced for $z = f(x, y)$ with $\frac{\partial z}{\partial x} > 0$ and $\frac{\partial z}{\partial y} < 0; x, y, z$ are real valued variables and f is a mapping $f \colon I\!R^2 \longrightarrow I\!R$. The plausible idea in behind is to replace the rather crude model of a ratio $\frac{x}{y}$ by the more flexible model $[x] \oslash [y]$ in ordinal variables; the latter has the same qualitative features as the former in that it increases strictly with x and decreases strictly with y. The other two of Huntington's equations are interpreted in the same manner. x and y are understood to be ordinal variables; z also has to be ordinal, this because z enters one of the other equations as independent variable. (According to Simon, an ordinal variable "can be replaced by any other that does not disturb the ordering by transposing items" [Simon, 1987, p. 2].)

In the next section this ordinal interpretation of Huntington's equations by Simon will be analyzed in detail. Inspired by Simon's credo of "sympathetic reading", I will consider an interpretation which is even more in favor of Huntington by weakening Simon's assumptions including the type of ordinality. From the mathematical analysis it will follow that even this refined model is impossible.

3. Analysis of the Refined Simon–Model

Let $z = f(x, y)$ with $f \colon I\!R^2 \longrightarrow I\!R$; here $I\!R$ is the set of real numbers ordered by $\leq$ and $I\!R^2$ denotes the Cartesian product $I\!R \times I\!R$. In his interpretations Simon asssumes for f that the partial derivatives exist with $\frac{\partial f}{\partial x} > 0$ and $\frac{\partial f}{\partial y} < 0$. In a first step we weaken this assumption by assuming only that f increases strictly with x and decreases strictly with y. Call a variable u *affine ordinal* if it is defined only up to transformations $u \longmapsto au + b$ with real numbers a, b with $a > 0$. If u is ordinal in the sense of Simon it is a fortiori affine ordinal, but not vice versa. To formulate our weakened type of ordinality let $\tau \colon I\!R^2 \longrightarrow I\!R^2$ be a *positive affine transformation* in two dimensions, that is

$\tau(x, y) = (ax + b, cy + d)$ with real numbers a, b, c, d and $a > 0, c > 0$.

The assumption that $z = f(x, y)$ models a relationship between affine ordinal variables then requires the following *invariance condition*:

$$f(x,y) \leq f(u,v) \text{ if and only if } f(\tau(x,y)) \leq f(\tau(u,v)) \qquad (\star)$$

for every positive affine transformation τ and for all real numbers x, y, u, v. (Note that for real numbers $z, \bar{z}$ it holds that $z \leq \bar{z}$ if and only if $az + b \leq a\bar{z} + b$ for all real numbers a, b with $a > 0$.)

It is in no way obvious how a function looks like which satisfies the invariance condition $(\star)$ and which is strictly increasing in x and strictly decreasing in y. The reader who wants to find an example of such a function will probably test various functions f which come to his mind. All these attempts must fail because it will follow from a general result below that no such function is possible. To be sure, this impossibility does not occur if only one independent variable is considered. Any strictly increasing bijection of $\mathbb{R}$, as e.g. $f(x) = x^3$, satisfies the invariance condition $(\star)$ (in one dimension). Also, the impossibility may disappear if the invariance condition $(\star)$ is not required to hold for all positive affine transformations τ. For example, if $(\star)$ is required only for transformations $\sigma(x,y) = (ax, cy)$ with $a > 0, c > 0$ then a possible relationship would be $z = f(x,y) = \frac{x}{y}(y \neq 0)$; or, if $(\star)$ is required only for transformations $\lambda(x,y) = (x + b, y + d)$ then a possible relationship would be $z = f(x,y) = e^{x-y}$.

Considering the invariance condition $(\star)$ I go one step further in generalizing Simon's ordinal interpretation. Any function $f \colon \mathbb{R}^2 \longrightarrow \mathbb{R}$ comes with an ordering $\preceq_f$ on $\mathbb{R}^2$ defined for $P, Q \in \mathbb{R}^2$ by

$$P \preceq_f Q \text{ if and only if } f(P) \leq f(Q).$$

An *ordering* on $\mathbb{R}^2$ means a binary relation $\preceq$ on $\mathbb{R}^2$ which has the following properties (P, Q, R arbitrary points in $\mathbb{R}^2$):

- Reflexivity, i.e. $P \preceq P$
- Transitivity, i.e. $P \preceq Q$ and $Q \preceq R$ imply $P \preceq R$
- Completeness, i.e. $P \preceq Q$ or $Q \preceq P$.

Let $\sim$ denote *indifference*, i.e. $P \sim Q$ means that $P \preceq Q$ and $Q \preceq P$. Obviously, the relation $\preceq_f$ defined by f has the three above properties; $P \sim_f Q$ means $f(P) = f(Q)$. (This shows that indifference classes of $\preceq$ may contain many points; for this reason an ordering is sometimes called a quasi–ordering.)

Though every function $f \colon \mathbb{R}^2 \longrightarrow \mathbb{R}$ induces an ordering $\preceq_f$ on $\mathbb{R}^2$, the reverse is not true, that is, there are oderings on $\mathbb{R}^2$ which cannot be represented by a function f as $\preceq_f$. Thus, our model becomes more general by considering instead of $f \colon \mathbb{R}^2 \longrightarrow \mathbb{R}$ any ordering $\preceq$ on $\mathbb{R}^2$. The invariance condition $(\star)$ for an arbitrary ordering $\preceq$ is, of course,

$$P \preceq Q \text{ if and only if } \tau(P) \preceq \tau(Q) \qquad (\star\star)$$

for all positive affine transformations τ and all points $P, Q \in \mathbb{R}^2$.

An ordering satisfying condition $(\star\star)$ will be referred to as an *invariant ordering*.

The weakening of the invariance condition from $(\star)$ to $(\star\star)$ also meets in part Koblitz' doubt concerning the measurement by numerical values. By "sympathetic reading" one might say that Huntington thinks of variables more in a qualitative than in a quantitative sense, e.g. if he observes a "rise" in political instability given by a sharp "increase" in political participation. (It is true, however, as pointed out by Koblitz, that Huntington implies also much stronger properties when he speaks of "levels", "ratios" and correlations of "0.50".) Generally speaking, a more qualitative model is obtained by employing an ordering $\preceq$ instead of a numerical function f and by employing invariance condition $(\star\star)$ instead of $(\star)$. In any case, we can check the possibility of Simon's interpretation by looking for functions $f\colon I\!\!R^2 \longrightarrow I\!\!R$ which induce an invariant ordering $\preceq_f$ and which are increasing in the first component and decreasing in the second component. The crucial step now is the following result which describes completely all invariant orderings on $I\!\!R^2$.

Theorem. There exist precisely three possibilities for an invariant ordering on $I\!\!R^2$, namely:

 i) The ordering is *trivial*, i.e. there holds indifference between all points of $I\!\!R^2$.

 ii) The ordering is *degenerated*, i.e. the ordering is solely determined by the ordering in only one of the components.

 iii) The ordering is *essentially lexicographic*, i.e. it is one of the two standard lexicographic orderings on $I\!\!R^2$ or a variant of them obtained by order reversal for components.

One verifies easily that a trivial or degenerated or essentially lexicographic ordering must be an invariant ordering. The interesting part of the Theorem is that the three cases are the only possible ones. The proof which will be given for this in the next section will provide also a geometric explanation of the cases i) – iii). (For the above Theorem see also [Krause, 1992a], where a different proof is presented; a generalization of the Theorem as well as an application to social welfare functions gives [Krause, 1995]; related results can be found in [Krause, 1987].)

Suppose now that $f\colon I\!\!R^2 \longrightarrow I\!\!R$ is a function for which $\preceq_f$ is an invariant ordering. By the Theorem, $\preceq_f$ must be of one of the types i) — iii). In case i) we must have $f(P) = f(Q)$ for all $P, Q \in I\!\!R^2$ and, hence, f must be a constant function on $I\!\!R^2$. In case ii) f can be dependent on one component only; e.g., if $\preceq_f$ is determined solely by the ordering in the first component then $f(x, y) = f(u, v)$ iff $x = u$, irrespective of y and v which means that f is solely a function of x. The case iii) is impossible for $\preceq_f$ since it is well–known that a lexicographic ordering cannot be represented by a real valued function (cf., e.g., Krantz *et al.*, 1971, p. 38–39). Thus, the only functions $f: \mathbb{R}^2 \longrightarrow \mathbb{R}$ for which $\preceq_f$ is an invariant ordering are the functions depending at most on one component. Since such a function cannot be increasing in the first component and decreasing in the second component we arrive at the conclusion that Simon's ordinal interpretation is impossible; this impossibility holds even for our weakened version of Simon's interpretation.

For the most general situation where the numerical function f is replaced by an arbitrary ordering, the Theorem may be viewed as providing three possibilities for a qualitative relationship between social mobilization (M), social frustration (F) and economic development (D). The trivial case means that nothing can be inferred from M and D for F. The degenerated case means, e.g., that an increase in M brings about an increase in F irrespective of what happens with D. The case of an essentially lexicographic ordering means that one of the variables M and F has priority over the other, the latter being not completely irrelevant but of secondary importance only. None of these possibilities, however, presents a proper relationship between three variables as imagined by Huntington or by Simon. Thus, in the most general situation considered so far, one faces de facto also an impossibility.

4. Proof of the Theorem

Let $\preceq$ be an arbitrary invariant ordering on $\mathbb{R}^2$. The proof of the Theorem proceeds in three steps.

First step. The set $C = \{P \in \mathbb{R}^2 \mid 0 \preceq P\}$ is a convex cone in $\mathbb{R}^2$. (For elementary notions related to convexity see, e.g., [Rockafellar, 1970].) Obviously, by the invariance condition $(\star\star)$, $P \in C$ and $\lambda > 0$ imply $\lambda P \in C$ (take $\tau(x, y) = (\lambda x, \lambda y)$). If $P, Q \in C$ then $P + 0 \preceq P + Q$ by $(\star\star)$ and $0 \preceq P + Q$ by using transitivity of $\preceq$. Hence C is a convex cone.

Second step. By the above the set $U = C \cap (-C)$ is a linear subspace of $\mathbb{R}^2$. There are three possibilities for the dimension of U, namely $\dim U = 2$ or $\dim U = 1$ or $\dim U = 0$. By completeness of $\preceq$, $C \cup (-C) = \mathbb{R}^2$ and it is geometrically plausible that the above three cases imply the following features of C:

case i: dim $U = 2$ case ii: dim $U = 1$ case iii: dim $U = 0$

$C = I\!R^2$ C a closed half space C a half–space without
                                                      the dotted half–line

This is plausible but needs to be proven.

case i) is obvious because $U = I\!R^2$ by dim $U = 2$ and hence $I\!R^2 = U \subseteq C$, that is $C = I\!R^2$.

In **case ii)** dim $U = 1$ and there exists $P_0 \in C$, $P_0 \notin U$. The ray $I\!R_+ P_0 = \{\lambda P_0 \mid \lambda \geq 0\}$ determined by P_0 belongs to C. Since $U \subseteq C$ and C is convex, the convex set generated by U and $I\!R_+ P_0$, which is a closed half–space H, is contained in C:

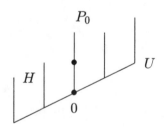

Suppose, there would be a $P_1 \in C$, $P_1 \notin H$. Then, as before, the convex set generated by $I\!R_+ P_1$ and U would belong to C; this would imply $I\!R^2 \subseteq C$, a contradiction to dim $U = 1$.

For **case iii)** let $S = \{P \in I\!R^2 \mid \| P \| = 1\}$ be the unit sphere with respect to the Euclidean norm $\| \cdot \|$. Define $I = C \cap S$.

By convexity of C, I is a connected segment on the unit sphere. From dim $U = 0$ it follows that $C \cap (-C) = \{0\}$ and, hence, I and $-I$ are disjoint segments of S. From $C \cup (-C) = I\!R^2$ it follows that $I \cup (-I) = S$. The latter together with $I \cap (-I) = \emptyset$ implies that I must be a half–sphere of S such that one endpoint P belongs to I whereas the other, Q, does not:

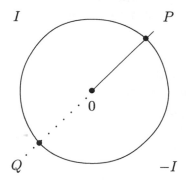

Because C consists of all rays determined by points of I, C must be a half–space which contains the ray $\mathbb{R}_+ P$ but contains no point of $\mathbb{R}_+ Q$.

Third step. Consider for case iii) a half–space C as in the second step with ray $\mathbb{R}_+ P$ belonging to C. Suppose that $P = (x, y)$ and $x \neq 0$ as well as $y \neq 0$. By condition ($\star\star$), all points (ax, by), for arbitrary $a > 0$ and $b > 0$ must belong to C:

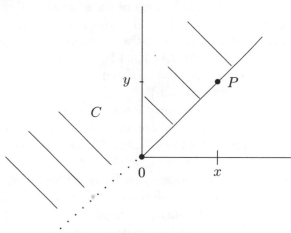

Then the whole orthant which contains P must belong to C, which is impossible. Therefore, we must have $x = 0$ or $y = 0$ which means that the ray $\mathbb{R}_+ P$ must be one of the half–axes of the Cartesian coordinate system. This shows that $\preceq$ must be essentially lexicographic. Similarly for case ii), the subspace U must be one of the axes $x = 0$ or $y = 0$. In the former case $\preceq$ depends solely on x, in the latter case solely on y. This finishes the proof of the theorem.

5. The Critical Capacity of Mathematics or You can't Always Get What you Want

On the occasion of the Huntington/Simon–Koblitz/Lang controversy we have discussed for a rather simple model only just one aspect of meaningfulness, namely that it should not matter how units and origins of measurement are choosen. This seemingly innocent condition, which was made precise by the invariance conditions $(\star)$ and $(\star\star)$ respectively, led to the result that the intended model is not possible. This is an unpleasant consequence which, however, could not be foreseen immediately. Some mathematical analysis was needed to make it clear that an invariant ordering which is neither trivial nor degenerated is possible in two variables only by giving one variable priority over the other. This little example shows that mathematics is not only useful in building up models but also in critically assessing the meaningfulness of models. Whereas the model designer wants to capture as much of "reality" as he can, the mathematician sometimes has to tell to him "You can't always get what you want." To be sure, impossibility need not always be the end of the dicsussion. If good reasons could be supplied that the variables involved do not require a full invariance with respect to changing units and origins of measurement then the road would be open to other conclusions. But this too needs a critical mathematical analysis — intuitive reasoning and the intention of producing a 'realistic' model alone are not sufficient. (As Karl H. Müller, however, pointed out to me, intuitive reasoning may serve to structure the flood of data before a proper model can be constructed.)

Having discussed the possibility for a rather simple model in three variables, one can imagine that for complex models employing many variables and equations the question of possibility becomes a much harder issue. At the same time, the question of possibility is ubiquitous if one thinks of the flood of mathematical models overflowing more and more disciplines. This development has been accelerated during the last decade by the increasing use of computers. There is, unfortunately, a run for producing models which display some mathematical apparatus and glance with up to date computer techniques without checking the internal consistency of the whole model. Checking consistency would be more serious than making pretentious claims about how 'realistic' these models are. Traditionally, mathematics serves other disciplines by what is called applied mathematics, which, however, uses mathematics often as a tool–box only. Traditionally, mathematicians don't bother whether this brings about dead ends, misuse or inconsistencies. (Koblitz and Lang, among others, being exceptions.) It appears to me that in the future there will be a role for mathematics and for mathematicians to assess critically the applications of mathematics made in various disciplines, especially in the social sciences. As Koblitz put it: "Just as the medical profession tries to combat medical quackery, members of the mathematical profession can take a stand against mathematical quackery." (Koblitz, 1981, p. 120). Points to have an eye on may be the propagandistic use of mathematics, the ad hocerism brought about by tool–box mathematics and, as explained in the present paper, the question if the model wanted is possible at all. Mathematics is needed as an X–ray apparatus to check all the influential models claimed to be 'realistic' or 'practically relevant' for the minor property of meaningfulness.

References

Huntington, S.P. (1968) *Political Order in Changing Societies*, Yale University Press, New Haven.

Koblitz, N. (1981) Mathematics as propaganda, in L.A. Steen (ed.), *Mathematics Tomorrow*, Springer Verlag, New York, pp. 111–120.

Koblitz, N. (1988a) A tale of three equations; or the emperors have no clothes, *The Mathematical Intelligencer* 10 (1), pp. 4–10.

Koblitz, N. (1988b) Reply to unclad emperors, *The Mathematical Intelligencer* 10 (1), pp. 14–16.

Koblitz, N. (1988c) Simon falls of the wall, *The Mathematical Intelligencer* 10 (2), pp. 11–12.

Krantz, D.H., Luce, R.D., Suppes, P., and Tversky, A. (1971) *Foundations of Measurement*, 1, Academic Press, New York.

Krause, U. (1987) Hierarchical structures in multicriteria decision making, in J. Jahn, W. Krabs (eds.), *Recent Advances and Historical Development of Vector Optimization*, Springer Verlag, Berlin, pp. 183–193.

Krause, U. (1992) Impossible relationships, in Bühler, W. et al. (eds.), *Operations Research Proceedings* 1990, Springer Verlag, Berlin.

Krause, U. (1995) Essentially lexicographic aggregation. *Social Choice and Welfare*, 12, pp. 233–244.

Lang, S. (1981) *The File*, Springer Verlag, New York.

Lang, S. (1987) *On a non–election to the National Academy of Sciences*, unpublished paper.

Narens, L. (1971) A general theory of ratio scalability with remarks about the measurement–theoretic concept of meaningfulness, *Theory and Decision* 13 (1), pp. 1–70.

Niiniluoto, I. (1987) *Truthlikeness*, Reidel, Dordrecht.

Rockafellar, R.T. (1970) *Convex Analysis*, Princeton University Press, Princeton.

Simon, H.A. (1987) *Some trivial but useful mathematics*, unpublished paper.

Simon, H.A. (1988a) Unclad emperors: A case of mistaken identity, *The Mathematical Intelligencer* 10 (1), pp. 11–14.

Simon, H.A. (1988b) The emperor still unclad, *The Mathematical Intelligencer* 10 (2), pp. 10–11.

Simon, H.A. (1988c) Final reply to Koblitz, *The Mathematical Intelligencer* 10 (2), p. 12.

THE WORLD AS A PROCESS

Simulations in the Natural and Social Sciences

STEPHAN HARTMANN
Fakultät für Physik
Universität Konstanz
Konstanz, Germany

Simulation techniques, especially those implemented on a computer, are frequently employed in natural as well as in social sciences with considerable success. There is mounting evidence that the "model-building era" (J. Niehans) that dominated the theoretical activities of the sciences for a long time is about to be succeeded or at least lastingly supplemented by the "simulation era". But what exactly are models? What is a simulation and what is the difference and the relation between a model and a simulation? These are some of the questions addressed in this article. I maintain that the most significant feature of a simulation is that it allows scientists to *imitate one process by another process*. "Process" here refers solely to a temporal sequence of states of a system. Given the observation that processes are dealt with by all sorts of scientists, it is apparent that simulations prove to be a powerful interdisciplinarily acknowledged tool. Accordingly, simulations are best suited to investigate the various research strategies in different sciences more carefully. To this end, I focus on the function of simulations in the research process. Finally, a somewhat detailed case-study from nuclear physics is presented which, in my view, illustrates elements of a typical simulation in physics.

1. Introduction

Major parts of current research in the natural and social sciences can no longer be imagined without simulations, especially those implemented on a computer, being a most effective methodological tool. Natural scientists simulate the formation and development of stars and whole galaxies [Kippenhahn and Weigert, 1991], the detailed dynamics of violent high-energy nuclear reactions [Blättel *et al.*, 1993] as well as aspects of the intricate process of the evolution of life [Eigen and Schuster, 1979], while their col-

R. Hegselmann et al. (eds.),
Modelling and Simulation in the Social Sciences from the Philosophy of Science Point of View, 77–100.
© 1996 *Kluwer Academic Publishers. Printed in the Netherlands.*

leagues in the social science departments simulate the outbreak of wars [Hermann and Hermann, 1972], the progression of an economy [Anderson *et al.*, 1988] and decision procedures in an organization [Simon, 1970] — to mention only a few. Recently, computer simulations even proved useful in moral philosophy[1]. In fact, there is almost no academic discipline without at least a little use for simulations.

Although simulations are therefore of considerable importance in science, philosophers of science have almost entirely ignored them. Only recently have whole articles and even conferences been devoted to their metatheoretical analysis[2]. There are, however, some interesting considerations on simulations scattered in the literature pre-dating the works cited above, such as the studies of M. Bunge (see Bunge (1969) and Bunge (1967), p. 266). Besides, a few scientists working actively with simulations made some more general remarks — mostly in the introductory part of their papers — on the scope and function of simulations [3].

But why should a philosopher of science be interested in simulations at all? I see three main reasons: Firstly, in order to formulate a "theory of science" (Giere 1988) one cannot ignore simulations, for they are too important a tool for today's scientists. Secondly, as F. Rohrlich and others [Humphreys, 1994] have emphasized, computer simulations provide "a qualitatively new and different methodology ... that ... lies somewhere intermediate between traditional theoretical physical science and its empirical methods of experimentation and observation" [Rohrlich, 1991, p. 507] for the sciences: numerical experimentation.[4] It is the task of the philosopher of science to elaborate this claim. Thirdly, since simulations are used by natural as well as by social scientists, they are best suited to compare critically the different respective methodological strategies.

It is thus worth asking how and why scientists use simulations, what *function* they have in the research process and what their respective advantages and disadvantages are. The purpose of this article is to provide a provisionary account to this task. I shall stress here that it is definitely *not* my aim to urge social scientists simply to copy the blessed methods of their colleagues in the, say, physics departments and everything will be

---

[1]See R. Hegselmann's contribution in this volume and references cited therein.

[2]See the recent work by P. Humphreys (1991; 1994; 1995) and F. Rohrlich (1991). R. Laymon (1991) discusses the role of idealizations and approximations in computer simulations.

[3]A collection of case-study simulations taken from the social and administrative sciences including some useful methodological analyses can be found in [Guetzkow *et al.*, 1972].

[4]I will come back to this in Sec. 3.

fine.[5] This is — I hold — not appropriate; first of all, every science has its own methods *sui generis*[6]. Instead, I suggest we first describe carefully, and thus take seriously, what working scientists actually do. In this context it is worth considering structures of typical problems, exemplary research strategies, the status of empirical tests etc. in the different sciences independently. Later on, ambitious methodologists might reach for "unified" models which fit many sciences. For our purpose, simulations are a good starting point to compare the research strategies of various sciences since the simulation method has become such a generally acknowledged tool.

The term "simulation" has many facets and is used with various meanings [Guetzkow *et al.*, 1972; Humphreys, 1991]. In the following I can only focus on some of them. More specifically, I shall not be concerned with what I call *experimental simulations*. In an experimental simulation a real physical (or biological) process is imitated by another real physical (or biological) process. As an example take the endeavours of M. Eigen and his collaborators to mimick processes which presumably occurred in the early stages of the evolution of life in a specially prepared reactor [Eigen and Schuster, 1979]. Theoretical simulations (that I shall call "simulations" in the remainder of the paper for the sake of brevity), on the other hand, are closely related to theoretical models, as will be pointed out in the next section. These simulations are usually carried out on a computer.

The remaining paper is organized as follows. Sec. 2 deals with the relation between theoretical models and simulations; in this context a definition of the term "simulation" is suggested and confronted with other proposals recently given in the literature. Sec. 3 focuses on the various functions of simulations in the everyday research process. I point out those functions of simulations which are more important for social scientists than for natural scientists and vice versa. In Sec. 4, a typical simulation from physics is presented and analyzed in order to identify some of the characteristic features of simulations in physics.[7] Finally, Sec. 5 summarizes our main results.

---

[5]I will leave unanswered the question as to whether scientists really need the advice of a philosopher at all, see [Hoyningen-Huene and Hirsch, 1988].

[6]It is a common complaint that current philosophy of science has been developed taking physics [Hausman, 1992], or even theoretical physics as the paradigmatic science [Cartwright *et al.*, 1995; Franklin, 1986; Galison, 1987; Hacking, 1983]. Although analyzing physics is surely worth doing, focusing too strongly on it is not without dangers. It is definitely not obvious that, for example, metatheoretical insights which were inspired by physics also make sense in other sciences. I would like to remind the reader of the various attempts to make economics fit Lakatos' methodology of scientific research programmes, see [de Marchi and Blaug, 1991]; for a critical discussion [Hausman, 1992], pp 192.

[7]I have chosen this physics case-study for two reasons. One is expertise (or better non-expertise), the other is that there are already many detailed social science simulations presented in this volume.

2. Models and Simulations

Models and simulations are apparently closely related. But what is the exact relation between both scientific tools? In order to clarify this question, I shall first make some general remarks about theoretical models[8] (Sec. 2.1). Subsequently, I propose a definition of the term "simulation" and confront it with another suggestion given in the literature (Sec. 2.2).

2.1. MODELS ...

In his "History of Economic Thought", J. Niehans claims that in 1894 the "era had began in which scientists interpreted their activity as model building."[9] In this very year, H. Hertz published his famous book "Principles of Mechanics". Therein the famous physicist writes:

> We make for ourselves internal images or symbols of the external objects, and we make them in such a way that the consequences of the images that are necessary in thought are always images of the consequences of the depicted objects that are necessary in nature ... Once we have succeeded in deriving from accumulated previous experience images with the required property, we can quickly develop from them, *as if from models*, the consequences that in the external world will occur only over an extended period or as a result of our own intervention.[10]

In fact, with the end of the late 19th century, model building began to dominate the (theoretical) activity in the field of physics: J.C. Maxwell used hydrodynamic analog models to derive the well known equations of electromagnetism and W. Thompson, later Lord Kelvin, stated that he could not understand a phenomenon until he had succeeded in constructing a (mechanical) model of the system under consideration [Thompson, 1884]. Presently, devising and exploring models forms an integral part of theoretical research [Giere, 1988]. Quite often, the term "model" is used — throughout the sciences — synonymously with "theory". By and large, scientists prefer "model", because — as I have spelled out elsewhere – it is safer to label one's thought products "models" instead of "theories" for they are most likely provisionary anyway, and the term "model" seems to acknowledge this right from the beginning [Hartmann, 1995].

---

[8]In the remainder I will drop the attribute "theoretical" for I am here only interested in this special sub-category of models. For a discussion of other types of models see [Giere, 1988; Leatherdale, 1974] and references cited therein.

[9][Niehans, 1990], p. 313

[10]Quoted from [Niehans, 1990] (translation from the German by J. Niehans), p. 313.

According to J. Niehans, it took some thirty years before the model method also conquered a social science, viz. economics.[11] Indeed, since the nineteen-thirties model building has been dominating economic theorizing. Other social sciences, such as sociology and psychology, followed somewhat later.

I now wish to explicate the concept of a model in fairly more detail. This is not an easy task, since the term "model" is used with many different meanings in the sciences and in philosophy. Nevertheless, it may be useful to have a precise definition. This is what the Logical Empiricists, such as R. Carnap (1939), R. Braithwaite (1964) and E. Nagel (1961), reached for. These philosophers identified a model in science with a model in mathematical model theory: A model is nothing but an *interpretation of the theory's calculus*[12].

It soon became clear that this definition of "model" is too narrow; it fails, *inter alia*, to illuminate what role models play in the actual research process. Why, then, are models important at all if they are only another interpretation of a given formalism?[13]

In the course of the general critique of various views of Logical Empiricism starting from the 1960's, philosophers of science such as P. Achinstein [Achinstein, 1968], M. Bunge [Bunge, 1973] and M. Hesse [Hesse, 1970] developed conceptions of models which are closer to the scientist's intuition of that concept. All of them stress elements of a typical model that have not been taken into account by the Logical Empiricists.

M. Hesse and P. Achinstein, on the one hand, emphasize the role of *analogies* in the procedure of developing a model, while leaving only insuf-

---

[11]Niehans mentions, however, that the classics, such as A. Smith and D. Ricardo, already used models at several stages of their work.

[12]Braithwaite (1964), p. 269. There are, of course, differences between the views of the above mentioned authors, especially between E. Nagel, who is more sensitive concerning the practice of science, and R. Carnap and R. Braithwaite. It is, however, not important for the remainder of this paper to discuss them here in detail, see Psillos (1995).

[13]This critique is elaborated in [Psillos, 1995]. It may, however, be unfair to include R. Carnap in this criticism for he did not intend to provide a reconstruction of the way scientists use the term "model". Carnap attempted to characterize the relation between syntax and semantics of a scientific theory. For this purpose, the model concept proves to be extremely helpful. Among the philosophers who boldly identify models in science with models in the sense of mathematical model theory is P. Suppes. After quoting A. Tarski ("A possible realization in which all valid sentences of a theory T are satisfied is called a model of T" (Suppes (1960), p. 287)) Suppes maintains:

> I claim that the concept of model in the sense of Tarski may be used without distortion and as a fundamental concept in all of the disciplines from which the above quotations are drawn (= physics, economics, psychology etc., *S.H.*). In this sense I would assert that the meaning of the concept of model is the same in mathematics and the empirical sciences. (Suppes (1960), p. 289)

For a forceful criticism of this see [Bunge, 1973], p. 111.

ficient space for general background theories [Psillos, 1995], such as Newtonian mechanics or quantum field theory, which often restrict scientist's freedom in the modeling process considerably.

In M. Bunge's approach, on the other hand, general background theories constitute an integral part of a model. According to Bunge, a model (or a special theory) consists of two components:

- A general theory,
- A special description of an object or system (*model object*).

The Billiard Ball Model of a gas illustrates this: In this case the general theory is Newtonian mechanics, the special description contains statements about the nature of a gas, e.g., that the molecules are point-like particles moving in a chaotic way in a given box. With the so characterized Billiard Ball Model it is now possible to derive the equation of state of an *ideal* gas[14]: $PV = RT$

There are many examples of a model à la Bunge in physics. It is harder to find cases in the social sciences, for there often does not exist a general theory[15]. This lack, though, does not prevent scientists from constructing models. I have argued elsewhere [Hartmann, 1995] that in Bunge's conception, the role models play in these cases is not recognized well-enough. In such cases models prove to be a favorite tool for theorists trying to obtain a (provisionary?) description of an object or system.[16]

For the purpose of this paper it suffices to characterize a model minimally as "a set of assumptions about some system" [Redhead, 1980, p. 146]. Some of these assumptions may be suggested by a general theory (such as symmetry principles), others serve merely as (idealized) descriptions of a special object or system.

It is useful to distinguish between *static* and *dynamic* models. A model is called *static*, if it only covers assumptions about systems at rest. A model is called *dynamic*, if it furthermore includes assumptions about the time-evolution of the system.

Although most systems — be they natural or social — evolve in time, it is nevertheless not generally unreasonable to construct a static model. The main reason for this is that it is commonly much easier to acquire a

---

[14]Here, p, V and T represent the pressure, volume, and temperature of the gas respectively, R is the gas constant.

[15]An exception is – as H. Lind [Lind, 1993] has pointed out – economics. Microeconomic theory / general equilibrium theory is, Lind maintains, *the* fundamental theory in economics. To study concrete systems in detail, special model assumptions have to be made. See also [Anderson *et al.*, 1988] for a critical assessment of this approach.

[16]Some of these aspects are, however, evaluated somewhere else in Bunge's œuvre, see [Bunge, 1967].

thorough understanding[17] of a system at rest. Unfortunately, it often does not make much sense to study static aspects for the considered system is inherently dynamic. This holds especially true in the social sciences.

2.2. ... AND SIMULATIONS

Simulations are closely related to dynamic models. More concretely, a simulation results when the equations of the underlying dynamic model are solved. This model is designed to imitate the time-evolution of a real system. To put it another way, *a simulation imitates one process by another process*. In this definition, the term "process" refers solely to some object or system whose state changes in time. [18] If the simulation is run on a computer, it is called a *computer simulation.*

I maintain that the definition given above is in agreement with the scientists' usage of that term. It emphasizes the function of a simulation to investigate real dynamic systems. In a recent article, P. Humphreys concentrates on another decisive feature of a computer simulation, viz. the possibility to explore otherwise untractable models. After critically examining several alternative definitions of a computer simulation, Humphreys suggests the following *working-definition*:

> A computer simulation is any computer-implemented method for exploring the properties of mathematical models where analytic methods are unavailable.[19]

Hence, having simulations as a tool it is not necessary any longer to make dubious approximations in order to obtain analytically solvable equations. In fact, most interesting, i.e. "most non-linear ODE's (= ordinary differential equations, *S.H.*) and almost all PDE's (= partial differential equations, *S.H.*) have no known analytic solution"[20].

---

[17]It is notoriously hard to explicate the notion "understanding". I certainly mean more than R. Carnap did when writing:

> An "intuitive understanding" ... is neither necessary nor possible. ... He (i.e. the modern physicist, *S.H.*) knows how to use the symbol 'ψ' in the calculus in order to derive predictions which we can test by observation. ... Thus the physicist, although he cannot give us a translation into everyday language, understands the symbol 'ψ' and the laws of quantum mechanics. He possesses that kind of understanding which alone is essential in the field of knowledge and science. ([Carnap, 1939], p. 69)

A adequate explication of "understanding" is certainly a *desideratum* of contemporary philosophy of science for this concept is of utmost importance in actual scientific practice.

[18]I should say that in using the notion "process" here, as well as in the title of this contribution, I do not intend to allude to the metaphysics of A.N. Whitehead (1978). I restrict myself to a methodological analysis of simulations in the various sciences.

[19][Humphreys, 1991], p. 501

[20][Humphreys, 1991], p. 499

I shall mention two objections to Humphrey's working-definition: Firstly, Humphrey's working-definition does not stress the dynamic character of the model in question. As far as I see, scientists reserve the term "simulation" exclusively for the exploration of *dynamic* models. Secondly, a computer simulation may also be helpful even if analytic methods are available. Visualizing the result of a simulation on a computer screen is just one advantage. This may increase our understanding of the system more than complicated formulas written down on a paper would ever do.

It is convenient to distinguish between *continuous* and *discrete* simulations. In a continuous simulation the underlying space-time structure as well as the set of possible states of the system is assumed to be continuous. The corresponding dynamic model is conveniently formulated in the language of differential equations [Rohrlich, 1991]. Discrete simulations are based on a discrete space-time structure right from the beginning [Wolfram, 1994]. Moreover, the set of possible states of the system is assumed to be discrete. The appropriate language for discrete simulations are cellular automata (CA) [Rohrlich, 1991]. Here, the state of a cell of the system at time t_{i+1} follows from the state of the neighboring cells at time t_i according to certain rules.[21]

It should be mentioned here that the numerical integration of a differential equation also uses a discrete space-time resolution. The aim is, however, to extrapolate to zero resolutions.

In recent years, the availability of computer simulations has supplemented the methodology of the natural sciences. Furthermore, the delay to the social sciences was much shorter than in the case of models. This volume documents the impressive work that has been done so far. There is no doubt that we are at the beginning of the "simulation era".

3. The Functions of Simulations

In this section I shall discuss the various functions of simulations in science. The following are — as far as I see — the main motives to run simulations:

1. Simulations as a technique: Investigate the detailed dynamics of a system
2. Simulations as a heuristic tool: Develop hypotheses, models and theories

---

[21] F. Rohrlich maintains that CA's "are necessarily of a phenomenological nature rather than of a fundamental one" ([Rohrlich, 1991], p. 516). I could not find any argument for this statement in Rohrlich's paper. Maybe Rohrlich considers a discrete space-time to be an approximation. However, it has been suggested that Nature itself is in fact discrete. Then, a special CA would indeed be fundamental, see also [Hedrich, 1990], chap. 7.3.

3. Simulations as a substitute for an experiment: Perform numerical experiments
4. Simulations as a tool for experimentalists: Support experiments
5. Simulations as a pedagogical tool: Gain understanding of a process

While discussing these functions in detail I shall pay special attention to the different weights of them in the natural and social sciences. Besides, I will point out some inherent drawbacks of computer simulations. It is interesting to note that both, the advantages as well as the disadvantages, result from the fact that simulations are run on powerful computers.

3.1. SIMULATIONS AS A TECHNIQUE

One major advantage of simulations is that they allow scientists to explore the detailed dynamics of a real process. In many cases it is not possible for pragmatic reasons to extract this information experimentally: the relevant time scale turns out to be either too large (e.g. for the evolution of galaxies) or too small (e.g. for nuclear reactions). For this purpose, simulations are often the only appropriate tool to learn something about a system's time-evolution. This applies particularly for very complex systems, systems that are composed of many interacting sub-systems. These sub-systems may, for example, be atoms (as in the case of solid state physics) or human beings (as in sociology). In these cases it is practically impossible to derive analytical solutions of the corresponding equations that have been formulated to describe real systems.

Furthermore, certain approximation schemes may wipe out effects that would otherwise occur in a full treatment of the model. Well known instances from chaos-theory illustrate this point. For example, the famous Russian physicist L.D. Landau failed to explain the phenomenon of turbulence because he cut off curls with very high frequencies in his harmonic-oscillator-treatment. We now know that a thorough description of turbulence can only be achieved when curls at all scales are taken into account. In many cases this can only be done by solving the corresponding equations "exactly" with the help of high-powered computers. I used quotation marks here to indicate that a numerical solution is not exact in the same sense as an analytical solution is. Indeed, in a computer-aided solution of, say, a differential equations space and time are always discretized. However, the corresponding lattice spacings can – in principle – be made arbitrarily small. At least, extrapolations to vanishing lattice spacings are possible.

The possibility to obtain very accurate solutions of equation in a simulation procedure has an interesting consequence: It allows a test of the underlying model or theory. I will explain this by distinguishing two cases:

1. Discrete simulations: In this case any difference between the simu-

lation and empirical data directly blames the model assumptions, i.e. the transition rules of the CA. This enables a critical assessment of these rules. Besides, there is no way to make approximations in a CA simulation that may be called into question for the deviation [Wolfram, 1994].

2. Continuous simulation: This case is more complicated. Let us therefore distinguish two other cases:

(a) There is no background theory: Then, any disagreement between data and theory can blame – in principle – any of the model assumptions. However, by "playing around" with the different assumptions etc. one may be in the position to detect the wrong one.[22]

(b) There is a background theory: Then, we have to distinguish between that background theory and the model assumptions (or the model object – to use Bunge's term). Now, any disagreement between theory and data can be either due to the theory or due to the model or due to both. Is it then possible to test the underlying theory? R. Laymon answers this question in the affirmative. In a recent article, Laymon suggests the following criterion for the confirmation (or disconfirmation) of a theory.

> A scientific theory is confirmed (or receives confirmation) if it can be shown that using more realistic idealizations will lead to more accurate predictions.
>
> A scientific theory is disconfirmed if it can be shown that using more realistic idealizations will not lead to more accurate predictions [Laymon, 1985, p. 155].

I take Laymon here to identify his "idealizations" with my "model assumptions". A theory is then, according to Laymon, confirmed when a better model object leads "to more accurate predictions". This sounds plausible. Take, e.g., the Billiard Ball Model of a gas mentioned in the last section. Idealizing the gas particles as point particles leads to the well-known equation of the ideal gas. Now, it is well known that molecules are *not* point-like particles. They have a finite volume V_0 and this volume affects the system's equation of state, too. A modification of the model object along this line leads to the van der Waals equation:

$$(p + \frac{a}{V^2})(V - b) = RT$$

with adjustable parameters a and b depending on the special system under investigation. This equation accounts for much more phenomena and leads "to more accurate predictions". Therefore, Newtonian Mechanics is confirmed.

---

[22]This sounds perhaps a bit naive for *aficionados* of the Duhem-Quine Thesis. However, as A. Franklin demonstrated convincingly, scientists usually have strategies (within a specified context) to detect wrong assumptions [Franklin, 1986].

But what exactly does it mean to make the model assumptions "more realistic"? It certainly does not mean to add more and more terms to the equations of the model which exhibit additional free parameters that can be adjusted suitably to experimental data.

There is, however, a strong temptation in science to complicate the underlying dynamic model in order to increase the simulation's empirical adequacy. I wish to confront this practice with N. Cartwright's fine discussion of models. In her book "How the Laws of Physics Lie" Cartwright claims that

> [t]he beauty and strength of contemporary physics lies in its ability to give simple treatments with simple models, where at least the behavior of the model can be understood and the equations can not only be written down but can even be solved in approximation [Cartwright, 1983, p. 145].

This romantic picture may change soon since high-powered computers facilitate the treatment of very complex models. That is the dilemma of using computers in science: People no longer spend that much time thinking about "simple treatments" but just complicate the model in order to increase its empirical adequacy. At this point it is worth noting that we need *independent* evidence for the terms involved: "A simulation is no better than the assumptions build into it." [Simon, 1969, p. 18]. Every term in the model has to be interpreted thoroughly. There is no understanding of a process without a detailed understanding of the individual contributions to the dynamic model. Curve fitting and adding more and more ad hoc terms simply doesn't do the job.[23]

There is still another (psychological) problem with many "realistic" simulations which fit all data well. They make us forget that – as always in science – idealizations and approximations were involved in deriving the model. A serious appraisal of computer simulation has to pay attention to this fact.

In closing this sub-section I shall mention that the term "simulation" is sometimes also used in the context of mathematical-statistical algorithms. A special example is the technique of so called Monte-Carlo-Simulations[24]. It has often been emphasized [Guetzkow *et al.*, 1972] that a Monte-Carlo-Simulation is in fact *not* a simulation at all. It is an effective numerical technique to tackle, say, complicated integrals. After all, this technique is often used in simulation procedures. In a way one can try to "save" the

---

[23]I hasten to add that in some cases curve fitting is also valuable, mainly if there is no better alternative in sight. It does not, however, facilitate real understanding.

[24]A description of this technique or method is given in [Binder and Heermann, 1988]. A special example, the treatment of the Ising model, is analyzed from a philosophy of science point of view in [Humphreys, 1994].

label "simulation" here: The problem of evaluating an integral with the Monte-Carlo method is *dynamized* in so far as one has to select successive random numbers, then calculate the value of the respective function, sum all the values up, do the same again for a new set of random numbers, ... and finally average over all those preliminary results.

3.2. SIMULATIONS AS A HEURISTIC TOOL

Simulations play an important role in the process of developing hypotheses, models or even new theories. Analyzing the results of very many runs of a simulation model with different parameters *may* suggest new and simple regularities that would not have been extracted from the model assumptions otherwise. Some of these hypotheses can, in turn, serve as basic assumptions of, say, an easier model. This happens quite often in the natural sciences. Simulating the dynamics inside hadrons at finite temperature starting from the fundamental model[25], quantum chromodynamics (QCD), reveals, for example, that there are certain phase transitions. Starting from this result, physicists intend to explore the consequences of these phenomena by modeling them in a simpler way.

The following quote illustrates the corresponding role computer simulations play in the social sciences:

> Even if the processes under study are not complex, simulation can give a better picture of them because of its greater likeness to them than other models, and this is useful not only for instructional purposes but also for research. Where the causal relationships are not well understood (that is, where theory is not well developed), sometimes the best one can do is attempt to imitate the change process itself in the hope of learning more about such relationships. Thus, the model becomes an aid of theory development. Guetzkow ... has given this as a chief reason for using simulations in the study of international relations. With such a model one can also try out a theory through manipulation of the model processes to see if results conform to real-world observations [Schultz and Sullivan, 1972, p. 10].

---

[25]I here use the somewhat queer notion "fundamental model" for two reasons: Firstly, QCD is a model in the sense of Bunge since it is a special realisation (interpretation) of a general quantum field theory [Wightman, 1986]. To make QCD a model it is necessary to specify the special fields (quarks and gluons), their respective symmetry groups (e.g. flavor $SU(3)$ and color $SU(3)$) and their interaction (gauged color $SU(3)$) [Hartmann, 1995]. Secondly, I added the specifier "fundamental" since it is often claimed that there is nothing more to be said about strong interactions than QCD does. However, this is not true in general, because – for example – Grand Unified Theories (GUTs) presumably modify the physics at high energies.

In this context "theory development" refers to the task of guessing suitable assumptions that may "imitate the change process itself". Confronting the outcome of a certain simulation with the "real-world" helps to critically assess the "theory" in question. The idea here is – as far as I see it – to establish theories by using the well known trial-and-error strategy (educated guesses). This sounds a bit blue-eyed; let's just try out as many assumptions as possible (remember: we have huge computers!) and as a result of this process we will finally get the aspired theory.

There are at least two problems with this approach: Firstly, as I mentioned already, one might not learn much from a simple reproduction of data with assumptions that are not satisfactorily understood.[26] Secondly, the social world is so complex that it is probably even with high-powered computers not possible to get close to "real-world observations". This raises again the problem of how to assess the model assumptions independently.

There is still another heuristic function of simulations that I shall mention here. S. Wolfram [Wolfram, 1994] maintained that the analysis of simulations may suggest an *ansatz* for an analytical solution of a given problem. Of course, there is no guarantee that this will lead somewhere; nevertheless, it may work. This aspect reflects the strong interaction between analytical and numerical endeavors - at least in the natural sciences.

3.3. SIMULATIONS AS A SUBSTITUTE FOR AN EXPERIMENT

Simulations may help scientists to explore situations that cannot (yet?) be investigated by experimental means. The performance of an experiment might be impossible for pragmatic, theoretical or ethical reasons. An example of a *pragmatically* impossible experiment is the study of the formation of galaxies; we simply cannot do much to manipulate galaxies. An example of a *theoretically* impossible experiment is the investigation of counterfactual situations. Theorists ask what happened, when some fundamental constants (such as the charge of the electron) had other values. These questions are also relevant for philosophers who study the so-called anthropic principle [Barrow and Tipler, 1986]. An example of an *ethically* impossible experiment is to ask for the long-term consequences of a raising of, say, the income tax by a factor of 1.5 or so. This question may well be of some interest for economists. However, it is hard to imagine that those experiments have a realistic chance of being performed in the future. In all these cases, an appropriate simulation is the best scientists can do.

In his article "Numerical Experimentation" P. Humphreys claims that "the computational methods of numerical experimentation constitutes a

---

[26]Compare, however, H.A. Simon's elaboration of the role of simulations of poorly understood systems in [Simon, 1969], pp. 19.

new kind of scientific method, intermediate in kind between empirical experimentation and analytic theory." [Humphreys, 1994, p. 103]. In fact, simulations help us to theoretically approach regions in a parameter space that are inaccessible by normal experiments. This novel possibility supports the thesis that a methodology is nothing fixed but something that evolves in a similar way as our scientific knowledge evolves.

Although social scientists have experience in performing laboratory experiments for several years[27], it is apparent that numerical experiments on a computer prove to be an appreciated complementary way to approach social systems. Now scientists can easily vary different kinds of parameters, visualize the effects and eliminate disturbing influences that unavoidably show up in real systems. It is precisely the lack of conclusive experiments that motivates social scientists to run computer simulations.

What is the difference between numerical experiments in the natural and social sciences? Methodologically there is no big difference. However, it is probably not too unfair to note that numerical experimentation is much more founded in the natural than in the social sciences. What reasons do we have to believe in numerical extrapolations? In the natural sciences models are (often) well confirmed in a certain parameter space and, furthermore, embedded in strong theories. Starting thus from such "solid ground" makes extrapolations in realms beyond experimental reach more trustworthy. In the social sciences, on the other hand, there often is no such "solid ground" to start with; this makes it much harder to trust numerical experiments.

I should stress again that it is important when running a simulation to have reason to believe in the details of the underlying model or theory which are independent of the results of complete simulations.

3.4. SIMULATIONS AS A TOOL FOR EXPERIMENTALISTS

I now discuss a function of simulations that is of significant importance for the natural sciences. Computer simulations nowadays constitute an essential tool to support real experiments. These are the dominant tasks:

- Inspire experiments
- Preselect possible systems and setups
- Analyze experiments

A simulation *inspires* experiments when, say, a new regularity or hypothesis has been found by analyzing the results of simulations for many different parameter sets. It is then worth confronting this hypothesis with a real experiment.

---

[27] For a survey of (laboratory) experiments in economics see [Smith, 1994].

A simulation helps to *preselect* possible systems and setups for pragmatic reasons. It is simply too difficult to find the parameters that demonstrate the effect experimenters are after most clearly in a real experiment, especially in cost (and time)-expensive high-energy physics [Barger *et al.*, 1986]. Then detailed simulations of possible experimental setups and arrangements are performed before the actual experiment is executed.

A simulation helps to *analyze* experiments when trivial or well-understood effects have to be subtracted in order make the actual effect visible. Simulations often prove to be useful for identifying these contributions. My example is again from high-energy physics. Physicists use standard methods to simulate so called "background"-processes. Subtracting their contribution from the experimental data (cross sections etc.) helps them to identify "nontrivial" contributions. In a way this is also done in the social sciences. Effects that are of entirely statistical origin can be identified (and then subtracted) by performing appropriate computer simulations.

3.5. SIMULATIONS AS A PEDAGOGICAL TOOL

Simulations prove to be extremely useful in instucting students. By "playing" with a simulation model and visualizing the results on a screen, students increase their understanding of the underlying processes and develop an intuition for what might happen in similar circumstances. Learnig things this way is both much cheaper and faster than performing real experiments (if this is possible at all!). Once again, all this only makes sense when we have good reasons to trust the underlying model.

With respect to this function there is no difference between the natural and social sciences.

4. Case-Study: The Simulation of Heavy-Ion Reactions

I now wish to present a somewhat detailed case-study from nuclear physics that shall illustrate some typical features of a simulation in physics.[28]

Nuclear physics has, so far, not attracted philosophers of science much. Particle physics, which separated from nuclear physics in the 1940's, seems to be much more attractive to the philosophical mind. The reason for this preference is, I suppose, that particle physics is directly concerned with the fundamental problems of the constitution of matter, questions which have

---

[28]Presenting this case-study I will be a bit more technical. It is often claimed, in philosophical analyses of scientific theories and in popular science books, that it is possible to *understand* the essentials by just over-reading the formulas. I think this is plainly wrong! Without at least a little grasp of the mathematical structure behind it is hopeless to gain an adequate understanding. Nevertheless, it is important to present the respective technicalities as simple as possible.

been worrying philosophers for a long time. Nuclear physics, on the other hand, is more applied and less fundamental and hence – one might infer – not worth looking at from a philosophical perspective.

Nevertheless, since physics is surely more than particle physics, a complete picture of science also has to include the efforts, say, in solid state physics and in nuclear physics. Besides, those branches of physics that deal with complex systems raise fascinating new philosophical questions.

In his recent book "Explaining Science", R. Giere also argues that nuclear physics is worth looking into:

> [N]uclear physics is itself a paradigm of twentieth century science. In its use of mathematical techniques, of computers and other advanced technology, as well as in its organization into research groups, nuclear physics resembles many other contemporary sciences [Giere, 1988, p. 180].

Giere carefully analyzes the development of relativistic models of nuclear structure (Dirac phenomenology) and confronts the upshot with his meta-scientific model, thus presenting an interesting example for a "naturalized" philosophy of science at work.

There are even more reasons why nuclear physics is challenging for philosophers of science:

- The models of nuclear physics are a colourful mix of fundamental theoretical principles, phenomenological assumptions and bold analogies. In this respect nuclear physics is, in my view, more representative for physics as a whole than, for example, particle physics.
- Nuclear physics is closely affiliated with other sub-disciplines of physics, such as particle physics (for high energy reactions), astrophysics (nucleosynthesis), solid state physics (nuclear solid state physics), statistical physics (as an underlying theory for the description of nuclear reactions) and quantum field theory (which is the "fundamental" background theory). It is thus a good starting point to examine the relations between the different branches of physics.
- Nuclear systems are special insofar as they are neither fundamental nor highly complex. The number of particles involved in a typical nuclear reaction is of the order of ten or hundred.

The following case-study is about heavy-ion reactions. After some introductory historical remarks (Sec. 4.1) I present the model (Sec. 4.2), discuss the detailed simulation process (Sec. 4.3) and point out what one can learn from this case-study concerning the function of simulations in physics (Sec. 4.4).

4.1. SOME HISTORICAL REMARKS

After E. Rutherford's crucial discovery in 1906 showing that there is a tiny but very heavy nucleus in the center of each atom, experimental physicists started systematically collecting a tremendous amount of data about masses, charges and other observables of numerous nuclei. These data sets could be structured in a variety of models such as the liquid drop model, the optical model, and the shell model, to mention only a few. All these models were designed – inspired by quantum mechanics – to help understanding static properties of nuclei, such as their masses and binding energies.

After some time physicists realized that the analysis of dynamic phenomena (nuclear collisions, scattering of electrons and photons on nuclei etc.) reveals even more about the nuclei in question. Consequently, experimentalists studied transition probabilities by exciting one nucleus in the field of another in peripheral reactions and later   once higher energies were available in the laboratory – direct nucleus-nucleus reactions. The corresponding observables turned out to be much more sensitive to the details of certain models. Thereby physicists learned more about the properties of nuclear matter, about matter under extreme conditions, and many other interesting topics.

The most modern of these experiments use heavy ions (up to uranium), systems that consist of several hundred protons and neutrons. Complementary to the "real" experiments in the laboratories (such as GSI near Darmstadt/Germany), theorists run extensive numerical simulations of the very processes.

4.2. THE MODEL

There are some models which describe the dynamical behaviour for low energies fairly well. One of them is the so called Boltzmann-Uehling-Uhlenbeck (BUU) model that I will sketch now.[29] For the sake of simplicity I will be concerned only with the non-relativistic version of this model.

The BUU model provides an equation for the phase space density $f(\vec{x}, \vec{p})$ of the nucleons, the constituents of the colliding nuclei. With the phase-space density in hand one can subsequently work out all interesting observables that can later be compared to experimental data.

Let us consider first the (highly idealized) case where no collisions between the nucleons occur. Now, a basic theorem of (classical) statistical mechanics ("Liouville's Theorem") demands that $f(\vec{x}, \vec{p})$ has to be a con-

---

[29] I follow the presentation given in the review [Blättel et al., 1993] where further details can be found. In this article, alternative models are shortly discussed as well.

stant in time:

$$\frac{df}{dt} = \sum_{i=1}^{3} \left[\frac{\partial f}{\partial p_i} \frac{dp_i}{dt} + \frac{\partial f}{\partial q_i} \frac{dq_i}{dt} \right] + \frac{\partial f}{\partial t} = 0$$

Including the general Hamilton equations of classical (!) mechanics (with a position and momentum dependent mean field potential $U(\vec{x}, \vec{p})$), one obtains the Vlasov equation:

$$\left[\partial_t + (\vec{\nabla}_p U)\vec{\nabla}_x - (\vec{\nabla}_x U)\vec{\nabla}_p \right] f(\vec{x}, \vec{p}) = 0$$

The Vlasov equation describes the time evolution of the phase space density in the presence of the mean field $U(\vec{x}, \vec{p})$. This mean field results from a self consistent treatment of the motion of the particles.

Thus far there are no collisions in the model. Besides, this treatment is completely classical. However, it is sometimes argued that quantum mechanical interactions alone are responsible for the generation of the mean field potential $U(\vec{x}, \vec{p})$. The rest can be understood using classical physics.

Now it is clear that in a real heavy-ion reaction there are also collisions between the individual nucleons. Formally, the right hand side of the last equation does not vanish once collisions are taken into account. Particles can be scattered into another phase-space cell or scattered out of one respectively. Considering only two-body collisions one gets the BUU equation:

$$\left[\partial_t + (\vec{\nabla}_p U)\vec{\nabla}_x - (\vec{\nabla}_x U)\vec{\nabla}_p \right] f(\vec{x}, \vec{p}) =$$

$$\frac{4}{(2\pi)^3} \int d^3 p_1 d^3 p' d\Omega \ v \ \frac{d\sigma}{d\Omega} \ \delta(\vec{p} + \vec{p}_1 - \vec{p}' - \vec{p}_1')$$

$$\left[\ f(\vec{x}, \vec{p}')f(\vec{x}, \vec{p}_1')(1 - f(\vec{x}, \vec{p}))(1 - f(\vec{x}, \vec{p}_1)) \right.$$

$$\left. -(f(\vec{x}, \vec{p})f(\vec{x}, \vec{p}_1)(1 - f(\vec{x}, \vec{p}'))(1 - f(\vec{x}, \vec{p}_1'))) \ \right]$$

This equation governs the time evolution of the phase space density $f(\vec{x}, \vec{p})$ in the presence of the mean field potential $U(\vec{x}, \vec{p})$ and two-particle collisions. The term on the right hand side is the so called *collision term*. $d\sigma/d\Omega$ is the corresponding two-particle collision cross section. We will come back to it below. The factors of the form $(1 - f(\vec{x}, \vec{p}))$ in the collision term take the Pauli principle into account, guaranteeing that no two particles occupy the same phase-space cell. This is in fact the only place where quantum mechanics explicitly enters the scene.

It is important to point out the different ingredients of the BUU-equation:

1. The gross structure of the BUU equation is a general demand of statistical physics that acts as a background theory.
2. The collision term is modeled in order to respect features of quantum mechanics.
3. Two components of the BUU model are fed in by other models:

   (a) The mean-field potential

$$U(\vec{x}, \vec{p}) = A \left(\frac{\rho(\vec{x})}{\rho_0} \right) + B \left(\frac{\rho(\vec{x})}{\rho_0} \right)^\sigma$$
$$+ 2 \frac{C}{\rho_0} \int d^3 p' \frac{f(\vec{x}, \vec{p}')}{1 + \left(\frac{\vec{p} - \vec{p}'}{\Lambda} \right)^2} \quad ,$$

   where A, B, C, Λ and σ are constants, is suggested by nuclear structure calculations. These parameters reflect *static* properties of the nuclei. It is therefore essential for the later simulation of heavy-ion reactions to have a solid description of the relevant nuclei at rest.

   (b) The "elementary" cross section $d\sigma/d\Omega$ is either taken from "fundamental" calculations in the framework of quantum field theory or it is just extracted from experimental data (data fits).

Thus, our (dynamic) model is a complicated combination of an inhomogenious set of theories and model assumptions: Simulations are an interplay between "real" experiments and theoretical considerations.

4.3. THE SIMULATION-PROCEDURE

The BUU equation describes the full dynamics of the model system. The equations are, of course, not analytically solvable. In order to solve them, theoretical physicists use what I call a *meta-simulation*. Applying the *test-particle method* one approximates the continuous phase space density $f(\vec{x}, \vec{p})$ by a phase space density of a large number of imaginary test-particles with phase-space coordinates $(\vec{x}_i(t), \vec{p}_i(t))$, i.e.

$$f(\vec{x}, \vec{p}) = \frac{1}{N} \sum_{i=1}^{NA} \delta(\vec{x} - \vec{x}_i(t)) \delta(\vec{p} - \vec{p}_i(t)) \quad .$$

N is the number of test particles per nucleon and A is the total number of nucleons described by $f(\vec{x}, \vec{p})$.

With this *ansatz*, the BUU equation can be solved numerically. One finally obtains a sequence of phase-space density profiles that imitate the real process in the laboratory.

The philosopher S. Toulmin stated that it is a typical feature of a model that it "suggests" how to extend it.

It is in fact a great virtue of a good model that it does suggest further questions, taking us beyond the phenomena from which we began, and tempts us to formulate hypotheses which turn out to be experimentally fertile [Toulmin, 1953, p. 38].

Like any good model, the BUU-model has also been extended in several directions:

1. Relativistic Formulation
   In order to study heavy-ion collisions with collision energies above, say, 1 GeV it is necessary to make the model consistent with relativity. The result is the RBUU (=relativistic BUU) model [Blättel *et al.*, 1993]. These investigations were initiated once experimentalists were able to produce heavy-ion beams with such high energies. Before this energy regime was experimentally accessible nobody thought seriously about developing a relativistic extension of the model.[30] This again stresses the main motivation for those studies: phenomenological success. Deeper theoretical understanding can only be gained by carefully analyzing and simulating thorough experimental studies.

2. Many Particle Scattering
   There is no reason why only two-particle collisions should show up. It is natural to also include three and more particle scattering effects [Blättel *et al.*, 1993]. This, however, makes the model much more complicated and – in a way – less intuitive.

3. Inclusion of Particle Production Mechanisms
   At very high energies new particles, such as strange mesons and other exotic particles, are produced in heavy-ion reactions. The different production rates strongly depend on the explicit structure of the interaction. Since electromagnetic interactions are well known it is especially interesting to study the production of electron-positron pairs and photons. So theorists hope to determine properties of the still (almost) unknown structure of the strong nuclear interactions.

4.4. DISCUSSION

Having introduced the BUU model and the procedure how to simulate heavy-ion reactions with its help, I shall now identify the motives to run simulations in the process of investigating the physics of heavy ions.

First of all, simulations of this kind are, so far, the best one can do as a theorist. It is impossible to derive cross sections and related observables

---

[30]See also R. Giere's insightful remarks in [Giere, 1988], p. 185.

from first principles. The systems are too complex and we do not yet know exactly how the strong force works. Because of the complexity of the problem it may be seriously doubted that one can ultimately learn something about the nature of the strong force. However, experiments are surely not compatible with every assumed force.

What one can learn from these studies is the relevance of certain physical processes. Is it possible to generate the spectra etc. by neglecting three-particle scatterings?

The following more general claims about simulations can be extracted from our discussion of the BUU model. Performing simulations is an intermediate step between theoretical and experimental work. They are inspired by experimental findings as well as by theoretical principles. On the other hand, simulations help both theoretical and experimental physicists.

Computer simulations (e.g. of the BUU type) help *theorists* to

1. develop an understanding of the relevant processes by calculating "macroscopic" properties (cross-sections etc.) from some assumed microscopic dynamics.
   In our example scientists study details of nuclear dynamics.
2. develop new models.
   In our example theorists try to discover regularities by varying the model parameters. These regularities may serve as an input ("model assumptions") in a modified model.
3. perform numerical experiments.
   In our example scientists do, for example, interpolate between known data regimes or extrapolate into regions that are experimentally inaccessible (such as the physics inside a supernova).

Computer simulations help *experimentalists* to

1. design new experiments.
   Experiments in nuclear and particle physics are so cost-intensive that it is common to perform detailed simulations in advance in order to determine the "best" experimental setup. Simulations help, e.g., to find the optimal detector location and to single out the best suited nuclear system for real experiments.
2. interpret real experiments.
   For given experimental data in high-energy physics (cross-sections etc.) researchers wish to know what part of the spectrum is simply "background". The contributions of those processes can be readily determined in computer simulations and then subtracted. What remains is the result of nontrivial dynamic interactions.

Summing up, this example demonstrates the involved interplay between fundamental theoretical concepts, model-inspired assumptions about the detailed dynamics and experimental inputs in a typical computer simulation in physics. An essential prerequisite for the phenomenological success of these simulations is definitely that the static aspects of nuclei are (theoretically) well under control.

5. Conclusions

In this paper, I proposed and explicated the following definition of a simulation: *A simulation imitates one process by another process.* The basis of a simulation is a dynamic model that specifies – besides some static properties – assumptions about the time evolution of the considered object or system. There are simulations that assume a continuous dynamics, and others that assume a discrete one. The first are formulated in the language of differential equations, the latter in the language of cellular automata (CA). Both types of simulations are applied in the natural as well as in the social sciences. So far, natural scientists (especially physicists) still prefer models based on differential equations, while social scientists frequently employ CA's that reflect the essential decision aspect right from the beginning.

Besides this difference there are many parallels between the various sciences concerning the question why simulations are an important tool at all. This is why the focus of this paper is on the *function* of simulations in the research process. Besides their pragmatic function to supply a description of the time evolution of a real system, simulations serve to perform numerical experiments that allow the extrapolation of data into experimentally inaccessible realms, they support experiments and provide useful heuristics to develop new models and – maybe – theories.

The final case-study from nuclear physics demonstrates that simulations in physics are often deeply grounded in models of the static aspects of the system whose time evolution is now in question. These static models are often well-established in physics, providing a solid basis for the dynamic part. The dynamic parts of the model assumptions is often severely constrained by the demands of statistical physics (such as Liouville's Theorem). However, there are still many free parameters that have to be (or can be) adjusted to experimental data.

Obviously, life is much harder for the social scientist. There are neither good descriptions of static aspects (if they make sense at all), nor are there generally accepted hypotheses for the details of the dynamics.

There is substantial evidence that computer simulations will become even more powerful in the foreseeable future. Progress in this field, at least progress in the descriptive power of computer simulations, is closely linked to progress in the development of new generations of high-powered computer systems. However, simultaneous progress in *understanding* natural and social phenomena can only be achieved when we use this mighty tool properly.

6. Acknowledgement

I wish to thank H. Carteret, P. Humphreys, K. Troitzsch and M. Weber for valuable comments on a draft of this paper and M. Stöckler for many helpful discussions and support. S. Wolfram provided me with his most recent writings on cellular automata. Thanks!

References

Achinstein, P. (1968) *Concepts of Science*. The John Hopkins Press, Baltimore.
Anderson, P.W., Arrow, K., and Pines, D. (eds.) (1988) *The Economy as an Evolving Complex System*. Addison-Wesley, Redwood City.
Barger, V., Gottschalk, T. and Halzen, F. (eds.) (1986) *Physics Simulations at High Energy*. World Scientific, Singapore.
Barrow, J., and Tipler, F. (1986) *The Anthropic Cosmological Principle*. Clarendon Press, Oxford.
Binder, K., and Heermann, D. (1988) *Monte Carlo Simulation in Statistical Physics*. Springer, Berlin.
Blättel, B., Koch, V., and Mosel, U. (1993) Transport-Theoretical Analysis of Relativistic Heavy-Ion Collisions. *Reports on Progress in Physics*, 56, p. 1.
Braithwaite, R.B. (1964) *Scientific Explanation*. Cambridge UP, Cambridge.
Bunge, M. (1967) *Scientific Research II*. Springer-Verlag, Berlin.
Bunge, M (1969) Analogy, Simulation, Representation. *Revue internationale de philosophie*, 87, pp. 16–33.
Bunge, M. (1973) *Method, Model, and Matter*. D. Reidel, Dordrecht.
Carnap, R. (1939) *Foundations of Logic and Mathematics. International Encyclopaedia of Unified Science 1, No. 3*. The University of Chicago Press, Chicago.
Cartwright, N. (1983) *How the Laws of Physics Lie*. Clarendon Press, Oxford.
Cartwright, N., Shomar, T., and Suárez, M. (1995) The Tool Box of Science: Tools for the Building of Models with a Superconductivity Example. In [Herfel *et al.*, 1995], pp. 137–149.
de Marchi, N., and Blaug, M. (1991) *Appraising Economic Theories*. Billing and Sons, Worcester.
Eigen, M., and Schuster, P. (1979) *The Hypercycle: A Principle of Natural Self-Organization*. Springer, Berlin.
Franklin, A. (1986) *The Neglect of Experiment*. Cambridge University Press, Cambridge.
Galison, P. (1987) *How Experiments End*. The University of Chicago Press, Chicago.
Giere, R. (1988) *Explaining Science*. The University of Chicago Press, Chicago.
Guetzkow, H., Kotler, P., and Schultz, R. (eds.) (1972) *Simulations in Social and Administrative Science: Overview and Case-Examples*. Prentice-Hall, Englewood Cliffs, N.J.
Hacking, I. (1983) *Representing and Intervening*. Cambridge UP, Cambridge.
Hartmann, S. (1995) Models as a Tool for Theory-Construction: Some Strategies of Preliminary Physics. In [Herfel *et al.*, 1995], pp. 49–67.
Hausman, D. (1992) *The Inexact and Separate Science Economics*. Cambridge University Press, Cambridge.
Hedrich, R. (1990) *Komplexe und fundamentale Strukturen. Grenzen des Reduktionismus*. B.I. Verlag, Mannheim.
Herfel, W., Krajewski, W., Niiniluoto, I. and Wójcicki, R. (eds.) (1995) *Theories and*

Models in Scientific Processes (= Poznań Studies in the Philosophy of the Sciences and the Humanities 44). Rodopi, Amsterdam.

Hermann, C., and Hermann, M. (1972) An Attempt to Simulate the Outbreak of World War I. In [Guetzkow *et al.*, 1972], pp. 340–363.

Hesse, M. (1970) *Models and Analogies in Science*. University of Notre Dame Press, Notre Dame.

Hoyningen-Huene, P., and Hirsch, G. (eds.) (1988) *Wozu Wissenschaftsphilosophie?* de Gruyter, Berlin.

Humphreys, P.W. (1991) Computer Simulations. In A. Fine, M. Forbes and L. Wessels (eds.), *PSA 1990*, 2, pp. 497–506, East Lansing.

Humphreys, P. (1994) Numerical Experimentation. In P. Humphreys (ed.), *Patrick Suppes: Scientific Philosopher*, 2, pp. 103–121, Dordrecht.

Humphreys, P. (1995) Computational Empiricism. *Foundations of Science*, 1, pp. 119–130.

Kippenhahn, R., and Weigert, A. (1991) *Stellar Structure and Evolution*. Springer, Berlin.

Laymon, R. (1985) Idealizations and the Testing of Theories by Experimentation. In P. Achinstein, O. Hannaway (eds.), *Observation, Experiment, and Hypothesis in Modern Physical Science*, pp. 127–146, Cambridge, Mass..

Laymon, R. (1991) Computer Simulations, Idealizations and Approximations. In A. Fine, M. Forbes and L. Wessels (eds.), *PSA 1990*, 2, pp. 519–534, East Lansing.

Leatherdale, W.H. (1974) *The Role of Analogy, Model and Metaphor*. North-Holland, Amsterdam.

Lind, H. (1993) A Note on Fundamental Theory and Idealization in Economics and Physics. *British Journal for the Philosophy of Science*, 44, pp. 493–503.

Nagel, E. (1961) *The Structure of Science*. Harcourt, Brace and World, New York.

Niehans, J. (1990) *History of Economic Thought*. The John Hopkins University Press, Baltimore.

Psillos, S. (1995) The Cognitive Interplay between Theories and Models: The Case of 19th Century Optics. In [Herfel *et al.*, 1995], pp. 105–133.

Redhead, M. (1980) Models in Physics. *British Journal for the Philosophy of Science*, 31, pp. 145–163.

Rohrlich, F. (1991) Computer Simulation in the Physical Sciences. In A. Fine, M. Forbes and L. Wessels (eds.), *PSA 1990*, 2, pp. 507–518, East Lansing.

Schultz, R. and Sullivan, E. (1972) Developments in Simulation in Social and Administrative Science. In [Guetzkow *et al.*, 1972], pp. 3–50.

Simon, H.A. (1981) *The Sciences of the Artificial*. MIT Press, Cambridge, Mass., 1969.

Simon, H.A. (1970) *Administrative Behavior. A Study of Decision-Making Process in Administrative Organization*. Macmillan, New York.

Smith, V. (1994) Economics in the Laboratory. *Journal of Economic Perspectives*, 8, pp. 113–131.

Suppes, P. (1969) A Comparison of the Meaning and Uses of Models in Mathematics and the Empirical Sciences. *Synthese*, 12, pp. 287–301. Reprinted in H. Freudenthal (ed.), *The Concept and the Role of the Model in Mathematics and Natural and Social Sciences*, pp. 163–177, Dordrecht, 1961, and in P. Suppes, *Studies in the Methodology and the Foundations of Science*, pp. 10–23, Dordrecht, 1969.

Thompson, W. (1884) *Notes of Lectures on Molecular Dynamics and the Wave Theory of Light*. Baltimore.

Toulmin, S. (1953) *The Philosophy of Science — An Introduction*. Harper, London.

Whitehead, A.N. (1978) *Process and Reality. An Essay in Cosmology*. Free Press, New York.

Wightman, A. (1986) Some Lessons of Renormalization Theory. In J. de Boer, E. Dal and O. Ulfbeck (eds.), *The Lesson of Quantum Theory*, pp. 201–226, Amsterdam.

Wolfram, S. (1994) *Cellular Automata and Complexity*. Addison-Wesley, Reading.

EVOLUTIONARY EXPLANATIONS FROM A PHILOSOPHY OF SCIENCE POINT OF VIEW

ULRICH MUELLER
Institut für Medizinische Soziologie
Philipps-Universität
Marburg, Germany

"Physics-envy is the curse of biology." Joel Cohen.

1. Introduction

In this article we will focus on the evolution of traits in living things. Traits can be physical properties as well as behavior patterns. Special attention will be given to the evolution of social behavior patterns among humans. With this, we mean a behavior that reacts to and is directed at the behavior other humans. We will concentrate on behavior patterns that are universal for all humans, and for which we may assume a well-developed genetical basis. The evolution of culturally transmitted behavior patterns, especially if we characterize them by features which are not universal for all humans may or may be not governed by the same mechanisms with govern the evolution of patterns which are universal for all humans. Examples would be languages, norms, bodies of technical or scientific knowledge. The mechanisms behind the evolution of culturally transmitted aspects of behavior patterns are not well understood within the theoretical framework of biological evolution (Pinker and Bloom 1990) so far. This may be caused by the fact that in studying social and cultural phenomena among humans, we still have not made the shift from typological to population thinking (Mayr 1975, p. 26 f.) which marked the beginning of modern evolutionary biology. But it might very well be that the evolution of exclusively culturally transmitted behavior, while not exempt from the laws of biological evolution, may also be shaped by some additional, hither too little understood mechanisms. Therefore, and also, because the evolutionary biology of humans is the basis of the evolutionary sociology of humans, this article will focus on evolutionary explanations in the sociobiology of human

R. Hegselmann et al. (eds.),
Modelling and Simulation in the Social Sciences from the Philosophy of Science Point of View, 101–122.
© 1996 *Kluwer Academic Publishers. Printed in the Netherlands.*

beings. The principal components of such explanations are populations, mutations, environmental change, competition for resources (among them sexual partners), selection, reproduction, heredity.

The principal law is that there is a quality standard for behavior displayed by the individuals of a population, namely adaptation, by which we mean the efficient use of environmental energy in order to survive and multiply/reproduce:

> If a is better adapted than b to some given environment T, then , in T, a will in the long run leave more offspring than b. (A)

The crucial criterion for adaptation is therefore survival and reproduction. Those individuals who solve this efficiency problem better than others have a higher chance of survival into the next time period and of leaving more offspring in the long run than other individuals within the same population. Since in all sexually reproducing species individuals must die, here differential reproduction is the criterion. We do not have to assume an inborn desire of all flesh to survive and reproduce in order to relate this quality standard to actual behavior. Single individuals may or may not have this desire. But, since children resemble their parents (the transmission mechanism being genetic or cultural), and since the offspring of individuals who have it tend to replace offspring of individuals who don't, any population after a while will consist of individuals, who whatever goes on in their minds, will behave such that they maximize their chances of survival and reproduction.

An evolutionary explanation is an application of that general law to specific behavior patterns in specific environments. But what is the status of this general law: is it an empirically testable hypothesis, or is it an axiom, which may be useful or useless, but defies falsification ?

Note that the reproduction mechanisms is not necessarily confined to the reproduction of genes by means of meiosis and the fertilization of an oocyte by a spermatocyte. All kinds of learning from parents will also contribute to reproductive performance. Parents are jealous to maintain exclusive access as teachers to their small children because they want to hand over their experiences and ways of seeing the world and doing things as they are jealous to hand down their genes to the children who grow up in their custody.

2.   Evolutionary Explanations and the Purpose of the Paper

Evolutionary explanations as defined here face the — old — objection of circularity. Question: *Who are the fittest?* Answer: *Those who survive/reproduce at best.* Question: *Who survives/reproduces at best?* Answer: *Those who are fittest.*

If applied to phenomena in modern human societies, evolutionary explanations also face the objection that they are self-immunizing against falsification: the human behavior potential evolved during the 5–8 Million years of humanization after the first australopithecines emerged as the savanna form of a common and widespread omnivorous ape (Kingdon 1993) when the given and man-made environmental conditions were radically different from those which prevailed during the last 150–200 generations since the Neolithic revolution, not to speak of the last 2–3 generations. Therefore no human behavior pattern in modern societies, which is definitively not fitness-maximizing, may be used as evidence against some evolutionary explanation of this behavior: it always can be argued that the behavior is not adapted to the novel environment which has evolved so fast that biological adaptation is lagging behind. On the other hand, since the environment has changed, it cannot be expected that humans display exactly the behavior they used to display on the savanna either.

Thus, whatever behavior we may observe in modern societies, it cannot be argued that this behavior is not well adapted. Therefore evolutionary explanations, perhaps useful for investigating hunter-gatherer societies, cannot contribute to the understanding of modern societies.

Indeed, central demographic features of modern societies are definitely non fitness-maximizing (overview of arguments in Low 1993), like the inverse relation between social status and fecundity, the high proportion of childlessness among the best-educated individuals, the course of demographic transition which universally started in the highest strata of the most advanced societies. These objections against the "novel environment" argument (Barkow *et al.* 1992) point to an broader weakness of the Theory of Evolution as outlined above: In a changing environment, especially if the change is not smooth, it is not possible to determine which traits constitute a successful adaptation. What might have been a fitness-maximizing behavior yesterday, need not be one today anymore, since it may have become colder (or hotter, wetter, drier etc.). Obviously, the future has always been opaque, and that a species went extinct after a swift change in the environment cannot serve as an argument against its well-adaptedness before the disaster. But how to determine present adaptedness if the direction of the selection pressure vector changes constantly in an ever unstable environment?

Even the idea of a selection of traits which actually diminish survival chances of the species — sometimes dubbed as "run-away-evolution" — would be perfectly compatible with evolutionary theory, namely when the evolution of certain traits itself triggers certain changes of relevant properties of the environment. One example would be sexual selection: a positive feedback loop between certain advertising highly visible properties of males (for example: the trail of peacocks) and female preferences. Provided that male properties are heritable, these properties might get fixed even if they are deleterious for male survival, because females cannot afford to have sons which do not meet female preferences — peahens lay more eggs for peacocks with larger trails (Petrie and Williams 1993) — even if these sons are better adapted to the physical environment (Bradbury and Andersson 1987). Sexual selection does not necessarily favor cheating: as long as the costs of advertising are higher for low quality males, honest signaling may be evolutionarily stable (Enquist 1985; Grafen 1990; Maynard Smith 1994), but advertising altogether may consume more resources than necessary for transmitting the information. Similarly, it can be argued that modern industrial societies, while exercising an overwhelming technical, economic, and therefore military dominance over all traditional forms of society, nevertheless damage the environment and fecundity to a degree that, once female life expectancy exceeds about 70 years, typically there is no industrial society in which the birth rate is not below the replacement level (Chesnais 1992).

There have been various attempts to overcome these principal objections against the status of the Theory of Evolution as a empirical theory by an axiomatization of the whole theory (for example Brandon 1978; Williams 1973) that is, treating (A) not as a empirically falsifiable law, but as an axiom, composed of primitive concepts. Frequently parallels are drawn between the law of evolution as in (A) and Newton's First Law of Motion

> "Definition IV: an impressed force is an action excited upon a body, in order to change its state, either of rest, or of uniform motion in a right line."
> "Law I: Every body continues in its state of rest, or of uniform motion in a right line, unless it is compelled to change that state by forces impressed upon it" (Newton 1686, quoted in Williams 1973).

Here again we might think of a circularity in this law: No change in state, if no impressed forces; and: if no change in state, then no impressed forces. Another example would be the law of energy conservation in closed systems (the first law of thermodynamics): if energy is conserved, the system must be closed, otherwise it must be open.

Another suitable example would be the principle of homo oeconomicus: that agents want to maximize utility: if agent do not as expected, their preferences must have dimensions which we failed to consider.

I think the suggestion of an axiomatization of these fundamental laws is misleading, that the doubts about the status of these laws as empirically testable laws are unfounded.

Being no physicist, I will concentrate upon the Law of Evolution, of which the principle of homo oeconomicus is all but an special case, an application to a human behavior, to which in its usual form we only have to add that the highest rank in the preference ordering of most people (as surveys in all societies show) is the survival of their children and of themselves. That some people may have different preferences is as little evidence against the law of evolution as the fact that some firms may not maximize profit but rather go bankrupt is evidence against be the principle of homo oeconomicus: selection got to work somewhere.

There are many well-written papers and essays about the status of the Theory of Evolution, which highlight many objections and the refutations of these objections (Vollmer 1987 presents both in a schematic overview) from various viewpoints. Of special interest is that fitness — the differential ability to survive/to succeed in the competition among conspecific — can be defined independently from its predicted effect — for example by energy use efficiency.

Therefore, I will look only into one aspect of the two main arguments against evolutionary explanations in the social sciences: (1) the alleged circularity of survival-of-the-fittest concept and (2) the alleged non-falsibility of the novel-environment concept; an aspect, which in my eyes has not received the attention which it deserves. I will argue that evolutionary explanations of stable results of evolution, which actually have decreased fitness, where the stable outcome of an evolutionary process has lead to suboptimal results, that is: apparent violations of the Fundamental Law of Natural Selection, in fact often offer prime opportunities for empirical testing of the Theory of Evolution. I will demonstrate this for the examples of frequency-dependent selection and fitness in novel environments.

3. Fitness in Frequency-Dependent Selection

Frequency dependent selection, while already in the 1930s investigated by R. Fisher in the special case of the evolution of the sex ratio, as a general phenomenon has received attention not before the 1970s. It is the field where evolutionary game theory has earned most of its merits (Maynard Smith 1982, 1989). Indeed, the evolution of most social behavior traits is subject to frequency dependent selection, which deals with traits whose adaptive value changes with its frequency within the population. Strangely enough, the ongoing discussion of the status of the Theory of Evolution in the Philosophy of Science has given this type of selection only little attention so far.

The precise formalization of the principle of Natural Selection is due to Ronald Fisher; his Fundamental Law of Natural Selection (1930) exists in several equivalent versions. The most widely quoted version is that the rate of evolution — the relative change of the frequency within the population of the gene under investigation — is proportional to the genetic variance in the population on which selection acts. From that follows that in any given population under a constant selection pressure the fitness change (the fitness gradient) must always be positive. It became also evident, that this version of the Fundamental Law does not strictly apply when fitness depends on two or more genetic loci, because the recombination of the alleles on these loci may interfere with the selection process (Moran 1962).

All these considerations apply only to frequency-independent selection — that is the selection of traits the evolution of which does not, in turn, change the selection pressure. Typically, such traits would be those which are to protect against physical hazards of the environment. If the average temperature in the environment decreases, developing a thicker fur would be such an example, because the thicker fur is useful for its owner, independent whether other members of the population also have developed thicker furs or not. Strictly speaking, frequency-independence requires that all other resources save the one under consideration are abundant, which is not the case in our example, because, if all population members with the old thin fur do not survive the winter, selection works more rapidly in favor of the new thick fur as long as this new trait is rare in the population. But while, here, this is only a secondary aspect of a game against nature, frequency-dependence becomes a crucial feature of selection, once either other members of the population are resources for which one has to compete, or competitors whose behavior can drastically alter the utility of a physical resource. For both cases, there are well known examples in the literature, the first one R. Fishers explanation of the sex ratio, the second the "Hawk"-"Dove" game described by Maynard Smith and Price (1973).

A human female can produce about twelve fertile egg-cells per year. Pregnancy lasts nine months. She may conceive again only a few months after childbearing, but for maximizing survival of subsequent children a much longer spacing between births in necessary - in extant hunter-gathers societies up to four years. A healthy male, on the other hand can produce one fertile ejaculation per day with no pregnancy costs for himself at all. If a couple can chose the sex ratio of their offspring, if the goal is maximizing number of grandchildren, why not raising more sons than daughters? On the other hand: Since female organisms are the bottleneck of reproduction (one man and ten women can produce 10 babies, ten men and one woman only one baby in one year), an uneven sex ratio in favor of more females would increase the reproduction potential and the growth rate of a given population — for example after epidemics or famines — and, thus, would enhance survival chances of the population as a whole. Following that logic, animal husbandry since its invention has decreased the proportion of male animals in a breeding population. Why has evolution not fixed a sex ratio of, say, 1 male to 10, 20. 30 females? Fishers argument goes like this:

1. Provided that it takes the same parental (mostly: maternal) investment to produce a viable son as to produce a viable daughter, there is no uneven sex ratio which cannot be beaten by another sex ratio. Let us assume that the sex ratio in a population is 1 male to 2 females, and let us assume that in this population, all individuals mate at random, and that each female has exactly three children. Let us assume that the population is large and that no one mates with a relative. Now let a mutant appear with an autosomal dominant gene, resulting in an offspring sex ratio of 2 males to 1 female. Because normal mothers have six grandchildren from each son, and three grandchildren from each daughter, average number of grandchildren per female is 12, while average number of grandchildren for a mutant female, as long as it is rare, is 15 (six from each of the two sons and three from the one daughter - see table 1).

TABLE 1. The evolutionary advantage of the rarer sex in a population with an uneven sex-ratio.

P-generation	1 normal couple	1 mutant couple
F1-generation	1 son, 2 daughters	2 sons, 1 daughter
F2-generation	4 grandsons,	10 grandsons,
	8 granddaughters	5 granddaughters
F3-generation	12 greatgrandsons,	30 greatgrandsons,
	24 greatgranddaughters	10 greatgranddaughters

Clearly, the mutant, having higher reproductive success, will invade the population. By exactly the same argument, however, a population with a offspring sex ratio of two males to one female cannot be safe against a invading mutant with a offspring sex ratio of one male to two females. In general, every population with an uneven offspring sex ratio can be invaded by a mutant with a higher proportion of the rarer sex. It follows that the even sex ratio strategy can invade every population with an uneven sex ratio of offspring.

2. The even sex ratio itself, however, cannot be invaded by any uneven mutant. In a large population with an equal number of young adults of each sex on the marriage market, sons, on the average, will not have more offspring than daughters.

Therefore, the even sex ratio is an unbeatable strategy: it can invade any alternative, but itself cannot be invaded by any alternative strategy. From whatever initial sex ratio, if we allow mutations, the sex ratio of offspring in a population will evolve toward an even one.

At any point on the respective trajectory taken by the population, direction and force of selection will depend on the present frequency of the relevant trait — the sex ratio of offspring. Relevant for our argument here is, that, *whenever the starting point for the trajectory of the evolution of the sex ratio is a sex ratio in favor of females, the course of evolution will decrease average fitness.* In the example, a two males / one female sex ratio of offspring with its initially 15 grandchildren will invade the one male to two female population with its 12 grandchildren per female, but the end point of evolution necessarily will be the even sex ratio with only 9 grandchildren per female.

Another apparent violation of the Fundamental Law of Selection is the Hawk-Dove game: Let us assume a population of animals who settle their competition for a particular resource peacefully. Once two individuals compete for a resource which can feed only one, they cast a perfect dice: the winner takes the resource, the loser leaves the site quickly and unharmed, free to try her luck elsewhere. Let the net value of the resource be 2, therefore the payoff to each one of both competitors applying this "Dove" strategy on the average (real doves are anything else than peaceful "Doves", an expression which here seems to be borrowed from military slang).

Now let a "Hawk" mutant appear, which immediately attacks a competitor with a dangerous weapon (teeth, horns, claws etc.). The "Dove" will retreat immediately, leaving with a zero payoff whenever such a pairing ("Hawk"-"Dove") occurs. Two "Hawks", however, will fight it out, until one of them - with even chances - is seriously wounded, which we may equate with a high negative payoff, say -200, the extra resources needed for healing and recovery. This translates into an average payoff to each contestants in an all-"Hawk" pairing of -99. The various payoffs are listed in the payoff matrix in table 2. If we keep population size and environment constant, and neglect all scarcity dynamics, it is obvious that no pure strategy can be stable. "Dove" can invade an all-"Hawk" population, because $0 > -99$, and "Hawk" can invade an all-"Dove" population because $2 > 1$.

TABLE 2. Payoff matrix of the "Hawk"-"Dove" game.

Strategy	played against...	
	Dove	Hawk
Dove	1	0
Hawk	2	-99

There is an equilibrium strategy at frequencies p "Hawks" and q "Doves" $(p + q = 1)$ at which payoffs to either strategy are equal: "Hawk" against "Hawk" plus "Hawk" against "Dove" equals "Dove" against "Hawk" plus "Dove" against "Dove":

$$-99p + 2q = 0p + 1q \tag{1}$$
$$\text{or } p = \frac{1}{100}$$
$$q = \frac{99}{100}$$

which is an equilibrium payoff of .99 to every member of the population. There is no other equilibrium and this one is stable: an increase in the population of "Doves" will offer the "Hawks" a higher payoff in the average contest, and an increase in the population of "Hawks" a higher payoff to the "Doves". Also, a probabilistic strategy of individuals choosing "Hawk" in 1 % and "Dove" in 99 % of all pairings, will be able to penetrate any other frequency distribution of strategies. The point for the subject of this article is, however, that, starting from any frequency distribution of strategies with "Doves" making up more than 99 % of the total population, evolution towards the evolutionarily stable strategy (ESS) as in (1) will lead to a decrease of average fitness in the population. Fitness is maximal in an all-"Dove" population. The "Hawk"-"Dove" game is just another example case of a social dilemma, a situation in which the social welfare optimum = Pareto optimum (a distribution of resources among members of a group, in which no one can be made better off without making someone else worse off), is not stable, or, conversely, where the equilibrium strategy, from which no player can unilaterally deviate without diminishing his individual payoff, is not pareto-optimal (Schulz et al. 1994).

In all kinds of frequency-dependent selection — and as we have seen, at a closer look, there is no selection which is absolutely frequency-independent — we therefore have to qualify the Fundamental Theorem of Natural Selection to the effect, that the predicted non-negativity of the fitness gradient applies to the situations where the initial population strategy was evolutionarily stable.

Frequency-dependence of selection, on the other hand, offers us the opportunity to test fitness maximization empirically much better than before: In a stable environment and a stationary population with frequency-independent selection, all what we can do is to declare the observed fitness of some fixed trait as maximal, because the few aberrant mutants which appear once in a while, show a lower fitness and appropriately go extinct. It is difficult, however, to determine whether the fixed trait does not incorporate some unknown impediment to even higher growth, to the effect that without the trait, an even higher fitness would be perfectly achievable. There is little more than the fact that there are no successful alternative mutants around which allows us to declare the fixed trait as fitness maximizing.

Now, in the case of a pareto-deficient evolutionarily stable strategy, as the result of frequency-dependent selection we are in a much better position for empirical tests of fitness maximization. We can empirically test invasion-proofness of any strategy and we can test the pareto-deficiency of the evolutionarily stable equilibrium — for example by temporarily removing "Hawks" and then "Doves" from the "Hawk"-"Dove" population. In principle, by measuring the payoffs of all strategies involved when played against each other, we can measure the average fitness of the population in any composition, and then predict evolutionarily stable equilibria. The payoffs may not be linear in frequencies (as we, for the sake of simplicity have assumed in (1)), but there is no principal obstacle for finding the appropriate model for this relation independently from the equilibrium strategy. Furthermore, we can determine whether a mutant strategy may invade a given population strategy, may the latter be pure or mixed. For this, we need not know the exact location of the new evolutionarily stable strategy or strategies in the state space (the set of all possible frequency distributions of strategies within the population).

If the invasion by a new mutant results in a internal evolutionarily stable strategy which comprises more than one pure strategy, and which is pareto-deficient, in comparison to the equilibrium we would have without the new mutant, then we have another way for an empirical test of the evolutionary explanation: we should then be able experimentally to increase the average fitness in the population by excluding all individuals playing the new strategy, or equivalently, by preventing all individuals from playing it.

4. Fitness in Novel Environments

The novel environment hypothesis refers to a fitness-suboptimality of a different kind: An overtly maladaptive behavior may be explained by demonstrating

a) that the behavior has a genetic basis which evolved over time in an environment to which it was well adapted;
b) that later on this environment has either changed much faster than the genetic endowment of this organism could — by the slow processes of mutation and subsequent selection — follow suit, or that the organism has recently left its original habitat;

c) that in the present environment the organism, by reacting to stimuli from the environment, which it is programmed to recognize as specific cues, displays the specific behavioral response which was well adapted to in the original environment, but is maladapted to in the novel one.

A fine, non-human example are moths circling around a candle light at night. The relevant feature of the environment they are adapted to is the light beams sent out by the moon or the stars. These beams, because of the great distance of their sources, virtually are parallel. Therefore, in order to maintain a straight flight course over the surface, a moth just needs to maintain a constant angle to the gravity vector (usually a 90 degrees angle) and a constant angle to the light of the moon or the brightest star (see figure 1, left). Enter homo sapiens with a candle. Maintaining in flight a constant angle to the beams of a point-source of light at finite distance necessarily results in a circular course, which, once you came close enough to the candle to perceive its light as the brightest around, by chance and natural imprecision, may bring you close enough to the heat such that your wings get burned (see figure 1, right).

Figure 1.   Moth flying at constant angle to parallel and to radial light beams

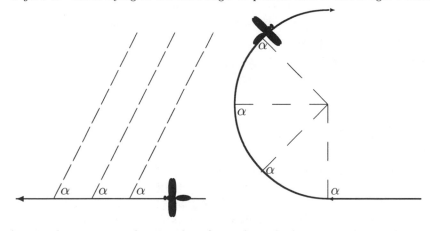

An application to the secular fecundity decline in industrial societies would go like this: Providing for own old age, people formerly accepted the burden of raising children, investing in their education and material endowment. Once, however, social security systems have been invented, which guarantee transfer payments and eventually a place in a nursing home with a much higher degree of certainty than children's gratitude and morals can guarantee, people prefer to spend their time with sex without procreation and with other hobbies, which provide fun without commitment. By refraining from having babies, however, people in rich industrial societies undermine the whole system of social security, because neither the capital stock in which the active population's contributions to the pension funds are invested nor the nursery homes function without real people, who, since they are not the children of the rich childless fun-maximizers, have to brought into the play from outside and might feel little obligations towards those old non-relatives who, stripped of power, have to rely on the benevolence of strangers.

Evolutionary explanations using the novel environment hypothesis, must make assumption which seldom can be tested, especially with regard to the cues from the environment. Crucial for the argument in this example would be the assumption, that in the Pleistocen, during the long process of hominization, living in rich environment with stable abundance of resources and an absence of hazards might have led early humans to stop having babies or at least to limit their own fecundity. On the other hand, once life became insecure and the environment unpredictable, people should have pushed their procreation. This assumption is very questionable, but is also difficult to falsify, since all pre-agrarian or pre-horticulturalist societies we can study today are all but the last remainders of societies which, by technologically superior competitors were forced into the least appealing and worst endowed habitats imaginable (deserts, rainy forests, tundras, and the like). Inferences to the kind of hunter-gatherer societies in which our ancestors lived for thousands of generations are very risky.

An even bigger obstacle to any novel environment explanation is that we must be able to demonstrate a genetic basis of the observed maladapted behavior. The stipulated inertia of this basis is the very core of the explanation.

Especially when it comes to complex social behavior, which in all observed human populations incorporates strong elements of cultural traditions and learning, genetics is still far away from identifying the exact genetic basis for, say reproduction centered behavior traits of humans - for example the desire for another baby. The general basis of the orientation in flight of moths is much easier to demonstrate, as well as the maladaptedness of getting one's wing burnt when circling around open light. Below-replacement-reproduction in industrial societies might be a well adapted behavior, when the goal, for example, is to maximize family-lineage survival chances rather than maximizing number of offspring in the first generation (Mueller 1992). Perhaps people in industrial societies not unreasonably get the cues from the environment that natural resources are becoming depleted at a high rate, that the carrying capacity is going to shrink dramatically within the next few generations. As long as the survival of the whole interbreeding group is not endangered, it might be difficult to contend maladaptation of a behavior displayed by the vast majority of all members of the group.

Furthermore, we must not only be able to demonstrate the genetic basis of the investigated behavior for which we seek an evolutionary explanation with the help of a "novel environment" hypothesis; we also have to show that this basis, while lagging behind the change of the environment, yet is adapting to the novel selection pressures. We must demonstrate evolution in progress, due to changing selection pressures in the novel environment. Here we are in a slightly better situation. We may still be unable to identify the genetic basis for some complex human social behavior, but we can demonstrate evolution in progress for several genetic traits in humans, due to the change in the environment of modern human, which they themselves have set in motion. In principle, on-going evolution of the genetic basis of modern humans is difficult to prove because of the astonishing genetic homogeneity of modern humans. The branching process of modern humans began just $\sim$ 8000 generations ago: the Eve female whose mitochondria (cell organelles which are passed exclusively from mother to children) we all carry with us, lived about 180.000 – 220.000 years ago in East Africa (Cann et al. 1987). In addition, human have multiplied enormously, and maintained migration rates which were sufficiently high to prevent a growth of genetic heterogeneity with eventual separation into different species (Cavalli-Sforza et al. 1993). In large populations with internal migration, adaptation to specific environments is very slow. But even in modern humans there are examples for ongoing evolution of the genetic basis. One example would be the trouble worldwide people have with their wisdom teeth. Without appropriate treatment, mortality from impacted third molars might be between 0,5 % and 1 % of total adult mortality. The still ongoing cranialization, the shift of skull architecture in favor of an even bigger space for the brain, especially its frontal lobe, goes parallel with a reduction and decline of teeth, chewing muscles and their bony anchorage, because better techniques of food-processing and increased nutritional value of food makes the big jaws and teeth of our forefathers more and more disposable. The level of specific mortality from impacted teeth, on the other hand, gives a hint of the relative evolutionary advantage of the bigger frontal brain.

For the topic of this section we may use two widespread behavior habits among modern populations of Northern European origin: sunbathing and consuming large quantities of milk, either fresh or processed milk. Both habits carry considerable health risks: Melanoma, an aggressive cancer of the skin, is associated with a history of excessive sun exposure, with blond, and pale individuals carrying a higher relative risk. It's incidence is growing fast among populations of Northern European origin at low latitudes (Southern United States, Australia, New Zealand, South Africa). Consumption of large quantities of milk fat is a major determinant of breast cancer and of coronary heart disease. Populations with a high proportion of fish meat and fish fat in their diet, but not of milk fat are exposed to a much smaller risk (Okayama *et al.* 1993).

Both habits have evolved as reactions to the same environmental challenge: the lack of sunlight at high latitudes.

Milk consumption as an adult requires the ability to absorb lactose (milk-sugar) which is the result of a culturally driven evolution. Because relevant general conclusions can be drawn from this example, some details will be given (A comprehensive overview which has been used for the following paragraph, is Durham 1991, chapter 5). Milk and many of its edible or drinkable derivatives are the embodiment of healthy nutrition. In the 1950 and 1960s when food aid was donated by industrial nations to developing countries all over the world, a substantial proportion of this aid came in the form of milk powder. It came as a surprise that a considerable proportion of white Americans and Europeans, and, even more important, the majority of populations in the world consists of individuals, who, after weaning, are lactose-malabsorbers. Those malabsorbers in diary societies quietly avoid milk, prefer milk-products which are low in lactose (like old cheese, curd etc.). Lactose is a disaccharide sugar, composed of two simple sugars, glucose and galactose. Lactose synthesis in the breasts of mammalian mothers has to be reversed exactly before lactose — in fact: its two components — can be absorbed through the membranous lining of the small intestine. The enzyme lactase (actually it is a whole class of them) which effects the hydrolysis of lactose back into glucose and galactose, is found in this membranous lining. Lactase is present in all mammalian newborns and sucklings; after weaning, its activity levels decrease rapidly. Given the fact that without lactase, any digestion of lactose is difficult (bacteries in the small bowl can break up small amounts of lactose), and drinking of large quantities of milk leads to fermentative diarrhea, the genetically controlled decline in lactose activity is a crucial, maybe the decisive factor triggering the onset of weaning. The evolutionary advantage of this early lactase decline in infants is obvious: it ensures that older children and adults do not drink the milk which has to go exclusively to small children. Among mammals, the presence of adults capable of digesting lactose in general will increase infant mortality. In fact, modern humans (and not even the domestic cat) are the only mammals with a substantial proportion of adult lactose absorbers.

Frequencies of lactose absorbers in human populations vary considerably by geographical location, history and technology (see table 3).

Several hypotheses were considered for explanation (Simoons 1978). The disease hypothesis: lactose malabsorbers have gone through an intestinal muccosa damaging disease; the induction hypotheses: continued milk consumption even after weaning induces the lactase enzymes in the mucous

TABLE 3. Indicators of Genetic and Cultural Diversity in Sixty Populations tested for Adult Lactose Absorption, 1966–1978, Categorized by Subsistence and Location

Category and population	Percentage of lactose absorbers	Approx. latitude	Total milk consumption (liters per person and year)	Cheese production (% of milk production)
Category A. Hunter-gatherers (traditionally lacking dairy animals): N=4				
1. Eskimos of Greenland	15.1%	62.2N	0.0	–
2. Twa' Pygmies of Rwanda	22.7	1.6S	0.0	–
3. !Kung Bushmen	2.5	19.6S	0.0	–
4. ǂhuâ Bushmen of Botswana	8.0	23.0S	0.0	–
Average	12.6 %[a]			
Category B. Nondairying agriculturalists: N=5				
5. Yoruba	9.0 %	6.3N	4.5	16.0 %
6. Ibo	20.0	4.4N	4.5	16.0
7. Children in Ghana	27.0	5.3N	0.6	12.0
8. Bantu of Zaire	1.9	4.2S	0.2	37.4
9. Hausa	23.5	12.0N	4.5	16.0
Average	15.5 %[a]			
Category C. Recently dairying agriculturalists: N=5				
10. Kenyans (mainly Bantu)	26.8 %	1.2S	67.5	0.3 %
11. Bantu of Zambia	0.0	15.3S	8.8	12.1
12. Bantu of South Africa	9.7	26.1S	92.1	10.2
13. Shi, Bantu of Lake Kivu area	3.6	1.6S	7.3	0.0
14. Ganda, other Bantu of Uganda	5.7	0.2N	29.8	0.0
Average	11.9 %[a]			
Category D. Milk-dependent pastoralists: N=5				
15. Arabs of Saudi Arabia	86.4 %	24.4N	NA[b]	
16. Hima pastoralists	90.9	1.0S	NA	NA
17. Tussi, in Uganda	88.2	0.2N	NA	NA
18. Tussi, in Congo	100.0	4.1S	NA	NA
19. Tussi in Rwanda	92.6	1.6S	NA	NA
Average	91.3 %[a]			

(continued)

living of the small bowl not to decline; the dietary inhibition hypothesis: consumption of spicy food or special stimulants like betel nuts, hashish or others consumed habitually in certain cultures prevents absorption of lac-

TABLE 3. Indicators of Genetic and Cultural Diversity in Sixty Populations tested for Adult Lactose Absorption, 1966–1978, Categorized by Subsistence and Location (continued)

Category and population	Percentage of lactose absorbers	Approx. latitude	Total milk consumption (liters per pers. & year)	Cheese production (% of milk production)
Category E. Dairying peoples of North Africa and the Mediterranean: N=5				
20. Jews in Israel	40.8 %	32.1N	203.6	40.1 %
21. Ashkenazic Jews	20.8	32.1N	203.6	40.1
22. N. Afrcan Sephardim	37.5	33.4N	28.5	10.7
23. Other Sephardim	27.8	37.0N	120.1	18.0
24. Iraqi Jews	15.8	33.2N	38.2	38.4
25. Other Oriental Jews	15.0	35.4N	71.1	33.8
26. Arab villager in Israel	19.4	32.1N	203.6	40.1
27. Syrian Arabs	5.0	33.3N	81.0	36.3
28. Jordanien Arabs	23.2	31.6N	15.8	54.3
29. Arabs (Jordan, Syria, etc.)	0.0	31.6N	15.8	54.3
30. Other Arabs	19.2	33.3N	81.0	36.3
31. Egyptian fellahin	7.1	30.0N	48.5	54.9
32. Greeks (mostly mainland)	52.1	37.6N	181.1	48.8
33. Greek Cretans	44.0	35.2N	181.1	48.8
34. Greek Cypriots	28.4	35.1N	141.2	56.4
35. Ethiopians/Eritreans	10.3	13.0N	21.8	5.4
Average	38.8 %[a]			
Category F. Dairying peoples of northern Europe (over 40°): N=12				
36. Danes	97.5 %	55.4N	1032.8	9.0%
37. Swedes	97.8	59.2N	393.1	24.4
38. Finns	85.1	60.1N	677.9	8.1
39. Northwest Europeans	87.3	51.3N	283.3	15.0
40. French	92.9	48.5N	580.3	21.2
41. Germans from Central Europe	85.5	52.3N	390.2	23.0
42. Dutch (living in Surinam)	85.7	52.2N	828.1	13.7
43. Poles (living in Canada)		52.2N	516.6	15.4
44. Czechs (living in Canada)	82.4	50.1N	391.5	20.2
45. Czechs (Bohemia, Moravia)	100.0	49.1N	391.5	20.2
46. Spaniards	85.3	40.2N	171.2	19.6
47. North Italians (Ligurians)	70.0	44.2N	210.9	37.6
Average	91.5 %[a]			

(continued)

TABLE 3. Indicators of Genetic and Cultural Diversity in Sixty Populations tested for Adult Lactose Absorption, 1966–1978, Categorized by Subsistence and Location (continued)

Category and population	Percentage of lactose absorbers	Approx. latitude	Total milk consumption (liters per pers. & year)	Cheese production (% of milk production)
Category G.Population Of "mixed" (dairying and nondairying) ancestry: N=13				
48. Iru	61.5 %	0.2N	29.8	0.0 %
49. Hutu	49.0	1.6S	7.3	0.0
50. Hutu/Tussi mixed persons	45.5	1.6S	7.3	0.0
51. Fulina/Hausa	33.3	12.0N	4.5	16.0
52. Yoruba/European mixed persons	55.8	6.3N	4.5	16.0
53. Nama Hottentots	50.0	26.0S	81.6	0.0
54. Eskimo/European mixed persons	62.0	69.2N	21.6	–
55. Yemen Jew/Arab mixed persons	55.6	32.1N	203.6	42.0
56. Skolt Lapps in Finland	39.8	69.5N	677.6	8.1
57. Mountain/Fisher Lapps	62.7	69.0N	677.6	8.1
58. Fisher Lapps in Finland	74.5	68.0N	677.6	8.1
59. Rehoboth Basters	35.0	26.1S	92.1	10.2
Average	62.0 %[a]			
OVERALL AVERAGE	62.0 %[c]			

a   Weighted by the sample size for each population in the category.
b   National-level statistics are not applicable to specialized pastoralists.
c   This figure must be used with caution because northern Europeans are greatly overrepresented in the subsample.
Source: Durham (1991) — Lactose absorption data from Simoons (1978).

tose; the genetic hypothesis: the observed differences are due to genetic difference between the groups. All hypotheses except the genetic one had to be abandoned, in the light of insurmountable counterevidence. There is considerable evidence, meanwhile, that adult lactose absorption is an autosomal (not located on the sex-chromosomes) dominant trait, and lactose malabsorption an autosomal recessive trait (Feldman and Cavalli-Sforza 1989). Cattle has been domesticated between 9.000 –11.000 years ago. Archeological evidence suggests, however, that initially, cattle was used as a source of meat only, and dairy did not emerge before 6.000 – 8.000 years ago, covering the short span of 250–300 human generations. Milk from other sources (sheep, goat, ren, camel, horse, donkey) came in use only later; none of these animals came even quantitatively close to cattle as a source of milk — especially fresh milk. Then milk was available beyond the strict limits of human milk, and the lactose malabsorption trait rather than be a necessary device for protecting survival of infants, excluded its bearer from a new and rich source of fat and proteins.

A look at table 3, however, shows that his new option was not sufficient a cause to make the malabsorption trait disappear. Human dairy technology, the art of turning fresh milk into more durable food which usually always entails breaking up lactose into carbohydrates which can be used also by lactose malabsorbers. A glance in the last column of table 3, especially in categories E, F and G, shows, that dairying people with substantial proportions of malabsorbers (for example those around the Mediterranean, the Arabian peninsula and in North Africa) consume only a fraction of the milk consumed in Europe and North America (there seem to be no data available for East and North Asian populations), but at the same time consume a much higher proportion of their milk production in processed form (cheese; curd etc.), we may - with all caveats - take this as an indicator for the proportion of milk which is consumed fresh.

Evidence has been gathered which shows that with some exceptions yet another selection pressure was needed beyond the general availability of bovine milk before the adult lactose absorption gene could make its way. Milk is not only a rich source of calcium, a crucial mineral for the skeleton, proper muscle function, blood coagulation and other vital processes, but its lactose can also serve as a substitute for vitamin D in facilitating the transport of calcium ions from the lumen of the small bowl through its walls — but only in lactose absorbers! Exposition to ultraviolet B light is necessary for synthesis of vitamin D in the skin. In high latitudes this exposition may be a rather scare resource, resulting in various diseases, for example, osteomalacia, hypertension and other rickets in children (this calcium absorption thesis was first proposed by Flatz and Rotthauwe 1973). Only where drinking fresh milk would bring the advantage of a superior calcium absorption, the lactose absorption gene spread. Incidentally — geographical areas where this is the case, are the same ones, where also skin depigmentation is advantageous: it lets more ultraviolet B light into the skin. What in the tropics is a hazard here is a welcome resource.

There seems to exist a counterexample to the calcium-resorption hypotheses: the not-dairying, dark skinned Eskimos (Inuit), among which, yet, rickets and osteomalacia are rare. At close inspection, however, the Eskimos provide another piece of support to the hypothesis. Traditional Eskimo diet contains large quantities of fish liver oil. Many fish, almost completely shielded by water against ultraviolet B light, have developed enzymatic ways of synthesizing vitamin D, which do not rely on light at all. Vitamin D, needed by all vertebrates, is enriched in the liver. In the Eskimo case, the nutrition culture and technology of a specific population found another way of solving the vitamin D deficiency problem; a solution which was so effective that it considerably reduced the selection pressure favoring depigmentation. The depigmentation of Nordic people around the Baltic sea (Scandinavians, Northern Germanic and Slavonic populations) is a hint that depigmentation has made its advance before the onset of dairy in this area. Indeed, it is thought that the spread of this trait (which seem to be controlled by some closely interacting genes) dates back between 10.000–35.000 years around the Baltic sea. The start of dairy could not revert the evolution of this feature in high latitudes.

In short: in the case of the milk-drinking Nordic populations with pale skin we have a perfect example of a genetic adaptation of a population to a self-made, novel environment. Widespread occurrence of rickets not only in Norway and Finland throughout history, but also in the crowded houses of the poor in Western and Central Europe, under the permanent smoke screen of early industrialization, points to the specific stress in the original environment; depigmentation and expanded milk consumption (with its negative by-consequences — for example the increased occurrence of breast cancer among those milk consuming populations) point to a successful adaptation. Modern food and transportation technology has eased, perhaps even eliminated the selection pressure for depigmentation in high latitudes. Indeed by migration and possibly by back-mutation, which today mostly will go unnoticed, pale skin, blond hair and blue eyes even in Scandinavia may be traits on the retreat. The new cultural habit of Northerners to expose their full body to the sun on southern beaches, a habit which relays on a possibly increased desire for sunlight among these populations (something black and brown skinned people are unknown to suffer from), makes them more vulnerable to cancer of the skin. Furthermore estrogens in the milk are suspected to cause infertility among male consumers of milk and milk products (Sharpe and Shakkeback 1993). Thus adaptation to the novel environment of widely availible vitamin D, cheap air travel and below-replacement birth rates in Northern countries, inviting migration from the South, may revert those evolutionary trends in the region towards lactose absorption, blond hair, blue eyes and pale skin. There are some cases which the calcium resorption hypothesis of lactose absorption cannot explain: some highly specialized milk-dependent pastoralists. (Category D in table 3). Their case shows that vitamin D deficiency — unlike an established dairy technology — is not a necessary condition for the spread of lactose absorption. The fact that the nutrition technology of these pastoralist did not become the dominant technology in their regions, on the other hand, speaks for the additional thrust provided by the vitamin D factor.

Shared food preferences are a very effective and universal mechanism among humans for creating groups identity: providing coherence within and marking differences between other groups. Given the equally universal tendency among humans to marry someone like themselves (homogameity) it can be expected that, in a diary society, immigrants from populations with a lower occurrence of the lactose absorption gene will find it easier to marry into the host population, if they carry the gene themselves. From a glance at the lower frequencies of this gene among Semitic populations (Jews living around the Mediterranean and in North Africa, and Arabs from various locations) in category E in table 3 it might be speculated that migrants form these populations to Northern Europe who have married someone of the host population who herself or himself is a lactose absorber, will possess this gene themselves at a higher proportion than migrants from their home population who have not. A remarkable feature of the evolution of pale skin in the populations around the Baltic sea is the contingency of the process, if we compare it with the Eskimo case. Indeed, had cattle breeding evolved earlier in the middle East and had the glaciers of the last Ice age receded later, depigmentation might not have evolved to the degree it has.No doubt, incidentally, that the Eskimo solution of the same problem of surviving in an environment with little sunlight - namely a diet based on fish — was superior to the Baltic solution — depigmentation and a bovine milk based diet — in terms of health hazards: the incidence of coronary heart disease among Eskimos in Greenland, living on a traditional fish-based diet, is 1/10 and less of the level found among Danes living in mainland Danemark (Kromann and Green 1980) or of Fins and Russians living in the Karelia region in Northern Finland or adjacent parts of Russia (Puska *et al.* 1993).

In this example: the spread of lactose absorption, we can in detail reconstruct the evolutionary path, we can identify lactose absorption as a new access to fat, protein and vitamin D, which offered new options as compared to the previous state (lactose malabsorption) in the populations around the Baltic sea, while it was on a global scale suboptimal, if compared with the Eskimo type of solution (a fish-based diet) in terms of the health hazards associated with a diary based diet.

From an evolutionary perspective, it is better to die from coronary heart disease or breast cancer than from rickets, which usually occurs many years earlier. But today, lactose absorption together with the features of blond hair, pale skin and blue eyes, has lost its evolutionary advantages even in those environments where it has evolved.

This explanation in all its parts can be falsified with according data: it may be shown, that there were alternative sources of vitamin D around the Baltic sea; that a large proportion of blond, pale and blue eyed persons are lactose malabsorbers; that all others things equal, dark pigmented populations like to sunbathe as blond and pale ones, and so forth. Also, it may be shown, that under original conditions, at high latitudes even today being a blond, pale, blue-eyed lactose absorber still may mean the difference between life and death.

5. Conclusion

It is true — and here the reader may recall the quotation from Joel Cohen, a demographer, at the beginning of this article — that evolutionary explanations cannot match physical explanations in their ability to predict future states of the processes under investigation. But that does not mean, that evolutionary explanations view the world only in the back-minor. We can make predictions which can empirically be tested. Therefore evolutionary explanations of human behavior as described here are neither circular nor, if they use the novel environment concept, necessarily self-immunizing. Frequency dependent selection is a field which seems to be especially fruitful for empirical tests of such explanations.

References

Barkow, J.H., Cosmides, L., and Tooby, J. (eds.) (1992) *The Adapted Mind. Evolutionary Psychology and the Generation of Culture.* Oxford University Press.

Bradbury, J.W., and Andersson, M.B. (eds.) (1987) *Sexual Selection: Testing the Alternatives.* New York: Wiley.

Brandon, R. (1978) Adaptation and Evolutionary Theory. *Studies in the History and Philosophy of Science 9*, pp. 181–206

Cann, R.L., Stoneking, M., and Wilson, A.C. (1987) Mitochondral DNA and Human Evolution. *Nature* 325 , pp. 31–36.

Cavalli-Sforza, L.L., Menozzi, P., and Piazza, A. (1993) Demic Expansion and Human Evolution. *Science* 259, pp. 639–646.

Chesnais, J.C. (1992) *The Demographic Transition. Stages, Patterns and Economic Implications.* Oxford: Claredon Press.

Durham, W.H. (1991) *Coevolution. Genes, Culture and Human Diversity.* Stanford University Press.

Enquist, M. (1985) Communication during Aggressive Interactions with Particular Reference to Variations in Choice of Behaviour. *Animal Behaviour* 33, pp. 1152–1161.

Feldman, M.W., and Cavalli-Sforza, L.L (1989) On the Theory of Evolution under Genetic and Cultural Transmission, with Application to the Lactose Absorption Problem. In Feldman, M. W. (ed.). *Mathematical Evolutionary Theory.* Princeton University Press.

Fisher, R. (1930) *The Genetical Theory of Natural Selection.* London: Oxford University Press (Reprint: New York: Dover 1958).

Flatz, G., and Rotthauwe, H.W. (1973) Lactose Nutrition and Natural Selection. *Lancet* 7820, pp. 76–77.

Grafen, A. (1990) Biological Signals as Handicaps. *Journal of Theoretical Biology* 144, pp. 517–546

Hasson, O. (1994) Cheating Signals. *Journal of Theoretical Biology* 167, pp. 223–238.

Kingdon, J. (1993) *Self-made Man.Human Evolution from Eden to Extinction.* New York:

Wiley.

Kromann, N., and Green, A. (1980) Epidemiological Studies in the Upernavik District, Greenland. *Acta Medica Scandinavica* 208, pp. 401–406

Low, B.S. (1993) Ecological Demography: A Synthetic Focus in Evolutionary Anthropology, *Evolutionary Anthropology* 1, pp. 177–187.

Maynard Smith, J. (1982) *Evolution and the Theory of Games.* Cambridge University Press.

Maynard Smith, J. (1989) *Evolutionary Genetics.* Oxford University Press.

Maynard Smith, J. (1994) Must Reliable Signals always be Costly? *Animal Behaviour* 47, pp. 1115–1120.

Maynard Smith. J.and Price, G.R. (1973) The Logic of Animal Conflict. *Nature* 246, pp. 15–18.

Mayr, E. (1975) *Evolution and the Diversity of Life.* Harvard University Press.

Moran, P.A.P (1962) *The Statistical Processes of Evolutionary Theory.* Oxford: Claredon.

Mueller, U. (1992) Birth Control as a Social Dilemma. In Haag, G., Mueller, U., Troitzsch, K.G. (eds.) *Economic Evolution and Demographic Change.* Springer, Heidelberg Berlin New York, 1992

Okayama, A., Ueshima, H., Marmot, M.G., Nakamura, M., Kita, Y., and Yamakawa, M. (1993) Changes in Total Serum Cholesterol and OtherRisk Factors for Cardiovascular Disease in Japan, 1980–1989. *International Journal of Epidemiology* 22, pp. 1038–1047.

Petrie, M., and Williams, A. (1993) Peahens lay more eggs for peacocks with larger trains. *Proceedings of the Royal Society London* B, pp. 1–5.

Pinker, S.and Bloom, P. (1990) Natural Language and Natural Selection. *Behavioral and Brain Sciences* 13, pp. 707–784.

Puska, P., Matilainen, T., Jousilahti, P., Korhonen, H., Vartiainen, E., Pukusajeva, S., Moisejeva, N., Uhanov, M., Kallio, I., and Artemjev, A. (1993) Cardiovascular Risk Factors in the Republic of Karelia, Russia, and in North Karleia, Finland. *International Journal of Epidemiology* 22, pp. 1048–1055.

Schulz, U., Albers, W., and Mueller, U. (eds.)(1994) *Social Dilemmas and Cooperation.* Springer.

Sharpe, R.M., and Shakkeback, N.E. (1993) Are oestrogens involved in failing sperms counts and disorders of the male reproductive tract? *Lancet* 341, pp. 1392–95.

Simoons, E.J. (1978) The Geographic Hypothesis and Lactose Malabsorption. *American Journal of Digestive Diseases* 23, pp. 963–980.

Vollmer, G. (1987) The Status of the Theory of Evolution in the Philosophy of Science. In Andersen, S., Peacocke (eds.) *Evolution and Creation.* Aarhus University Press.

Williams, M. (1973) The Logical Status of Natural Selection and Other Evolutionary Controversies. In M. Bunge (ed.) *The Methodological Unity of Science.* Dordrecht.

SOME THOUGHTS ON THE METHODOLOGICAL STATUS OF THE DARMSTADT MICRO MACRO SIMULATOR (DMMS)

HANS-DIETER HEIKE
Technical University of Darmstadt Institute of Statistics and Econometrics Darmstadt, Germany

1. Introduction

In this paper we attempt to evaluate the methodological status of the DMMS. We try to answer the question whether there is either a modest basis for an objective theory appraisal and theory choice or whether we have to be content with a subjective and conventional evaluation or whether perhaps methodological anarchism is the pure source of innovative science.

In order to give a tentative answer we have to delineate structure and application of the DMMS and at least some part of the model selection procedure. The evaluation is made on the basis of a subjective sample of methodological approaches reaching from logical empiricism to the positions of Kuhn, Feyerabend and Lakatos.

2. Structure and Applications of the DMMS

An overview of the structure of the DMMS is given by the presentation of the software architecture in Figure 1.

Basic components are the micro- and the macrosimulator, the linking driver, a relational database, an import interface with consistency check and software facilities such as the program and database generator and the runtime monitor (Heike *et al.* 1993).

Microsimulation models transfer a representative sample of decision units, e.g. households and enterprises, from one period to the following period. This transfer is performed with behavioral hypotheses, institution-

R. Hegselmann et al. (eds.),
Modelling and Simulation in the Social Sciences from the Philosophy of Science Point of View, 123–139.
© 1996 *Kluwer Academic Publishers. Printed in the Netherlands.*

Figure 1. DMMS software architecture

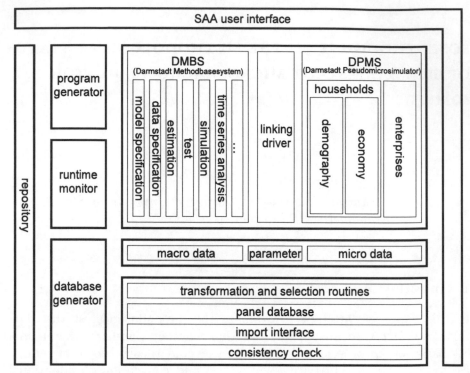

al regulations, technical relations and definitions that can be written in various forms as equations, tables, algorithms, programming sequences etc. A vector of attributes is computed for each agent by way of this procedure, so that one obtains a sequence of simulated samples (Heike and Kaufmann, A. 1987, Heike *et al.* 1988 und 1988a).

Figure 2. Principles of microsimulation

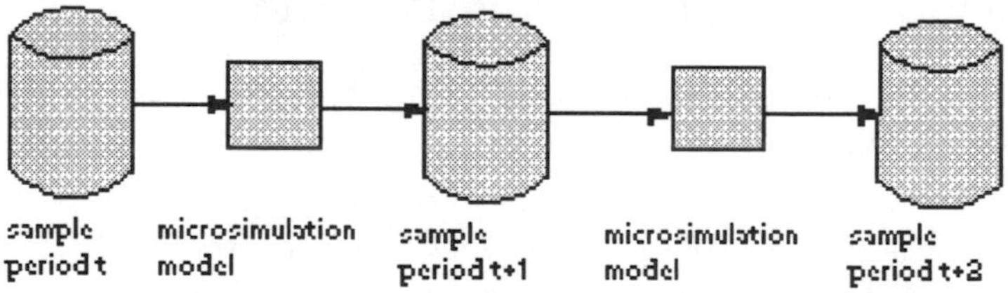

The variables of the microsimulation model of the household sector which are updated from period to period can be seen in Table 1.

The updating is performed sequentially in the way shown in Figure 3.

TABLE 1. Variables of the micro simulation model of the household sector

Area	Process of households and persons
Demographic Area	Death Birth Marriage Partnership Divorce Household mobilities
Education	School attendance Vocational guidance Retraining Additional education Private education and further education
Living situation	Rent or purchase of dwelling or house and maintenance Change of dwelling Purchase and maintenance of furniture
Transport and communication	Purchase and maintenance or rent of relevant goods and services
Participation in employment and employment income	Entry into occupational life Exit out of occupational life Change of occupation Change of enterprise Change of position Determination of working hours

(continued on next page)

The first implementation of the micro household sector was performed with data of the Income and Consumption Sample (EVS), which includes about 50000 households and 200 data per person. Simulations show satisfactory results, that is well documented by forecasting error calculations (Heike *et al.* 1988a).

The database is changed now from the EVS to the German Socio Economic Panel (GSOEP) provided by the Deutsches Institut für Wirtschaftsforschung (DIW), Berlin. This new database includes yearly samples of about 6000 households with roughly 1300 attributes (Heike *et al.* 1994).

The micro simulation model of the business sector (see Figure 4) resembles that of the household sector regarding the simulation technique; a

TABLE 1. Variables of the micro simulation model of the household sector (continued)

	(continued from previous page)
Area	Process of households and persons
Social transfers	Pensions Unemployment payments Dwelling subsidies Social aid Other social transfers Transfers from other private households
Taxes and social security	Wage and income tax Church-rate Capital tax Other taxes Legal and private payment for social security Transfers to other private households
Standard of living	Purchase of food and luxury goods Clothes and goods of household keeping
Leisure and entertainment	Purchase and maintenance or rent of relevant goods and services
Non-market activities	Cultivation of an own garden Illegal employment and relevant receipts and payments Free of charge services
Wealth	Saving Changing type of wealth Transfer of wealth

sample is written forth from period to period in order to show the structural development of the enterprise sector. The data include profit and loss accounts and balance sheets of enterprises (Güldner 1992).

However, the economic development covering several periods can only be simulated if the mainly cross-section oriented microsimulation models are linked to time series based macromodels. By this linking procedure medium to longer term influences of economic variables can be taken into account. For instance price, wage and income development have to be simulated in a macromodel and are then to be transferred through linkage. Furthermore secondary effects and feedback effects of political programs cannot be simulated without the linking procedure.

Figure 3.   Updating in a single simulation step

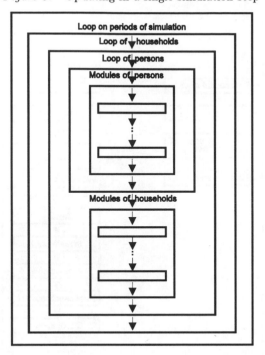

According to the total design of the DMMS the macrosimulation models used are not general equilibrium models but models that permit disequilibrium, i. e. non-market clearing at given prices resp. inconsistency of the hypothetical demand and supply plans. The macrosystems used are either traditional macroeconometric models built around the Hicksian IS/LM-scheme or rationing models.

Such a macromodel can be defined as a system of K nonlinear simultaneous equations by

$$\Phi_{kt}(Y_t, X_t, B_t) \quad = \quad u_{kt}$$

Y_t :   vector of endogenous variables $k \; = \; 1,2,\ldots,K$
X_t :   vector of exogenous variables $t \; = \; 1,2,\ldots,T$
B_t :   vector of unknown parameters
u_{kt} :   random variable

One of the main applications of macromodels is forecasting the vector of endogenous variables Y_t. This can be performed in the following way. First the parameter vector B_k has to be estimated, e. g. by using the method of nonlinear three-stage least squares or the full information maximum likelihood method. Then we have to solve the system for the endogenous

Figure 4.   Micro simulation model of the business sector

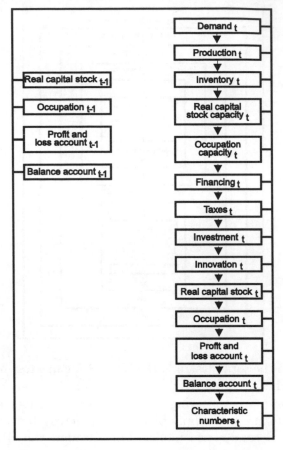

variables. Suppose that the structural equations can be written as

$$\Phi(Y_p, X_p, B) = U_p \qquad p : \text{prediction period}$$

If $\hat{B}$ is the estimator of B and we can solve the system for $\hat{Y}_p$ we have the predictor

$$\hat{Y}_p = \tau(X_p, \hat{B})$$

3. Model Selection

Model selection in DMMS covers a wide variety of procedures from the selection of behavioral relations and methods of estimating demographic probabilities to the solution algorithms of nonlinear models and the convergence algorithms of the linking driver.

3.1. A VARIANT OF THE STANDARD ECONOMETRIC PROCEDURE

The econometric modelling taken as representative of the DMMS will be
discussed in some detail. I start with a variant of the standard procedure
of which a rough flow chart is shown in Figure 5.

Figure 5.   Flow chart of the standard econometric procedure (variant)

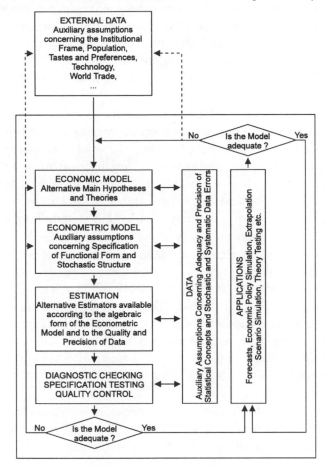

Step one determines the data frame, frequently including instutitional
regulations, population, preferences, technology and world trade. Assumptions concerning this frame have to be kept in mind as auxiliary hypotheses.

Step two symbolizes the role of economic theory which generally explains
how agents decide in an economy and how those agents interact through
markets. Theory includes therefore objective functions, decision variables,
expectations, constraints, the available information and the actual choices.
Usually the step from micro to macro level is made by introducing representative agents and ignoring the aggregational problems. On the macro

level additional constraints such as those of national accounting have to be met. The nature of theory implies that for each decision a number of economic models are available. Mostly these theoretical models are reduced forms of the decision making process and usually they provide relations between variables and signs of coefficients. Theory has little to say about functional form and lag length. Theory based on a priori and empirical knowledge therefore shows which variables are important in decision making and how data are generated. Consequently the data generating process is not characterized by general interconnectedness of data but by variables which directly influence economic agents in specific situations. This role theory can play so much the better the more it is corroborated in severe tests.

Step three includes collecting and screening of data. Data and generated variables should represent theoretical variables as close as possible. Data should meet i. a. the constraints of national accounting. Modelling, estimating and testing are restricted by quality and availability of data. Unobservable data, such as expectations, require special treatment. These have to be modelled within steps two and four. Assumptions concerning precision and adequacy of data may be handled as auxiliary hypotheses.

Step four. Since theory mostly does not specify functional form, lag length and stochastic structure, these features have to be specified by the econometric model. A variety of functional forms is available and various lag lengths will be tested. As concerns the stochastic structure most frequently a white noise error term is added. This error term is representing omitted variables and also measurement and approximation errors. The white noise property is regarded as theoretically and statistically convenient since e.g. its absence invalidates a number of statistical tests and omitted variables and misspecifications of the functional form can lead to nonrandom disturbances.

In view of the state of the art in economics and econometrics, econometric models are at best approximations of actual structures frequently in form of linear hyperplanes and they differ with respect to robustness against changes of the environment and exogenous variables. Summarizing the auxiliary assumptions of step four include i. a.: functional form, lag length and functional form of the error term.

Within step five alternative estimators are used dependent on the algebraic form of the econometric model, the available data and nonsample information.

The resulting econometric structures have to be checked diagnostically in various ways and by a number of specification tests within step six. Refutations first lead to modifications of the auxiliary assumptions. If these finally cannot be refuted by the above mentioned procedures and tests, then the main hypothesis is checked in various ways, i. a. by using methods of stochastic simulations and later on by means of real time forecasts.

This iterative model selection procedure is guided by methodological rules which comprise the determination of significance levels, the acceptance of limits to the repetitive process of testing after which contradictory evidence or the failure of refutation is accepted and the consideration of pretest bias which arises because of unavoidable β-errors (type II errors) in an iterative regression strategy. Since little is known concerning the sampling properties of pretest estimators the results of the described search strategy should be evaluated cautiously and interpreted in the light of further information such as theoretical knowledge and forecasting experiences. Nevertheless it should be kept in mind that pretesting also provides additional evidence in the course of specification.

3.2. ALTERNATIVE APPROACHES

In recent years alternatives to the traditional model selection procedure have been proposed and those of Hendry, Leamer and Sims together with cointegration and error correction will be outlined in the following subsections.

3.2.1. *Hendry's General to Specific Modelling*
In a series of papers Hendry — partially with coauthours — (Hendry 1979; Hendry and Richard 1982) proposes a general to specific modelling strategy which is mainly applicable to dynamic time series models. Starting with static long-run equilibrium relationships these equations are augmented by lags to model the disequilibria which are supposed to generate the data. These overparametrized models are then reduced by a sequence of simplification tests in accordance with a number of characteristics discussed by Hendry. Hendry criticises the bottom-up approach which is contrary to his general to specific strategy and frequently used in practical econometrics because it might not lead to the best model. He argues that the tests could be flawed, because they depend on assumptions which are tested later in this sequence and moreover the significance levels of the sequential tests are unknown. It has to be kept in mind however that also the Hendry procedure has its weaknesses e.g. the initial specification might omit relevant variables.

3.2.2. *Leamer's Bayes Oriented Approach*
Leamer's contribution to econometric methodology consists of a taxonomy of specification searches and a Bayes-oriented search strategy (Leamer 1978). According to him six types of specification searches have to be distinguished: hypothesis-testing search, interpretive search, simplification

search, proxy variable search, data selection search and post-data model construction.

Since specification searches as used in econometrics invalidate the traditional models of statistical influence he proposed a Bayesian approach. He shows e.g. that in case of two competing models such as

$$Z = Y_1 B_1 + U_1$$
$$Z = Z_2 B_2 + U_2$$

the choice between model 1 and model 2 depends on a posterior odds ratio (Leamer 1978, chapter 4). This ratio is a function of the marginal probability density functions of the sample observations being determined by prior densities of the parameters. Since actually there are unsurmountable difficulties to specify these a priori densities in a reasonable way Leamer proposes to subdivide the explanatory variables in focus variables and doubtful variables. For all combinations of the doubtful variables the coefficients of the focus variables are calculated with Leamer's Search program in order to get the extreme values of these coefficients. By way of this procedure the sensitivity of the focus coefficients can be analyzed under various specifications (Leamer 1978, chapter 5). The traditional approach could be supplemented by this procedure. Research in this direction seems to be rewarding.

3.2.3. *Sims' VAR Approach*

Sims distrusts the theoretical constraints on simultaneous econometric systems and introduces an equation system known as vector autoregressive model (VAR model) (Sims 1980; Sims 1988; Sims 1993).

$$
\begin{aligned}
Y_t &= A(L)U_t \\
Y_t &: \quad M \times 1 \text{ vector} \\
U_t &: \quad M \times 1 \text{ vector} \\
A(L) &: \quad \text{Parameter matrix in Lag operator } L
\end{aligned}
$$

which can be rewritten if $A(L)^{-1}$ exists as

$$A(L)^{-1}Y_t = B(L)Y_t = U_t$$

which is the case if the above system contains only stable difference equations. This system is estimated and applied in forecasting, testing for causality and identifying policy effects. It should be remembered however that only a small number of variables can be accounted for in view of the proliferation of parameters, that a VARMA (vector autoregressive process with moving average error process) process might be a better representation if these variables are generated by a stationary process, that the order of the process is generally unknown and that many economic time series show nonstationarities that cannot be removed by transformation. Furthermore a recent comprehensive analysis of forecasting performance covering twenty-two years shows that the group consensus forecasts of NBER-ASA (National Bureau of Economic Research, Cambridge, Massachusetts, - American Statistical Association) of survey ranks first in each of the periods covered, Michigan forecasts (econometric model forecasts) were second best, time series forecasts share mostly the third and fourth rank. Within the NBER-ASA survey that includes about hundred individual forecasters those who used well known econometric service bureaus ranked first according to forecasting performance (Zarnowitz and Braun 1993).

3.2.4. The Error Correction (EC) and Cointegration Approach

The error correction model is a generalization of lag models such as the partial adjustment model. Error correction models combine a long run relationship between an endogenous variable y and an explanatory x-variable with a short run relationship between the deviations of y from its long run trend and deviation of the x-variable from its long run trend as is demonstrated by a simple example which we start with the following dynamic model:

$$Y_t = a_1 + a_2 x_t + a_3 x_{t-1} + a_n y_{t-1} + u_t$$

and the long run relationship:

$$Y = b_1 + b_2 x \quad \text{implying the long run constraints}$$
$$\frac{a_2 + a_3}{1 - a_4} = b_2 \quad \text{and obtain}$$
$$\Delta Y_t = a_1 + a_2 \Delta x_t - (1 - a_4)(y_{t-1} - b_2 x_{t-1}) + u_t$$

It can be shown that the OLS-estimator of the coefficients is consistent if the variables are integrated I(1) and that the OLS-estimator of the coefficients are asymptotically normal distributed so that standard inferential procedures can be applied, if the y and x-variables are cointegrated and x is weakly exogenous. (Engle and Granger 1987; Park and Phillips 1988; Hansen 1993, S. 128 ff). Hence the ECM-cointegration approach warrants that

a) standard inferential procedures can be applied if cointegration and weak exogenity are taken for granted and

b) long run relationships may be explicitly accounted for in dynamic relationships.

This is a flexible dynamic approach which should be tested in further applications. One has to keep in mind however, that in reality no long run equilibria exist but only long run changing trends.

4. Methodological Positions

Structure, application and model selection of the DMMS are to be judged in the light of methodological positions ranging from rational theory choice to irrational method dadaism. An outline of a sample of methodologies out of this spectrum may be helpful to specify the basic elements and critical points of objective and subjective canons of theory appraisal.

The main tenets of logical empiricism (Carnap 1959; Carnap and Jeffrey 1971; Nagel 1961) are: Theories in natural and social sciences are of hypothetico-deductive nature. Only theories as a whole can be subjected to testing. Theoretical terms get meaning indirectly by testing the theories. Theory may be judged according to their degrees of confirmation. Confirmation depends on the number of favourable test outcomes, the precision of observation and measurement, the variety of supporting evidence and new test implications. The most highly confirmed theory is considered to be the most probable. It is realized that a highly confirmed theory need not be the true one. Additional criteria such as logical consistency, simplicity, generality, extensibility are recommended.

The position of logical empiricism has been criticized on a variety of reasons. One of the main critics was Popper and one of his central points of criticism was, that by the rules of logical empiricism the problem of induction was not solved and that on behalf of this no inductive logic had been achieved and the problem of choice was not solved.

According to Popper the proper method of science is to work with strong conjectures that forbid much, have high empirical content and could be subjected to severe tests. Those theories that survive have high empirical content but low probability (Popper 1965, 1976, 1984). Popper's approach too has been sharply criticized. In order to apply falsificationism a conditional test of a theory must be possible in which the conclusion follows from initial conditions and general laws. But in economics it is generally impossible to enumerate all initial conditions, furthermore general laws (specified numerically) are not available in economics and are also not provided by econometrics, so that clean tests are rare. Furthermore is has been pointed out that without accepting so-called ad hoc theory changes large of parts of the economic theory soon would be questioned if not eliminated by disconfirming evidence. Finally it has been argued that the epistemological position of Popper rules out the consideration of supporting evidence (Caldwell 1982, chapter 12; Caldwell 1991).

The objectivity of theory appraisal inherent in confirmationism and falsificationism was completely denied by Kuhn, Feyerabend and others. According to Kuhn (Kuhn 1970, 1977) traditional criteria such as precision, content, fruitfulness, fertility may be used within normal science which is dominated by one paradigm; yet also in normal science the specific weight of these criteria varies between scientists. But in time of scientific revolutions when paradigms are changing, theory choice is in no way guided by such criteria. Competing theories are incommensurable and so standard criteria are not sufficient for an objective choice. Norms and values are decisive and additionally social and political theories and other subjective factors. Sociological studies may be helpful for discovering the development of science.

Feyerabend carries the criticism of Kuhn to extremes (Feyerabend 1983). According to his view theories generally are incommensurable e.g. because the content of concepts is changing, they are irrefutable, furthermore, because all data are theory-dependent and there does not exist a neutral observation language. Experience shows, according to his view, that there is no theory which is consistent with all known facts and that scientific knowledge is the result of working with incommensurable theories. Therefore, no objective knowledge exists and consequently the methodological proposal is methodological anarchism. This captures the actual history of scientific development according to Feyerabend.

Lakatos (Lakatos 1978) believes that his methodology avoids the subjective relativism of Kuhn and the anarchism and dadaism of Feyerabend. According to his methodology of scientific research programs, that program has to be chosen which over time has progressive problemshifts, that means which continues to predict new facts some of which are corroborated. He is of the opinion that his program comprises elements of falsificationism and the growth of knowledge literature since refutation and corroboration as well as rational reconstruction of science are adequately included. There is no immediate rationality but long run objectivity in theory appraisal according to his position. But he never explains when and under what conditions one should accept a scientific program and refute the competing ones.

This is only a small sample of alternative methodological programs which does not include positions such as structuralism, holism or evolutionary approaches but I hope it will suffice to support a preliminary evaluation of the methodological status of the DMMS.

5. Tentative Methodological Evaluation of the DMMS

The general purpose of the DMMS is to serve as a frame supporting explanation, forecasting, policy formulation and evaluation, simulations of alternative monetary and fiscal actions and eventually testing and development of theories.

Applications such as forecasting, policy simulation and evaluation should be made only after intensive systematic testing to get a model that is an adequate and robust approximation of reality, where robustness means little model sensitivity against changes of external data and exogenous variables. The question is whether the standard procedure of model selection presented for econometric model choice supports this goal in a sufficient objective manner or not.

Within the described iterative research strategy we have to work with variables which can be measured as far as possible independently from theory so that these data can be used as objective instances in the process of refutation and corroboration. Numerous economic concepts are defined in the context of theories, there exist e.g. different concepts of the national product and when using a theory one has to decide which one should be applied. But since the content of these concepts regularly is defined with sufficient precision and can be measured independently of theory empirical statements including these may be used in an objective manner as test instances. It should be remembered furthermore that those data mostly are gathered and sometimes also defined independently of current research by official statistic institutions. Concepts may also have different meaning in competing theories, but this difference can be identifed and taken into account.

Using these data as test instances we have to consider by which rules it should be decided whether a model is refuted or corroborated in the light of empirical evidence. There is no simple answer available since models mostly are stochastic, complex, nonlinear and formulated under ceteris paribus conditions. The stochastic nature of models means that we have to agree upon a significance level which defines the regions of acceptance and refutation taking into account possible losses. The complexity of models implies that in view of the great number of assumptions accompanying the main hypothesis resp. the economic model an immediate refutation is not possible. If a forecast e. g. gives contradictory evidence to a model as a whole mostly it is not clear which part of the model is false so that a battery of tests concerning central and auxiliary hypotheses has to be performed. The consequence of nonlinearity is a more or less strong dependence of simulation solutions upon the initial conditions. From the c. p. clause it follows that if after severe testing a model finally is not refuted we cannot be sure that this is only a preliminary success not only because of the probability nature of our forecasts but also because of possible changes of preconditions such as assumptions reflecting the institutional frame, population, tastes and preferences, technology a. s. o. and also because of possible relations between variations of exogenous variables of our models and the econometric structures. This is a critical point because it is extremely difficult to completely internalize these relations in order to close the models. We have to realize as actual state of the discipline that our models are imaging only a limited space-time interval and have to be adapted from time to time.

Finally we have to acknowledge that owing to the stochastic nature of our models β-errors are possible which lead to pretest bias in the sequence of testing and estimating and invalidate standard inferential procedures. Consequently a further methodological norm has to be introduced which informally incorporates the pretest bias into our procedure and leads to a cautious interpretation of our results.

Summarizing we may conclude that the objectivity of our knowledge depends on the truth and precision of evidence and on the procedures of testing, refuting and corroborating. The objectivity is restricted as a consequence of the stochastic, complex and nonlinear nature of our model and the unavoidable use of *ceteris paribus*-clauses. This forces us to use methodological rules which lead to preliminary, time consuming and cautiously to be interpreted results of our test procedures. That implies that for a more or less long time competing theories and models are used simultaneously and that we cannot be sure that by way of the described selection process we eventually approximate reality. But experience shows that the above discussed program establishes theories that are modestly successful in forecasting, explaining and policy application. Finally those theories are discarded that are not corroborated in a sequence of trials; e.g. macroeconomic equilibrium models are no longer applied as guide to economic policy in situations of heavy long-run unemployment. One may tell an unemployed person once that his unemployment is the consequence of his free decision between labour and leisuretime but you cannot tell this story all unemployed persons all the time. We are not sure whether we are approximating truth but in any case corroborating test outcomes are a necessary precondition on the way to truth.

References

Blaug, M. (1980) *The Methodology of Economics or how Economists explain*. Cambridge, London, New York, Sydney.

Boland, L.A. (1982) *The Foundations of Economic Method. London*. Boston, Sydney.

Bryant, R.C., Henderson, D.W., *et al.* (eds.) (1988) *Empirical Macroeconomics for Interdependent Economies*. Washington.

Bryant, R.C., Holtham, F., and Hooper, P. (1988) Consensus and Diversity in the Model Simulations. Bryant, R. C., Henderson, D. W., *et al.* (eds.), *Empirical Macroeconomics for Interdependent Economies*. Washington, pp. 27–62.

Caldwell, B.J. (1982) *Beyond Positivism, Economic Methodology in the Twentieth Century*. London, Boston, Sydney.

Caldwell, B. J. (1991) Clarifiying Popper. *Journal of Economic Literature*, XXIX, pp. 1–33.

Carnap, R.(1959) The Elimination of Metaphysics through Logical Analysis of Language. Ayer, A. J. (ed.), *Logical Positivism*. Glencoe, Ill., pp. 60–81.

Carnap, R., and Jeffrey, R.C. (eds.) (1971) *Studies in Inductive Logic and Probability*.

Berkeley.

Darnell, A.C., and Evans, J.L. (1990) *The Limits of Econometrics*. Aldershot. Brookfield.

Engle, R.F., and Granger, C.W.J. (1987) Cointegrations and Error Correction. Representation, Estimation, Testing. *Econometrica*, 55, pp. 251–276.

Feyerabend, P.K. (1983) *Wider den Methodenzwang. Skizze einer anarchistischen Erkenntnistheorie*. Frankfurt.

Granger, C.W.J. (1992) Evaluating economic theory. *Journal of Econometrics*, 51, pp. 3–5.

Greenwald, B.C., and Stiglitz, J.E. (1988) Examining Alternative Macroeconomic Theories. *Brookings Papers on Economic Activity*, pp. 207–260.

Güldner, M. (1992) *Das Nachfragemodul im Mikrosimulator Unternehmenssektor — Modellkonstruktion, Probleme und Lösungsansätze*. Dissertation, Technical University Darmstadt

Hansen, G. (1993) *Quantitative Wirtschaftsforschung*, München.

Heike, H.-D., Beckmann, K., Fleck, C., and Ritz, H. (1994) The Darmstadt Micro Macro Simulator: GSOEP — Consistency Check and Data Modelling. *Vierteljahreshefte zur Wirtschaftsforschung*, Heft 1/2, pp. 139–144.

Heike, H.-D., Beckmann, K., Kaufmann, A. and Sauerbier, T. (1994) Der Darmstädter Mikro–Makro-Simulator: Modellierung, Software Architektur und Optimierung. Faulbaum, F. (ed.), *SoftStat'93, Advances in Statistical Software 4*. Stuttgart, New York, pp. 161–170.

Heike, H.-D., Beckmann, K., and Ritz, H. (1993) *The Development of a Micro Macro Simulator in a 4GL Environment*. Paper presented at the IARIW-Conference on "Microsimulation and Public Policy", Canberra, Australia. Paper No. 80 of the Institute of Economics, Technical University Darmstadt.

Heike, H.-D., Hellwig, O. and Kaufmann, A. (1988) Der Darmstädter Pseudomikrosimulator, Modellansatz und Realisierung". *Angewandte Informatik*, pp. 9–17.

Heike, H.-D., Hellwig, O. and Kaufmann, A. (1988a) Das Darmstädter Mikrosimulationsmodell — Überblick und erste Ergebnisse. *Allgemeines Statistisches Archiv*, 72, pp. 109–129.

Heike, H.-D., and Kade, G. (1968) Methodologische Probleme makroökonomischer Theorien. *Konjunkturpolitik*, 14, pp. 291–374.

Heike, H.-D., and Kaufmann, A. (1987) Charakterisierung und Vergleich des Sfb 3 und des Darmstädter Mikrosimulators. *Angewandte Informatik*, pp. 9–17.

Heike, H.-D. and Krupp, H.-J. (1972) Die Ökonometrie verbindet Theorie und Empirie. Molitor, R. (ed.) *Kontaktstudium Ökonomie und Gesellschaft*, Frankfurt, pp. 128–133.

Hendry, D.F. (1979) Predictive Failure and Econometric Modelling in Macroeconomics: The Transactions Demand for Money. Ormerod, P., ed., *Economic Modelling* London pp. 217–242.

Hendry, D.H., and Richard, J.F. (1982) On the Formulation of Empirical Models in Dynamic Econometrics. *Journal of Econometrics*, 20, pp. 3–33.

Kaufmann, A. (1988) *Systematische Entwicklung von Mikrosimulationssoftware*. Dissertation, Technical University Darmstadt.

Kastrop, C. (1993) *Rationale Ökonomik, Überlegungen zu den Kriterien der Ökonomischen Theoriendynamik*, Berlin.

Kuhn, T.S. (1970) *The Structure of Scientific Revolution*. 2nd. Enlarged Edition. Chicago.

Kuhn, T.S. (1977) *The Essential Tension*, Chicago.

Lakatos, I., (1978) *The Methodology of Scientific Research Programmes*. Vol. 1 of His Philosophical Papers. Edited by Worrall, J. and Currie, F., Cambridge.

Leamer, E.E. (1978) *Specification Searches. Ad hoc Inference with Nonexperimental Data*. New York, Toronto.

Nagel, E. (1961) *The Structure of Science: Problems in the Logic of Scientific Explanation*. New York.

Park, Y. and Phillips, P. C. B. (1988) Statistical Inference in Regression with Integrated

Processes; Part 1. *Econometric Theory*, 4, pp. 468–497.

Popper, K.R. (1965) *Conjectures and Refutations*, 2nd ed., London.

Popper, K.R. (1976) *Logik der Forschung*, 6. Auflage, Tübingen.

Popper, K.R. (1984) *Objektive Erkenntnis*, 4. Auflage, Hamburg.

Redman, D.A. (1991) *Economics and the Philosophy of Science*. New York, Oxford, Toronto, Melbourne.

Sims, C.A. (1993) A Nine-Variable Probabilistic Macroeconomic Forecasting Model. Stock, J. H. and Watson, M. W. ed., *Business Cycles, Indicators and Forecasting*. Chicago, London, pp. 179–212.

Sims, C.A. (1988) Identifying Policy Effects. Bryant, R. C., Henderson, D. W., *et al.* (eds.), *Empirical Macroeconomics for Interdependent Economies*. Washington, pp. 305–321.

Sims, C.A. (1980) Macroeconomics and Reality. *Econometrica*, 48, pp. 1–48.

Stock, J.H., and Watson, M.W. (eds.) (1993) *Business Cycles, Indicators, and Forecasting*. Chicago, London.

Zarnowitz, V., and Braun, P., (1993) Twenty-two Years of the NBER-ASA Quarterly Economic Outlook Surveys: Aspects and Comparisons of Forecasting Performance. Stock, J. H., Watson, M. W. ed., *Business Cycles, Indicators and Forecasting*. Chicago, London, pp. 11–93.

ON THE MEASUREMENT OF ACTION

WOLFGANG BALZER
Institut für Philosophie, Logik und Wissenschaftstheorie
Universität München
München, Germany

Most social theories have actions among their objects. This may not seem obvious for pure macro-theories [1] and for purely numerical mathematical models [2] where actions can not be made out. Even in these cases, however, actions do play a central role. They form an underlying level from which a theory's terms either are obtained by statistical procedures of aggregation or are obtained as numerical representations. In the case of pure macro-theories the determination of action does not appear as a major methodological difficulty because variation of action apparently is cancelled out by aggregation. In this paper, I have nothing to say about theories of the latter kind which does not mean that they are without problems. Also, I will not discuss the problems of 'representationalist' theories in which actions are represented by numerical terms. What I have to say here will however become pertinent to such cases once the representation or aggregation is no longer taken for granted, and itself becomes the object of theoretical investigation so that representation or aggregation itself explicitly occurs in the theoretical model. In the present paper, I concentrate on the main case which is most important methodologically and perhaps also by the number of its occurrences in science. In this case a social theory has actions among its objects. In the following I will refer to theories of this kind as *primary* social theories.

1. A Theory's Objects and Their Functioning

Each theory's models are made up from at least two kinds of things: objects and relations. From a standpoint external to the respective theory the objects may be simple or complex, observational or theoretical, idealized or

---

[1] See Allen (1968) and the discussion in Schlicht (1985).
[2] Like Debreu (1959) in micro economics or Taylor (1976) in game theory.

R. Hegselmann et al. (eds.),
Modelling and Simulation in the Social Sciences from the Philosophy of Science Point of View, 141–156.
© 1996 *Kluwer Academic Publishers. Printed in the Netherlands.*

realistic. From within the theory however, its objects are 'atomic', simple, unspecific. They get their specificity exclusively in terms of classifications, properties and relations to other objects which are stated and made explicit by the theory itself. A theory's relations ('terms', 'functions'), on the other hand, relate nothing but the theory's objects, and perhaps numbers or similar purely[3] mathematical stuff. This characterization of objects and relations seems obvious enough; it is the basis of model theory, and is sufficiently documented in the literature both theoretically and in terms of examples. [4]

In applying an empirical theory to a real system, data have to be collected which form the basis of various procedures of check, or fit, with the theory's models or hypotheses. The specific way or ways in which data and models are fitted is of no concern here. What is important is the format of data. Data usually have the form of atomic sentences, $R(a_1, ..., a_n)$, where R is (a symbol for) an n-ary relation, and $a_1, ..., a_n$ are (names of) objects.[5] A necessary condition for an atomic sentence to become, or express, a datum, is that the theory's practitioners have agreed on the particular objects $a_1, ..., a_n$ and on the nature or meaning of the relation R. This, in turn, usually implies that practitioners have agreed on special ways of determining the objects and the relation. They have procedures in order to refer to each object by its name in a definite way and procedures to find out whether relation R obtains among n given objects or not. These procedures may be called methods of determination. In order to collect data for a theory, it is necessary to have methods of determination for at least some of the theory's objects, and thus to determine at least some of the objects.

It is important not to be misled by ordinary language according to which it is a *property* (like length, distance, colour, etc.) which is measured or determined. Using this terminology one might reason as follows. Actions are not properties, so actions are not measured, so there is no problem of measuring actions.[6] The falsity of this way of reasoning can be seen through two lines of argument. First, on a foundational level we must use precise language and cannot systematically use the sloppy language of the working scientist. Speaking precisely, a property is a unary relation, and the expression 'to determine (measure) a property' is a shorthand for 'to determine (measure) whether a given object has a property'. In any actual measurement the object is given, and what is measured is a 'degree' or

---

[3]To be more precise, 'purely' mathematical entitites are those for which there exist standard interpretations in a commonly accepted mathematical theory. This excludes things like events in probability theory from being seen as purely mathematical.

[4]See, for instance Balzer *et al.* (1987).

[5]For reasons of simplicity, I suppress 'negative' data having the form of negated atomic sentences.

[6]I am indebted to H.Westmeyer at this point.

'expression' or 'quantity' in which this object has a property. To measure 'work satisfaction', for instance, means to determine for a concrete person, or a well specified statistical aggregate, the 'degree' to which the person or the statistical representative of the aggregate is satisfied, and this is true even for the simplest, nominal scale type. Each measurement in social science involves objects in this sense. The fact that these objects, like 'work satisfactions', often are not explicitly introduced as forming part of a theoretical picture certainly is interesting and needs explanation, but it cannot be used to argue that such objects are irrelevant. We have to accept that every measurement involves objects. In social science, these objects often are of a highly abstract kind, as indicated by 'intelligence', 'satisfaction' or 'role'. This leads to abstract, formal or mathematical models the interpretation of which is difficult. As stated in the introduction I will not go into the analysis of such theories here but simply claim that such analysis ultimately leads to a primary social theory. In primary social theories among the objects there are actions, and these are involved in measurements. In standard terminology, the determination of an action may be regarded as measurement on a nominal scale: the 'nominal value' of the action is measured.

A second line of arguing that actions are measured is by analogy to the natural sciences. In measuring length, for instance, it is said in a sloppy way that we determine a 'property'. However, any real measurement of length involves a real object whose length is measured, and what is determined is the 'degree' or 'measure' in which the property 'length' is realized or 'possessed' by that object. Strictly speaking it would be rather strange to say that 'the property length' has been determined, for the combined efforts of all mankind are not sufficient to do so. Measurement only determines some few, concrete 'lengths'. In the natural sciences the objects involved in measurement usually are 'simple': material bodies, well specified events, or simple constructs of such. They usually are explicitly mentioned as 'parts' of, say, physical theories.

In the natural sciences the determination of objects often is easy; it is achieved by means of other, more elementary theories, by means of experiment, and sometimes by direct observation. Even in the natural sciences, however, we meet clear cut cases in which some of a theory's objects are determined by using the theory itself. In Daltonian chemical theory, for instance, molecular weight — which is a theoretical concept in this theory — may be used in order to determine a particular chemical substance.[7] In primary social theories, on the other hand, the complex nature of human actions raises problems of determination and measurement of a special kind

---

[7]See Lauth (1989).

which have been the subject of considerable thought and writing.

2. The Structuralist View of Measurement

One way of avoiding sweeping discussion here is to focus on a precise notion
of determination or measurement. The term measurement is sometimes
used in a rather narrow meaning according to which an assignment of real
numbers to objects is one essential ingredient of measurement. I will use the
term in a broader way — to be described below — such that measurement
can take place without numerical representation and covers all kinds of
determination. This broader usage is backed by instances both from the
natural sciences (e.g. yes–no experiments in quantum physics which do
not involve numbers) as well as from the social sciences (e.g. qualitative,
'nominal' , scales as instruments of measurement).

In order to bring out the essential point of my account of measurement
it will be helpful to begin with the received view of measurement in its
most extreme form. [8] Of course, there are many other forms documented
in a rich literature, [9] some of which are less extreme. The extreme form is
exemplified by ordinary length measurement. In order to measure a length,
i.e. the distance between two marked points on a rigid body, a number of
unit lengths (millimeters, for example) are concatenated such that the resulting object has a length equal to that to be measured. The concatenated
object simply is that part of the measuring rod (or rope) equal in length to
the given distance, and equality is checked by means of incidences of end
points. [10] If n is the minimal number of units thus needed to construct a
concatenated object of length equal to that of the given object a then n
is called the length, in general: the *measure*, of a according to the procedure described, and denoted by $\phi(a)$: $\phi(a) = n$. In this way, to each object
a from a given domain D a number $\phi(a)$ is assigned the determination of
which refers to concatenation of objects, $\circ$, comparison of objects (includ-
ing equality), $\preceq$, a set $U \subseteq D$ of unit objects, and the set $\mathbb{R}$ of real numbers.
The whole array of objects and their measures obtained in this way then
may be represented as a structure [11] x

$$x = \langle D, \mathbb{R}, U, \preceq, \circ, \phi \rangle \tag{1}$$

where D and U are non-empty sets, $U \subseteq D$, $\mathbb{R}$ is the set of real numbers,
$\preceq \subseteq D \times D, \circ \subseteq D \times D \times D$, and $\phi : D \to \mathbb{R}$. ϕ is called a *scale*.[12] In

---

[8]Compare Balzer (1992) for a more extensive discussion of the following.

[9]For instance, Krantz *et al.* (1971), Narens (1985), Pfanzagl (1968).

[10]See Balzer *et al.* (1980) for a detailed analysis of this example.

[11]I refer to Balzer *et al.* (1987) and (1985) for that notion. Compare Balzer (1992) for
a detailed account of the structure described in the following.

[12]On some more abstract level, a scale is an equivalence class of such ϕ' s.

the 'standard' cases used most frequently in social science according to the degree in which ϕ is determined in a structure x, ϕ is called a nominal, ordinal, interval or ratio scale. [13] Under suitable axioms for x the scale for distance mentioned will be a ratio scale. [14]

The point of this conceptual reconstruction is this. As a matter of empirical fact, if the method of comparison of an object with a suitable concatenation of units is repeatedly applied (twice, for simplicity) to an object a the resulting numbers $\phi(a)$ and $\phi^*(a)$ are the same or nearly the same if the units are very small in comparison to the 'resolution' of the means of checking incidences. Repeated mesaurements of the length, say, of a table in millimeters will yield the same value, perhaps up to a margin of $1 \ mm$. Such repetition may be represented by two structures x, x^* of the form 1 above which both capture 'the same' system of objects D with the same units U and relations $\preceq, \circ$. The empirical identity of $\phi(a)$ and $\phi^*(a)$ (for $a \in D \cap D^*$) may be regarded as expressing a property of the 'empirical' components $D, U, \preceq, \circ$ in terms of which the process of measurement is described. If it is possible to formulate this property of $\langle D, U, \preceq, \circ \rangle$ in an explicit way then in case of proper measurement that property should imply the observed identity of the ϕ-values. Let me write $\mathcal{B}(D, U, \preceq, \circ)$ to express that $\langle D, U, \preceq, \circ \rangle$ has the described property, and $\mathcal{A}(D, U, \preceq, \circ, \phi)$ to express that the ϕ-values are obtained in the way described. The statement of identity then takes the following form:

For all ϕ, ϕ^* :   (2)

$\qquad$ if $\mathcal{B}(\mathcal{D}, \mathcal{U}, \preceq, \circ)$ and $\mathcal{A}(\mathcal{D}, \mathcal{U}, \preceq, \circ, \phi)$ and $\mathcal{A}(\mathcal{D}, \mathcal{U}, \preceq, \circ, \phi^*)$ then $\phi = \phi^*$.

Informally: if the empirical objects and relations have the property $\mathcal{B}$, and if ϕ and ϕ^* are two scales representing numerically $\preceq$ and $\circ$ as described by means of formula $\mathcal{A}$ then the two scales must be identical. Still differently: there is at most one scale ϕ which can be added to the empirical objects and relations which is obtained as described by $\mathcal{A}$, provided the empirical entities have property $\mathcal{B}$.

In most applications uniqueness in (2) is replaced by 'uniqueness up to a given kind of invariance'. For instance, if uniqueness obtains up to a positive factor, ϕ represents a ratio scale, if it obtains up to bijections, ϕ represents a nominal scale. [15] Formally, strict uniqueness always can be achieved by introducing an equivalence relation $\sim$ which identifies the different, admitted 'variants' of ϕ. Replacing '=' in (2) by '$\sim$' , uniqueness up

---

[13] A full, abstract theory of such scale types is given in Narens (1985).

[14] See Balzer (1992).

[15] The latter case means that the only property of ϕ determined by $\mathcal{A}$ and $\mathcal{B}$ is that ϕ separates the objects in some weak form: objects differing under $\preceq$ are represented by different numbers.

to $\sim$ then is covered.

For all ϕ, ϕ^* : (3)

   if $\mathcal{B}(\mathcal{D},\mathcal{U}, \preceq, \circ)$ and $\mathcal{A}(\mathcal{D},\mathcal{U}, \preceq, \circ, \phi)$ and $\mathcal{A}(\mathcal{D},\mathcal{U}, \preceq, \circ, \phi^*)$ then $\phi \sim \phi^*$.

This may be called the standard form of the uniqueness condition in which $\sim$ is an equivalence relation which is externally defined on the set of all possible functions ϕ.

Properties $\mathcal{B}$ of the kind discussed here have been extensively studied. [16] In the case of length measurement there is a particularly simple one. $\mathcal{B}(D, U, \preceq, \circ)$ holds iff 1) $\emptyset \neq U \subseteq D$, 2) $\preceq \subseteq D \times D$ is transitive, reflexive and connected, 3) $\circ$ is a partial funcion from $D \times D$ to D, and is associative, 4) for all $a, b, c \in D$ for which $a \circ b, a \circ c, b \circ c$ are defined:[17] $a \prec a \circ b$, and $(a \preceq b$ iff $a \circ c \preceq b \circ c$ iff $c \circ a \preceq c \circ b)$, 5) for all[18] $b, b' \in U$: $b \equiv b'$, 6) for all $a \in D$ there exist $n \in \mathbb{N}$ and $b_1, ..., b_n \in U$ such that $b_1 \circ ... \circ b_n$ is defined and $b_1 \circ ... \circ b_n \equiv a$. In this setting, $\mathcal{A}(D, U, \preceq, \circ, \phi)$ can be defined to hold iff 1) $\langle D, U, \preceq, \circ \rangle$ satisifes 1) - 3) of the previous definition of $\mathcal{B}$, 2) $\phi : D \to \mathbb{R}$, 3) for all $a, b \in D$: $(a \preceq b$ iff $\phi(a) \leq \phi(b))$ and (if $a \circ b$ is defined then $\phi(a \circ b) = \phi(a) + \phi(b))$, 4) for all $a \in D$: $\phi(a)$ is the minimal number n such that there exist $b_1, ..., b_n \in U$ for which $b_1 \circ ... \circ b_n$ is defined and $b_1 \circ ... \circ b_n \equiv a$. For these definitions of $\mathcal{B}$ and $\mathcal{A}$, (2) above can be easily proved to hold. [19]

To put things more informally, (2) states that the measured values $\phi(a)$, or the 'quantity' or 'scale' ϕ, are uniquely determined by the property $\mathcal{B}$ of objects, concatenation and comparison, and by the way — expressed in the definition of $\mathcal{A}$ — in which these values are produced. This uniqueness is intended to model the empirical fact of identical results under repetition and at the same time expresses that this identity is not accidental but is the systematic consequence of the objects (plus $\preceq$ and $\circ$) having property $\mathcal{B}$.

Generalizing this special case, in the received view of measurement structures of the form $\langle D, \mathbb{R}, R_1, ..., R_n, \phi \rangle$ are considered, modelling a domain D of objects together with 'empirical' relations $R_1, ..., R_n$ among these objects and a 'representing function' or 'scale' $\phi : D \to \mathbb{R}$ assigning to each object a its measure $\phi(a)$. The general condition for such a system to capture some way of measurement is that, for given $D, R_1, ..., R_n$, a representing function ϕ exists and is uniquely determined (up to a certain, externally fixed degree of variation).

---

[16] See Krantz et al. (1971).
[17] $x \prec y$ means $x \preceq y$ and $x \neq y$.
[18] $x \equiv y$ means $x \preceq y$ and $y \preceq x$.
[19] See Balzer (1992).

The *structuralist* view of measurement extends this account to cases where the distinction between empirical relations and numerical representations is dropped. According to this view, and allowing for objects of different sorts, a *measuring model* (for ϕ) in general is a structure of the form

$$\langle D_1, ..., D_k, A_1, ..., A_m, R_1, ..., R_n, \phi \rangle$$

where $D_1, ..., D_k$ are sets of objects, $A_1, ..., A_m$ are sets of purely mathematical things, $R_1, ..., R_n$ and ϕ are relations 'over' [20] $D_1, ..., D_k, A_1, ..., A_m$, and ϕ is uniquely determined by some suitable property $\mathcal{B}$ of structures of the form $\langle D_1, ..., R_n \rangle$. The class of structures having property $\mathcal{B}$ I call a *method of measurement* and I will denote it by $\hat{\mathcal{B}}$.

Uniqueness usually is relaxed to 'uniqueness up to transformations of type g' where g is externally given, see (3) above. However, condition (3) still is idealized because it does not explicitly account for inexact measurement, that is, for different scale values to occur in repeated measurements of the same object. Such inexactness being unavoidable for real-valued scales, further relaxation of (2) is needed in order to obtain methods of measurement describing 'real-life' methods as used in science. A more general notion allowing for inexactness is this.

$\hat{\mathcal{B}}$ is a $\langle W, \delta \rangle$-*method of measurement* iff 1) $\hat{\mathcal{B}}$ is a class of structures of the above form $\langle D_1, ..., D_k, A_1, ..., A_m, R_1, ..., R_n, \phi \rangle$ characterized by a property $\mathcal{B}$, 2) W is a uniform space [21] on the set of all ϕ occurring in elements of $\hat{\mathcal{B}}$, and $\delta \in W$, 3) in each structure $x = \langle D_1, ..., R_n, \phi \rangle \in \hat{\mathcal{B}}$, ϕ is uniquely determined up to δ, that is

For all $x = \langle D_1, ..., R_n, \phi \rangle$ $\hfill (4)$

and all ϕ^*, if $x \in \hat{\mathcal{B}}$ and $\langle D_1, ..., R_n, \phi^* \rangle \in \hat{\mathcal{B}}$ then $\langle \phi, \phi^* \rangle \in \delta$.

Condition (4) is weaker than (3) in the following sense. The part '$\langle \phi, \phi^* \rangle \in \delta$' in (4) is not necessarily transitive, whereas transitivity holds for the corresponding part $\phi \sim \phi'$ in (3). Clearly, if we let δ go to 'zero' ('zero' in a uniform space corresponds to the diagonal $\{\langle x, x \rangle / x \in W\}$), and if the uniform space is a Hausdorff space then requirement (4) approaches that of unique determination. If δ is sufficiently small a $\langle W, \delta \rangle$-method of measurement determines ϕ almost uniquely. In practice this is the best one can achieve because data and hypotheses (here: those characterized by $\mathcal{B}$) fit only approximately.

---

[20] See Balzer (1992) for a precise definition of this 'over' along the lines of Bourbaki (1968).

[21] Compare, for instance, Balzer *et al.* (1987), Chap.7, for a precise definition. Each $\delta \in W$ is a degree of uniformity which intuitively represents a fixed 'distance' between members of W. More formally, $\langle x, y \rangle \in \delta \in W$ expresses that x and y have at most distance δ, or are similar to each other in degree δ.

Note that ϕ here is not required to be a numerical function. In general, ϕ can be an arbitrary relation or function, and among the functions I include zero-place functions, that is, constants. [22] A constant γ simply is (denotes) one particular object from one of the base-sets [23] $D_1, ..., D_k$. Thus, in the case of a primary social theory special actions might be represented as constants and in this way get into the position of a ϕ in (4) which is to be measured. An 'ordinary' nominal scale for a system of actions, on the other hand, would be treated by a function ϕ from D_i to the reals where D_i denotes a set of actions or action types.

3. Actions, Tokens and Types

Actions are dealt with at two levels, that of tokens and of types. Roughly, an action-type is given in terms of language, by a verbal phrase or some more complex description from which a proper sentence is obtained when a subject and possibly other items are adduced. An action token, on the other hand, is a real event located in space-time and possibly other 'dimensions'. In experimental psychology, for instance, observed behavior is classified according to a system of categories. Each such category then represents one action-type. A corresponding action-token of such a type is given by the concrete behavior of a particular person which is observed as, and then classified as, being of the special type.

There is a long and ongoing discussion in philosophy of the nature of action-tokens which up to now has led to the following insights. First, actions are a special kind of events, namely events involving one or more persons as actors. Events, in turn, at least according to one kind of metaphysics are the basic stuff the world is made of, and are at the same time somehow elusive since they may acquire endless complexity and reflexivity. Second, events and therefore also actions are parasitic on language. Events cannot exist if they do not have a possible expression in language. Even such an innocent event like the collision of a planet and a comet acquires part of its eventhood only because we can verbally describe it. Without humans and their languages there would be no way to identify such an event. Third, actions are very sensitive to their descriptions. The action of i's beating j is not the same as that of i's beating j in a play though there may be no material difference expressible in terms of material movements. The difference comes in by the different intentions indicated by the

---

[22]It has to be mentioned that standard structuralist accounts in Balzer *et al.* (1987), Stegmüller (1986), or Balzer (1985) do not allow for constants. This is not so, however, for deeper reasons but only for reasons of simplicity. A fully general account is given in Balzer *et al.* (1993).

[23]More generally, it seems to be necessary also to allow for numerical constants, i.e. elements of $A_1, ..., A_m$.

two descriptions. 'i beats j' indicates an intention of i to hurt or to cause pain while 'i beats j in a play' does not. This is a typical case: the identification of an action-token depends on its description, and on the actor's intention indicated by the description. A change of intention may change an action without visible trace in the material world 'outside' the intending person. Other examples are 'i's shooting a thief in his home at night' versus 'i's mistakenly shooting a friend coming in without ringing', or 'i's acquiring the belief that Venus is the morning star' versus 'i's acquiring the belief that Venus is the evening star'. This sensitivity is hightended in the case of joint action, that is, action which can be performed only by several persons jointly (carrying a table, destroying a town). The deployment of troops along the frontier may be described as an agression by one side and as a prevention (which is a different kind of action) by the other side. I abstain from further analysis of the dependence of action-tokens on language and the ontological implications of this fact. It is sufficient to state that, ontologically, action-tokens are very complex and that they depend on language and propositional attitudes.

Action-tokens must not be confused with 'behavior', movements, or other, more 'observational' items. Of course it is always possible to *postulate* a distinction between an 'observational' and a 'theoretical' level, and to insist that measurement should be restricted to observable items. However, the problem of social science is that there is no 'observational level' prior to that of action. This is more or less implied by the discussion of the previous paragraph. If intentions are essential to the identification of action the descriptions of 'mere behavior' will fail to discrimiate between different actions if 'behavior' means 'behavior which can be described without reference to intentions'. Otherwise, if 'behavior' is admitted to include intentions, the distinction between action and 'observable behavior' vanishes. It has to be stressed here that I am discussing foundational issues, and that in spite of the foundational, and in my view irreducible problem stated, 'locally', in a sufficiently narrow context of research a distinction between interpreted and uninterpreted behavior may be very useful.

In-between tokens and types, cognitive psychologists have developed a useful account according to which an action roughly is constituted by two components: a schema and a goal.[24] In order to perform an action the actor must have internalized a certain schema which may be more or less complete. Often, the schema is incomplete, comprising only first steps, and gets concretized in the course of execution which may be contingent on external factors. Such completion, concretization or just continuation is possible because of the presence of the goal. The goal and the actor's

[24]See, for instance, Aebli (1980).

commitment to it together yield the actor's intention.[25]

Each of the three ways of approaching action, by types, tokens, or goal directed schemata yields a different perspective for measurement. I will discuss these in turn.

4. Measuring Action Tokens

Consider the condition of uniqueness (4) above when applied to a case in which one would like to determine an action-token a up to a given degree δ of approximation:

$$\text{For all } x, a, a^*, \text{ if } \langle x, a \rangle \in \hat{B} \text{ and } \langle x, a^* \rangle \in \hat{B} \text{ then } \langle a, a^* \rangle \in \delta. \tag{5}$$

Here $\langle x, a \rangle$ and $\langle x, a^* \rangle$ are structures in whose base sets a and a^* occur, respectively. So $\langle x, a \rangle$ has the form $\langle D_1, ..., D_k, A_1, ..., A_m, R_1, ..., R_n \rangle$ where $a \in D_i$ for some $i \leq k$. $\hat{B}$ is a class of such structures characterized by suitable assumptions, and δ is a degree of uniformity from a uniform space defined on some suitable superset of D_i. The base set (D_i) to which a and a^* belong may be interpreted as a set of action-tokens without any problem, and also there is no problem of assuming that a and a^* denote action-tokens occurring in D_i.

Suppose, for instance, that a is a symbol denoting the action of my writing down just this sentence here and now, that is, a denotes the action: Balzer, at August 29, 1994, 17.35, writes down the sentence 'Suppose, for instance ... of Warsaw airport' at the check in of Warsaw airport. This particular symbol certainly is specific enough to guarantee a unique denotation; there is only one such person in the hall writing this sentence.[26] D_i is a set of action-tokens and a^* denotes an arbitrary element of this set.

Now how would $\hat{B}$ have to look like in order to assure that (5) is true for the example under consideration? To fix the ideas, suppose that a^* denotes another action, say, Szienkiewicz's smoking a cigarette at the same time and place. (5) then implies that $\langle x, a^* \rangle$ cannot satisfy the axioms for $\hat{B}$ for a^* is very different from a, that is, not: $\langle a, a^* \rangle \in \delta$. By (5), $\langle x, a^* \rangle \in \hat{B}$ must imply that a^* is approximately identical with a: $\langle a, a^* \rangle \in \delta$. This seems to be possible only if $\hat{B}$ characterizes the content or meaning of a, and it is hard to see how this might be achieved without using in the definition of $\hat{B}$ syntactical and semantical properties of the natural language in which the actions collected in D_i are described. On the account of Sec.1, however, this

---

[25]There is no space to review the different approaches to action in this family, but it may be noted that similar ideas have come up recently in AI, see for example Cohen and Levesque (1990).

[26]Of course, in other situations problems of uniqueness of denotation may arise, and may be difficult to be overcome.

implies that the structures in $\hat{B}$ cover a substantial part of natural language. Though I do not want to rule out in principle that such structures can be defined – witness Montague's grammar – to do so seems an heroic task and rather utopic.

The example indicates that actions in general, i.e. actions which can be described in terms of natural language, can be determined only in a theoretical setting which includes ordinary language descriptions of the actions to be determined. A 'theory' comprising such descriptions for a large array of actions would amount to a general theory of human action and interaction – which is not available. I conclude that the measurement of action-tokens in general, that is, when the domain and structure of actions considered is not substantially restricted, may be too difficult in order to be practically applicable.

The situation changes when the actions occurring in the intended systems are of a very specific kind which can be theoretically grasped by a simple type of structures and axioms. The intended systems might, for instance, only contain actions which are economic exchanges between few individuals. In this context the variety of actions can be captured in terms of a few notions like quantity, value, kind of commodity, person etc. which may be used to establish simple criteria of identity for actions of such kind. The example is generic. In general, whenever there is a theory about a very limited class of actions this theory can be used in order to define a class $\hat{B}$ of measuring models for particular such actions, conceived as tokens. Thus, in the frame of exchange economics[27] it is possible[28] to capture a specific exchange-token by means of a method of measurement; in my theory of power [Balzer, 1992a] it is possible to do this for a token of an exertion of power; and in Downs' economic theory of democracy [Downs, 1957] it is possible to do it for a token of voting.

5. Measuring Action-Types

When switching from tokens to action-types things become simpler. Action-types are typically represented by expressions of a language, and such expressions naturally form a system in which one expression may be explained and determined in terms of others. The condition of uniqueness now reads

$$\text{For all } x, \tau, \tau^*, \text{ if } \langle x, \tau \rangle \in \hat{B} \text{ and } \langle x, \tau^* \rangle \in \hat{B} \text{ then } \langle \tau, \tau^* \rangle \in \delta. \qquad (6)$$

where τ and τ^* are action-types which we may simply imagine as appropriate expressions of a language. In the most simple case, τ is a verbal

---

[27]See Balzer (1982) for simple models of such a theory.

[28]Possibility here means that the respective theory has to be appropriately (re-)formulated and/or specialized.

phrase, like 'to go', 'to scream', 'to beat someone' etc. The actor or actors involved in the actions of such a type are not essential for the type's identity. Different actors can perform the same action-type.

In linguistics and political science, classifications of action-types have been developed [29] in which one type may well be characterized by its 'position' in such a classification. For instance, in the domain of international politics, there is a special vocabulary which is used to describe the parties' actions on a high level of abstraction. A system proposed by Brecher consists of action-types like: threat of attack, accusation of planned attack, demand to attack,..., blockade, embargo,...,mobilization of reserves, movement of forces closer to frontier,...,challenge to legitimacy by other actor(s), challenge to legitimacy of incumbent elite by international organization etc. [Brecher, 1977]. In such a system, the types have specific 'positions' or scale values on a nominal scale, and it is not difficult to introduce natural theoretical vocabulary to express these. In such a setting a method of measurement $\hat{B}$ for one particular action-type can then be defined without major problem. The practical problem with this account is that in order to determine one action-type, other action-types must be determined beforehand, namely those to which one has to refer when describing the 'position' of the former in the whole system of action-types.

It is more efficient to use properties of the actions observed in order to determine their types. This comes closer to the use of an ordinary theory whose models have the general form described above. The action-types may be taken as objects, i.e. as elements occurring in one base set, while their properties are expressed by relations $R_1, ..., R_n$.

Attempts at the determination of action-types along these lines in fact are found in science, though they are not (yet?) formulated as sharply and theoretically precise as would be necessary in order to lay the ground for respective *theories* or methods of measurement in the above sense. Let me indicate two clear cut examples.

The first method is exemplified by the work of Westmeyer on interaction in dyads. [30] Interactions, for instance between mother and child, are taped during some period of time and then coded. The goal is to cut the stream of behavior into a sequence of single 'atomic' actions of distinct action-types. This typification is performed by collaborators which obtain some training to do their job. They are made familiar with the possible types of actions, i.e. those which the researcher is willing to consider theoretically. These are labelled by means of suitable expressions in natural language, characterized by properties or relations to other types, and by reference to suitable examples. Some evaluation of sequences of actions is actually

[29]Compare, for instance, Ballmer and Brennenstuhl (1981).
[30]See Westmeyer (1989) and the references given in that paper.

performed in the presence and with the aid of the researcher, and borderline cases are discussed. All this is necessary in order to obtain some satisfactory degree of similarity of evaluation by the different observers needed in order to encode the mass of observational material: the interobserver agreement.

Even from this very informal and brief description it should be clear that the procedure has already gone a long way towards a method of measurement, and given the relatively short period of existence of experimental psychology ('short' in comparison to natural science) there is every reason to expect that accounts of that kind over time will grow into full fledged methods of measurement for action-types.

As a second example let me consider methods of content analysis as used in empirical social research, political science, and in communication theory. [31] In political science, the method is applied to a very large number of texts, typically newspaper articles, over a longer period of time. The texts are coded by searching for key words, and deciding, on every token found, whether the text expresses some event, or opinion from a range of previously fixed categories. The parallels to the first example are obvious. Again, a theoretical system of event-types or opinion-types is made up, and the evaluators have to find out which of those are referred to in the text. Again, previous training is needed in order to get acquainted with the possible types, and with practical rules for deciding in ambiguous cases. Again, much attention and methodological thought has to be spent on the possibility that different evaluators can come to a different classification of the same event, and on the achievement of interobserver agreement.

6. Measuring Goal Directed Schemata

While the two previous approaches to the measurement of action remain well inside the boundaries of the meta-theoretical notion of measuring a constant, new aspects come up when action is represented by goal directed schemata. In the previous cases there was no difficulty in regarding action-types and action-tokens as objects. They were seen as objects of the kind which occur in a base set of a corresponding structure. When actions are represented as schemata this seems no longer possible for a schema itself is naturally represented by a structure and not by an unstructured object. Thus the entities s, s^* whose approximative identity is expressed in the condition of uniqueness

$$\text{For all } x, s, s^*, \text{ if } \langle x, s \rangle \in \hat{\mathcal{B}} \text{ and } \langle x, s^* \rangle \in \hat{\mathcal{B}} \text{ then } \langle s, s^* \rangle \in \delta. \tag{7}$$

[31] A good example from political science is McClelland (1968). Some general description is found, for instance in Friedrichs (1980), Sec. 5.10.

should naturally be treated as structures and the question is how these are related to the structures $\langle x, s \rangle$ of which they are parts.

One might think of circumventing the problem by concentrating on the goal which belongs to each action, and by trying to identify an action just by means of its goal. This does not work, however, because the same goal may be achieved by means of different actions which shows that the schema-part of the action is essential.

At the abstract level, two general remarks can be made about the possibility of determining a schema or a plan according to (7). First, if the entities s, s^* occurring in (7), i.e. the representations of actions, are themselves structures we have to make sure that this can be accomodated by the structuralist format described in Sec.1. Now the structure $\langle x, s \rangle$ of which s is a part may contain a relation R_i the elements of which are structures of the kind of s. This can be arranged when the elements of R_i are appropriately typified over the base sets occurring in x. That this is possible in principle can be shown by a simple example. Suppose that x has one base-set D and that the description of a relation R 'over' D specifies that the elements of R are elements of the set $\mathbf{Po}(D) \times \mathbf{Po}(D \times D)$. Among these elements there are proper set theoretic structures of the form $\langle A, F \rangle$ where $A \subseteq D$ and $F \subseteq A \times A$.

Second, one may think about the precise relation between one such structure s and the 'rest' of the overall structure which is denoted by x in (7). One natural way to fix s in terms of x consists of s's being defined or constructed from other 'parts' of x such that the definition is made explicit in $\mathcal{B}$. However, this would mean that the measuring model determines s in a rather trivial way: in order to apply the corresponding method of measurement, s has to be constructed or defined in the way this is described by $\mathcal{B}$. Less trivial but of substantially higher complexity is an approach in which x 'contains' a whole set of structures of the form of s, one of which, namely s, is distinguished. In the light of the real, processual character of schemata perhaps an intermediate possibility will be most realistic according to which x contains some description or representation of a method for constructing and realizing schemata out of given 'building blocks' much in the way this seems to happen in real persons.

7. Conclusion

I have shown that the structuralist account of measurement can be applied to primary social theories, and that actions can be in principle determined or measured in this sense. Despite the ontological complexity of actions the measurement of action-types and of actions of a theoretically restricted domain is feasible. This shows that the 'hermeneutic', interpretative features which necessarily occur in the collection of data about actions are not a fundamental obstacle to the empirical foundation of the social sciences.

In the social sciences we find empirical methods which already are rather close to the ideal-types of structuralist methods of determination for action-types, and there is every reason to expect that these methods will further develop towards the ideal notion when resources comparable to those present for the natural sciences will be available for social science.

8. Acknowledgement

I am indebted to H. Westmeyer for stimulating comments on an earlier draft.

References

Aebli, H. (1980) *Denken: Das Ordnen des Tuns*, 1, 2, Stuttgart: Klett-Cotta

Allen, R.G.D. (1968) *Macro-Economic Theory: A Mathematical Treatment*, London: Macmillan.

Ballmer, T. and Brennenstuhl, W. (1981) *Speech Act Classification*, Berlin etc.: Springer.

Balzer, W. (1982) A Logical Reconstruction of Pure Exchange Economics, *Erkenntnis* 17, pp. 23–46.

Balzer, W. (1985) *Theorie und Messung*, Berlin etc.: Springer.

Balzer, W. (1992) The Structuralist View of Measurement: An Extension of Received Measurement Theories. C.Wade Savage and Ph.Ehrlich (eds.), *Philosophical and Foundational Issues in Measurement Theory*, Hillsdale NJ: Erlbaum, pp. 93–117.

Balzer,W. (1992a) A Theory of Power in Small Groups, in H.Westmeyer (ed.), *The Structuralist Program in Psychology*, Seattle etc.: Hogrefe and Huber, pp. 191–210.

Balzer, W. and Kamlah, A. (1980) Geometry by Ropes and Rods, *Erkenntnis* 15, pp. 245–67.

Balzer, W., Moulines, C.U., and Sneed, J.D. (1987) *An Architectonic for Science*, Dordrecht: Reidel.

Balzer, W., Lauth, B., and Zoubek, G. (1993) A Model for Science Kinematics, *Studia Logica* 52, pp. 519–48.

Brecher, M. (1977) Toward a Theory of International Crisis Behavior, *International Studies Quarterly* 21, pp. 39–73.

Bourbaki, N. (1968) *Theory of Sets*, Paris: Hermann.

Cohen, P.R., and Levesque, H.J. (1990) Intention is Choice with Commitment, *Artificial Intelligence* 42, pp. 213–261.

Downs, A., (1957) *An Economic Theory of Democracy*, New York: Harper and Row.

Debreu, G. (1959) *Theory of Value*, New York: Wiley.

Friedrichs, J. (1980) *Methoden empirischer Sozialforschung*, 13th edition, Opladen: Westdeutscher Verlag.

Krantz, D.H., Luce, R.D., Suppes, P., and Tversky, A. (1971) *Foundations of Measurement*, 1, 2, 1989, New York: Academic Press.

Lauth, B. (1989) Reference Problems in Stoichiometry, *Erkenntnis* 30, pp. 339–62.

McClelland, C.A. (1968) Access to Berlin: The Quantity and Variety of Events, 1948–1963, in J.D.Singer (eds.), *Quantitative International Politics: Insights and Evidence*, New York: Free Press, pp. 159–186.

Narens, L. (1985) *Abstract Measurement Theory*, Cambridge: Cambridge University Press.

Pfanzagl, J. (1968) *Theory of Measurement*, Würzburg: Physica Verlag.

Schlicht, E. (1985) *Isolation and Aggregation in Economics*, Berlin etc.: Springer.
Stegmüller, W. (1986) *Theorie und Erfahrung*, Dritter Teilband, Berlin etc.: Springer.
Taylor, M. (1976) *Anarchy and Cooperation*, London: Wiley.
Westmeyer, H. (1989) The Theory of Behavior Interaction: A Structuralist Construction
    of a Theory and a Reconstruction of its Theoretical Environment, in: H.Westmeyer
    (ed.), *Psychological Theories from a Structuralist Point of View*, Berlin etc.: Springer,
    pp. 145–85.

STRUCTURALIST MODELS, IDEALIZATION, AND APPROXIMATION

C. ULISES MOULINES
Institut für Philosophie, Logik und Wissenschaftstheorie
Universität München
München, Germany

1. Introduction

For many decades since the constitution of philosophy of science as a more or less autonomous discipline at the beginning of this century, philosophers of science had not paid much attention to the essentially approximative nature of empirical science, i.e. to the fact that there is no such thing as an *exact empirical* theory (unless it is completely trivial). This fact was mostly seen as a rather unimportant, accidental feature of empirical knowledge — a kind of "noise" we could forget about when trying to unveil the essential structure of science. Since the mid—seventies this evaluation of inaccuracy in science has gradually changed. Philosophers of science of different schools and using different methods have become increasingly concerned with the issue, quite independently of their idiosyncratic epistemological views. In particular, two notions have become the focus of much attention: idealization and approximation. They are seen as quite central for an adequate explication of inaccuracy in science.

The background premise of the present paper is that the issue of inaccuracy, and in particular the concepts of idealization and approximation, may be most adequately analyzed by using the formal tools of a particular metatheory of science: the so-called "structuralist view of theories", or "structuralism" for short.

Structuralism was one of the first formal metatheories of science to take approximation very seriously. Its first systematic contribution to this topic is dated 1976 [Moulines, 1976]. Since then, more results have been obtained in the application of structuralist methods to this area. A whole chapter of

157

R. Hegselmann et al. (eds.),
Modelling and Simulation in the Social Sciences from the Philosophy of Science Point of View, 157–167.
© 1996 *Kluwer Academic Publishers. Printed in the Netherlands.*

structuralism's standard reference work, *An Architectonic for Science* [1] , is devoted to approximation. Indeed, one of the central tenets of structuralism in its present version is that the concept of approximation essentially belongs to the systematic explication of the notion of an empirical theory. The reason for this tenet is clear: no scientific theory fits the "facts" it is supposed to systematize in a completely accurate way; almost no theory is related to other theories in an exact way; generally, no theory works well unless a certain degree of idealization and approximation with respect to its "outer world" is admitted. In sum, any "serious" empirical theory can be effectively applied or related to other theories *only if* some fuzziness is allowed. Approximation and idealization are necessary conditions of success in empirical science.

Now, the degree of inaccuracy with which a theory works may be expressed either numerically or qualitatively, according to the type of theory and application. In so-called "mathematized" empirical theories, that is, in those theories systematically using metrical concepts, there will be a tendency to express their degree of inaccuracy partially, though *not completely*, in numerical terms, or, to be more precise, in terms of comparisons between real numbers expressing values of magnitudes; whereas in so-called "qualitative" theories, that is, in those theories only using classificatory or comparative concepts, the degree of inaccuracy assumed will be given, explicitly or more frequently implicitly, by topological, non-quantitative comparisons. In the quantitative case, we usually express the accepted inaccuracy by saying that the difference between the theoretically and the empirically determined values of a given magnitude is less than an accepted real number ε . In the qualitative case, we say, or rather suggest, that the empirically determined extension of a predicate lies within the accepted fuzziness of its theoretically determined extension. At any rate, the use of accepted inaccuracies is almost universal.

2. Idealization vs. Approximation

Now, it has been assumed in the foregoing introductory remarks that approximation and idealization are *more or less* the same thing. And there may be some plausible reasons for making this assumption. It might be argued that the only real difference between idealization and approximation consists in a different way of describing the same phenomenon: theoretical inaccuracy. The difference would be a matter of nuance, or perhaps even just terminology. "Idealization" would be the coarse description for a phenomenon which would be regarded as an approximation in a fine-tuned

---

[1]See Balzer *et al.* (1987). Since this work will often be cited in this paper it will be abbreviated as "Architectonic".

description. For example, saying that we have *idealized* the earth as a perfect sphere would be a rough-and-ready way to express that describing the earth as a sphere is a good *approximation* of the geodesical values we would obtain if we measured the distance from the center of the earth to its surface point by point. Or saying that we *idealize* the weekly market in a small Maya village taking it as an example of a pure exchange economy might be understood as an abbreviation of the claim that the exchange of goods predicted by theoretical economics is a good approximation of the exchange patterns observed in that remote village. If this were all we might sensibly say about the difference between idealization and approximation, then there certainly would not be much reason to keep both notions apart in a systematic treatment.

However, though this way of speaking about idealization and approximation is indeed plausible, it seems to me that it does not reflect the only sensible uses one finds in the literature. It appears convenient to make a difference between two substantial notions in this context, which we might reasonably label "idealization" and "approximation", respectively; they are intimately related to each other but nevertheless quite distinct. The structuralist methodology provides, it seems to me, the right tools to make the distinction precise and justified.

The explication I would like to propose runs along the following lines. The first point to note is that idealization should be considered the more general, approximation the more particular notion. In a sense, every approximation is (or rather: rests on) a specific kind of idealization, though the converse is not true: not every kind of idealization may be translated into a form of approximation.

Idealization, under its most general interpretation, may just be characterized as any attempt to relate a given theory, as a conceptual unit, to something outside itself — be it the "empirical reality" or the "data", or perhaps also other theories as different conceptual units. To make such a relationship fruitful, one must "idealize" — that is, in general terms, one must bring the theory's *ideal conceptual structures* in touch with this "something" outside itself, which is always *alien* to the theory. This operation does not work adequately unless one is prepared to do some *force* to one of the two terms of the relationship (or to both at the same time). This force is "idealization" as different from approximation. On the other hand, approximation in its more specific sense (as not *mere* idealization) may be viewed as a process of comparison of structures which are already idealized (in the more general sense). Approximation is a symmetric relationship whereas non-approximative idealization is asymmetric.

These metaphors may be given a more controllable meaning by using technical notions of the structuralist approach. According to structuralism, one of the essential components of a theory's identity is the class of its potential models, M_p. Roughly speaking, M_p represents a theory's *conceptual framework*. Now, idealization in general has to do with the use of the elements of M_p to say something about things outside the theory. If we idealize a physical system so as to make it amenable to treatment by theory T, what we are doing is to conceptualize that system as a potential model of T. For example, when we say that, in order to apply Newton's gravitational theory to the planetary system, we have to idealize the latter as a set of particles moving on smooth paths, what we are doing is to "convert" or "reconstruct" that system into a potential model of Newton's theory. Whether or not it will *also* be an *actual* model, that is, a system satisfying Newton's law of gravitation, is a further question. To answer it, we generally need not only idealization in its general sense, but also its specific kind called "approximation". We may be successful in idealizing but fail in approximating. That is, we may be able to show that, in a given context, it is reasonable or plausible to deal with the planetary system *as if* it would just be a set of particles moving along smooth paths, and we may be able to provide an exact mathematical description for this idealization — in terms of ellipses, for example. But we may fail to show that this mathematical description is close enough to the mathematical description that should come out from applying the law of gravitation.

Under the general setting of idealization and approximation suggested here, many more interesting things could be said about idealization in general, and more particularly about idealization which does *not* involve approximation. However, this essay will concentrate on approximation as a particular form of idealization. The results offered here are a continuation and revision of work already done on approximation within the structuralist program (especially in the works mentioned in the footnotes above). The main new result is a reformulation of the approximative version of the so-called "central empirical claim" of a theory according to structuralism.

3. The structuralist apparatus to deal with approximation

For obvious reasons, the essentials of the structuralist apparatus to represent theories have to be summarized quite radically here. A theory, according to structuralism, is to be represented as an ordered net of structures called "theory- elements". Each theory-element consists of a core K and a domain of applications I, and the "(central) empirical claim" of each theory-element is just that K can be effectively applied to I. Each core K is a complex entity constituted, in turn, by a number of different components: the class M_p of potential models, the class M_{pp} of partial potential models (the relative "non-theoretical" structures or "data"), the class M of actual models, and some other components which will be left out of consideration now for the sake of simplification because they are not essential to the argument. M_p and M_{pp} are related by a many-one function r ("restriction"), which applies, to each potential model, its corresponding non-theoretical structure in M_{pp}. [2]

The theory's empirical claim may then be represented synthetically as follows:

$$I \subseteq r(M).$$

Now, the metatheoretical problem we have to face is that, in any "really existing" empirical theory, the proposition expressed by this formula will normally be false (unless it is trivially true, which sometimes also happens). The reason is, of course, that no interesting empirical theory applies exactly to the whole of its domain of intended applications. There will always be some "noise". But, though false, the empirical claim will be, in more or less "well-functioning" theories, "approximately true" — or, at least, this is what we would like intuitively to be able to say. That is, intuitively we may want to say that, although "$I \subseteq r(M)$" is false, "$I \widetilde{\subseteq} r(M)$" is true. Then, our metatheoretical problem becomes the problem of looking for a satisfactory explication of the dash "$\sim$" in that formula.

The most basic intuition we wish to explicate is essentially a metatheoretical claim of the sort: "potential (or partial) model x is an approximation of potential (or partial) model y in the theory-element T". How can we render this idea precise?

Potential models of a theory-element T are all elements of the same class M_p of T. The particular form the elements of this class may take does not matter on the general level. They need not be structures of metrical concepts, nor do we need to suppose that the comparison rests on some special measurements methods. All we need is a well-defined class M_p (that is a class of structures determined by the same structural axioms) and an approximation relation between any two elements of that class. There is a well-known method for defining such a relation in topology, namely to introduce the concept of a *uniform structure*, or "*uniformity*" for short. Thus, let us explicate model-theoretical approximation by defining a uniformity on M_p.

Definition 1: If M_p is a class of potential models, then U is a uniformity on M_p iff

1. $\emptyset \neq U \subseteq \wp(M_p \times M_p)$
2. $\forall u_1, u_2 : u_1 \in U \wedge u_1 \subseteq u_2 \to u_2 \in U$

---

[2]For more details on all this, see *Architectonic*, especially the first chapters.

3. $\forall u_1, u_2 : u_1 \in U \wedge u_2 \in U \rightarrow u_1 \cap u_2 \in U$

4. $\forall u \in U : \Delta(M_p) \subseteq u$

5. $\forall u \in U : u^{-1} \in U$

6. $\forall u_1 \in U \exists u_2 \in U : u_2 \circ u_2 \subseteq u_1$

where "$\wp$" denotes the power-set operator, $\Delta(M_p)$ is the set of all pairs of identical elements of M_p, "$\cdots^{-1}$" denotes the converse of a relation, and "o" denotes the composition of relations.

The elements u_i of U we call "blurs". They are sets of pairs of potential models. Intuitively speaking, each u_i represents a certain degree of "proximity" between different structures or models; or, seen from another viewpoint, each u_i represents a "cloud" of a certain degree of fuzziness around a fixed model.

A detailed exposition of the rationale for the conditions of Definition 1 as interpreted in the framework of empirical theories may be found in *Architectonic*, Ch.VII. A model— theoretical uniformity U is, according to structuralism, the base to deal with approximation, be it quantitative or qualitative. However, this notion is not sufficient by itself to deal with the specific problems of *empirical* theories. Uniformities in this general sense have to be endowed with more restrictive conditions to deal especially with the following issues, which according to structuralism are quite basic when analyzing empirical theories and empirical approximation:

1. The relationship between T-theoretical and T-non-theoretical concepts: when considering approximation, we would like to expect that approximation at the T-theoretical level (M_p) is closely related to approximation at the T-non-theoretical (or "data") level (M_{pp}).
2. "Really admissible" approximation: not all "blurs" $u_i \in U$ will be acceptable to express genuine empirical approximation (otherwise we would trivialize the idea itself). Some further constraints should be put on a subset A of admissible blurs in U.
3. The precise approximate version of the empirical claim: different alternatives to apply the dash $\sim$ in "$I \subseteq r(M)$" are conceivable, and we should choose the most plausible one.

Chapter VII of *Architectonic* includes some formal or quasi— formal solutions to the above problems. Though I still think that the solutions proposed there are on the right track, some important shortcomings have been detected later on. Better proposals to handle these problems could and should be made. In recent times, my collaborator Reinhold Straub and I myself have been working on these problems. Our results have already been published [Moulines and Straub, 1994].

The most important issue among the above-mentioned problems is the last one: the adequate formulation of the approximative empirical claim. The final part of this essay is devoted to it.

Let us presuppose we already have an adequate explication of what an admissible blur is. A blur u is admissible if it is a member of a particular subclass $A \subseteq U$ which satisfies certain conditions in addition to those of a uniformity. Then, we may express any approximative model—theoretical statement by using only such admissible blurs. They provide the adequate explication for the approximation operator "$\sim$". For example, if we want to say that potential model x_1 is an admissible approximation of actual model x_2, we write:

$$\exists u \in A : \langle x_1, x_2 \rangle \in u,$$

and this, in turn, may be abbreviated into: $x_1 \sim x_2$.

(The relation $\sim$ is reflexive and symmetric but not transitive, which is what we intuitively should expect.) Analogously, to express that a partial potential model $y \in M_{pp}$ is an admissible approximation of an intended application $i \in I$, we write

$$\exists v \in r\,[A] :< y, i >\in v.$$

(Note that r may also be defined for pairs of structures in a natural way: $r(< x_1, x_2 >) =:< r(x_1), r(x_2) > .$)
Let's abbreviate also the foregoing formula into: $y \sim i$.

The relation $\sim$ may be generalized in a natural way into a relation between classes of models:

$$X_1 \sim X_2 \text{ iff: } \forall x_1 \in X_1 \exists x_2 \in X_2(x_1 \sim x_2) \wedge \forall x_2 \in X_2 \exists x_1 \in X_1(x_1 \sim x_2).$$

4. The approximate empirical claim

With this notation in mind, let us now see what the central empirical claim in its approximative version might look like. Several possibilities are open to us: We can blur the set of intended applications I, the class of actual models M, its restriction $r(M)$ and finally the restriction function r (or, equivalently, both I and $r(M)$ at the same time). We may render these four possibilities formally precise as follows:

$$\tilde{I} \subseteq r(M), \text{ i.e., } \exists X(X \sim I \wedge X \subseteq r(M)), \tag{EC1}$$

$$I \subseteq (\widetilde{r(M)}), \text{ i.e., } \exists X(X \sim r(M) \wedge I \subseteq X), \tag{EC2}$$

$$I \subseteq r(\tilde{M}), \text{ i.e., } \exists X(X \sim M \wedge I \subseteq r(X)), \tag{EC3}$$

$$\tilde{I} \subseteq (\widetilde{r(M)}), \text{ i.e., } \exists X, Y(X \sim I \wedge Y \sim r(M) \wedge X \subseteq Y). \tag{EC4}$$

We could consider a further possibility of "double blurring":

$$\tilde{I} \subseteq r(\tilde{M}) \tag{EC5}$$

But for reasons which will become apparent below, this does not mean an essential departure from (EC4). In *Architectonic*, (Ch.VII.2.3), the claim is made that (EC4) should be preferred for methodological reasons. However, some further reflections have shown that this is actually *not* the most plausible choice to express the approximate empirical claim of a theory.

To reduce the number of alternatives, we should follow two principles: a logical one, which says that we should not distinguish between set-theoretically equivalent formulations, and an epistemological one, which says that we should not blur the same components of the claim twice. A first step towards the reduction of alternative formulations can be made by noticing that (EC1) and (EC2) are equivalent. This is settled in the following proposition.[3]

Theorem 1: The following formulations of the approximate empirical claim are equivalent:

$$\tilde{I} \subseteq r(M) \tag{EC1}$$

$$\forall i \in I \exists m \in r(M) : i \sim m \tag{EC*}$$

$$I \subseteq (\widetilde{r(M)}) \tag{EC2}$$

So (EC1) and (EC2) are just two possibilities to express (EC*) in a more compact way. The third formulation $I \subseteq r(\tilde{M})$, is stronger than these two.

Theorem 2:

(EC3) $I \subseteq r(\tilde{M})$ implies (EC^*) $\forall i \in I \exists m \in r(M) : i \sim m$

Now, it appears that, for reasons of metatheoretical adequacy, the stronger claim (EC3) should be preferred to the weaker (EC*) in order to represent the general situation in empirical science. To understand why, the intuitive difference between (EC3) and (EC*) should be borne in mind. (EC3) means that we first blur the models in all their components (i.e. including the T-theoretical concepts) and then project this blurring into the purely T-non-theoretical structures in order to construct the approximate empirical claim. (EC*), on the other hand, means that we only blur on the T-non-theoretical level, not caring at all about the choice of the T-theoretical components of the models. This means that we would approximate the intended applications by choosing completely "crazy" T-theoretical terms which, by themselves, are not in an admissible approximative relationship. Such a choice, however, would not be plausible from

---

[3]The proofs of the theorems to follow may be found in the article by Straub and myself mentioned above [Moulines and Straub, 1994].

the point of view of normal scientific practice, especially in those theories where there is a chance that, from a diachronic perspective, some of the former T-theoretical terms later become T-non-theoretical — a possibility structuralism explicitly envisages and for which some historical examples may be given. For these reasons, it appears that (EC3) is a more adequate representation of the empirical claim than (EC*).

Now, I come to the last point: the comparison between (EC3) and (EC4). Here, too, some reflection shows that (EC3) is more adequate than (EC4). To see this, let me first make a general remark about blurring twice a given structure in the empirical claim: any formulation of the empirical claim which implies blurring twice the same component should be ruled out for epistemological and methodological reasons. Otherwise, approximation would become too lax. Indeed, for any two potential models x and y, if we find that $x \not\sim y$, we should conclude that x and y are too far apart, that none does approximate the other even if there is a third model z with $x \sim z$ and $z \sim y$, which is actually what double blurring implies. So blurring models twice runs against the idea of a boundary of the globally admissible inaccuracy.

Prima facie, this doesn't seem to be the case in (EC4), i.e. $\tilde{I} \subseteq (\widetilde{r(M)})$. This formula seems to express that it is allowed to blur the intended application and (the restriction of) an actual model. But as the following proposition shows, (EC4) just says that there is a "middle" model between i and $r(m)$, i.e. the intended applications are blurred twice in order to attain (a restriction of) an actual model.

Theorem 3: Define for M_{pp} : $x \approx y$ iff $\exists z(x \sim z \& z \sim y)$. Then $\tilde{I} \subseteq (\widetilde{r(M)})$ iff $I \subseteq (\widetilde{\widetilde{r(M)}})$.

For quite the same reasons, any other kind of double blurring should be ruled out as well, e.g. (EC5), which is stronger than (EC4), as can easily be seen.

So with regard to the admissible inaccuracy in the sense of closeness of intended applications and actual models, (EC3) seems to be the best explication of the empirical claim.

5. Conclusions

What do the general framework and the formal results presented in this paper mean for the practitioner of an empirical discipline, especially of the social sciences? To put it in a nutshell, the idea of structural approximation allows to understand the very nature of inaccuracy in the theory/data relationship. But let's try to give a more detailed answer. If the structuralist framework in general and its formal treatment of approximation in particular are assumed then the following insights come out naturally:

1. You shouldn't worry too much if your theory "doesn't fit the facts", since "misfit" (i.e. inaccuracy) is essential to any kind of empirical knowledge, both in the natural and social sciences. You should rather think about how big the misfit might be which still is useful for your purposes. (See also point 4 below.)

2. You should distinguish between proper idealization and approximation. You idealize whenever you conceive something "out there" in terms of your theory. In structuralist terms, this means that the first thing you should do is to construct a potential model of the form

$$x =< D_1, \cdots, D_m, R_1, \cdots, R_n >,$$

where D_i are basic sets of objects and R_j relations over those sets (including metrical functions).

You can do this even if you still don't know what the best formulation of the proper axioms ("fundamental laws") of your theory is. You can go a long way in idealizing, and even approximating, without being quite clear about the laws of the theory (i.e. without settling the actual models). This possibility is especially significant for theories which are still not very well developed or whose laws have not been well tested.

3. Once you have your potential models representing some phenomena, you may start blurring them, that is, taking "balls" of similar potential models. The balls should be defined some way or other in terms of (some of) the theory's basic domains and relations. This is what it means to construct a uniformity over M_p. Each ball is a blurred representation of a given (potential) phenomenon. For the general strategy, it does not matter much if you work with metricized concepts or with purely qualitative descriptions, so long as the whole array of balls really constitutes a uniformity.

4. Then you decide how big the still acceptable balls are to be. If you only work with metricized concepts you set some ε ; if you also have to take qualitative descriptions of phenomena into account you settle what you take to be an "acceptable similarity" between different qualitative descriptions. This is what it means to construct the set of admissible blurs A. Normally, for a given theory, there will be not just one ε or one similarity degree for a given primitive concept, i.e., normally A will *not* be a singleton. The acceptable blurring will depend on each particular model representing each particular phenomenon. Your blurring will usually be more liberal in a large-scale application than in a small-scale one; it will be narrower in a life-or-death issue (say, in applications involving the safety of human beings) than in other is-

sues, etc. But anyway, you should try to be clear about the maximum degree of fuzziness you are still ready to accept in each application.

5. If your theory allows for a clear distinction between concepts describing the "data" and those belonging to the theoretical "superstructure" (as it will normally be the case if your theory has been properly reconstructed along structuralist lines), then the content of the theorems 1–3 above and their methodological interpretation allow for the following recommendations:

> 5a. When trying to apply your theory to a given empirical domain I (representing your data basis), you should *not* blur it - i.e. you should take your data basis as if it were exactly given.

> 5b. You should blur at the level of the actual models (not of their T-non-theoretical restrictions), and each model should be taken thereby as a kind of holistic unit, i.e. you may blur the T-theoretical as well as the T-non-theoretical concepts, in different ways, depending on the particular case you are dealing with; in sum, blurring has to go from the top downwards and not conversely.

> 5c. Never blur twice, even if it seems to you that you are blurring at different levels: this would be a dirty trick to immunize your theory beyond reasonable limits.

References

Balzer, W., Moulines, C.U., and Sneed, J.D. (1987) *An Architectonic for Science - the Structuralist Program*, Dordrecht, Chap.VII.

Moulines, C.U. (1976) Approximate Application of Empirical Theories: A General Explication. *Erkenntnis*, 10/II, pp. 201–227.

Moulines, C.U., and Straub, R. (1994) Approximation and Idealization from the Structuralist Point of View. Kuokkanen, M. (ed.) *Structuralism, Idealization and Approximation*, Amsterdam, Atlanta: Rodopi, pp. 25–47.

A CONCEPT OF EXPLANATION FOR SOCIAL INTERACTION MODELS

HANS WESTMEYER
Department of Psychology
Free University of Berlin
Berlin, Germany

In this paper, I will recommend one particular concept of causal explanation, the concept of aleatoric explanation of Humphreys (1989), and discuss this concept in detail in the context of one particular theory of social interaction, the theory of behavior interaction in small groups [Westmeyer, 1989], which is an element of a particular class of social interaction theories the basic element of which is formulated by David Magnusson (1980) in his general interactionistic behavior concept.

1. Humphreys' Concept of Causal Explanation

The concept of aleatoric explanation of Humphreys is not very well known in the social sciences in which the concept of deductive-nomological explanation and its variants [Stegmüller, 1983] still prevail. Nevertheless, Humphreys' concept is ideally suited especially in the case of social interaction theories used to explain specific social episodes. Humphreys explicates the concept of causal explanation of specific events as shown in Table 1.

This concept will be applied to a certain class of social interaction theories which can be characterized as concretizations of the general interactionistic behavior concept formulated by the Scandinavian psychologist David Magnusson.

2. A General Interactionistic Behavior Concept

Magnusson's conception consists of four principal assumptions M1 to M4 (1980, p. 20 f.):

(M1) For each individual there is a certain population of possible behaviors. To some extent this population is different for different individuals.

169

R. Hegselmann et al. (eds.),
Modelling and Simulation in the Social Sciences from the Philosophy of Science Point of View, 169–181.
© 1996 *Kluwer Academic Publishers. Printed in the Netherlands.*

TABLE 1. The canonical form for causal explanations of specific events (cf. Humphreys 1989, pp. 286f.)

Request for explanation
What is the explanation of Y in S at t?
Appropriate explanation
Y in S at t [occurs, was present] because of $\boldsymbol{F}$ despite $\boldsymbol{I}$.

Notes

"Y"	is a term referring to a property or change in property.
"S"	is a term referring to a system.
"t"	is a term referring to a time.
"$\boldsymbol{F}$"	is a (nonempty) list of terms referring to contributing causes of Y.
"$\boldsymbol{I}$"	is a (possibly empty) list of terms referring to counteracting causes of Y.

Frame of reference

—   For something to be a cause, it must invariantly produce its effect.

—   Probabilistic causes produce changes in the value of the chance of the effect: a contributing cause of Y produces an increase, a counteracting cause of Y a decrease in the value of the chance of Y.

—   Explanations of this kind are conjunctive:
   If $\boldsymbol{F}$ is a proper subset of $\boldsymbol{F}'$ and $\boldsymbol{I} = \boldsymbol{I}'$, then the explanation of Y by $\boldsymbol{F}'$ is superior to that given by $\boldsymbol{F}$.
   If $\boldsymbol{F} = \boldsymbol{F}'$ and $\boldsymbol{I}$ is a proper subset of $\boldsymbol{I}'$, then again $\boldsymbol{F}'$ gives a superior explanation, in the sense that the account is more complete.

... Initially, no probability is attached to a certain behavior act.

(M2) The probability of a certain actual behavior is conditional and determined by the situational context: (a) by the situational frame of reference ..., and (b) by the continuously changing situational conditions functioning as sources of each specific behavioral act. ... The key factor determining the probability of each specific behavioral act is the individual's own interpretation of the total situation and of the situational cues. The factors working under (a) have been designated *across situation factors*, and those working under (b) *within situation factors*.

(M3) As soon as a person occurs in a certain situation, ... two things happen, which have implications for studying behavior across different situations: (a) a restricted sample of possible and probable behaviors is defined, the samples apparently differing between the situations ..., and (b) a probability of single acts of behavior is determined to some degree ...

(M4) Within the frame of reference given by a certain situation, the probabilities for a certain behavior change with changes in the flow of situational cues, as a result of changes in the physical conditions, as a result of other person's behavior, as a result of one's own actions, or as a result

of a combination of two or three of these factors.

The most important aspects of this theoretical approach are illustrated, in the case of dyadic interaction, in Figure 1.

Figure 1.   Illustration of important aspects of Magnusson's interactionistic behavior concept in the case of dyadic interaction

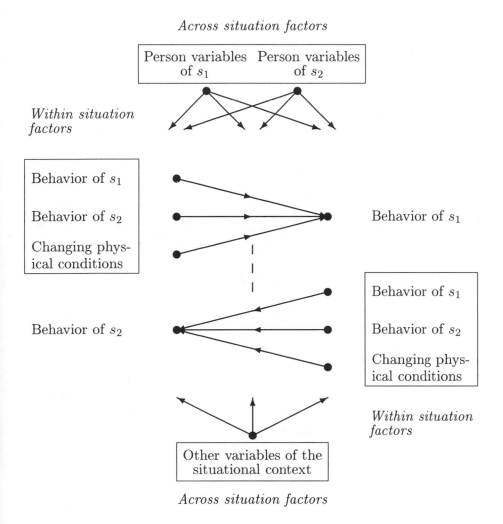

There are three kinds of within situation factors: Changing aspects of the physical conditions, the behavior of the interaction partner s_2, and the behavior of the target person s_1 her- or himself. The across or between situation factors that characterize the situational frame of reference in which the intrasituational dynamic interaction between s_1 and s_2 takes place and

that do not change during the interaction can be subdivided in two classes: The set of person variables of s_1 and s_2, and the set of other variables
of the situational context including the stable aspects of the physical situation, e.g., home setting, work setting, school setting, therapeutic setting,
and so on.

Within this theoretical framework, intrasituational interaction between
two or more persons can be studied by itself, or as a function of certain
across situation factors. The second alternative is illustrated by studies
dealing with comparisons of marital interaction in distressed versus non-
distressed couples (cf. Gottman, 1979; Hahlweg *et al.*, 1982), with the influence of natural gender and sex-role orientation of interacting partners on
their interactional involvement and their satisfaction with this involvement
(cf. Ickes, 1981, 1985; Westmeyer, 1992), or with differences in interactive
patterns resulting from interactions of the same persons, but in different
interactional settings.

I would like to go into the first alternative, the intensive examination
of social interaction as a function of within situation factors, more closely
by introducing the theory of behavior interaction of which a very elaborated structuralist construction already exists including a structuralist
reconstruction of its methodological environment [Westmeyer, 1989].

3. A Theory of Behavior Interaction

The theory of behavior interaction, a performance theory, deals with intrasituational dynamic behavioral interactions in small groups of two, three
or, at most, four persons. Intended applications of the theory are, for example, behavior interactions between mother and child in the natural setting
of their home (cf. Westmeyer *et al.*, 1984, 1988), behavior interactions in
dyadic face-to-face encounters between persons which never met before (cf.
Nell, 1982; Westmeyer, 1987, 1990), behavior interactions in father-mother-
child triads within specific social situations (cf. Nell *et al.*, 1988; Westmeyer *et al.*, 1987), and behavior interactions between client and therapist in
therapeutic settings. Actually the theory has already been tested for these
intended applications and found confirmed in several empirical studies.

The theory requires that the behavior of interacting persons is described
by behavior categories; the relative frequency of these behavior categories
is the central parameter chosen to characterize interactional episodes; behavior categories are differentiated in controlled versus uncontrolled, in
controlling versus uncontrolling ones; and frequency courses, i.e., increase,
decrease, or no change of frequencies, make up the events which are analyzed and explained by the theory.

Table 2 shows a simplified version of the non-theoretical concept of the theory of behavior interaction and its basic assumption.

TABLE 2. Non-theoretical concepts and basic assumption of the Theory of Behavior Interaction

---

Non-theoretical concepts
c_j is a behavior category of one interaction partner
c is a behavior category of another interaction partner

Category relations	*Frequency courses*
facilitating cat. rel.	increase of the frequency of c:
(c_j facilitates c,	
c is facilitated by c_j):	
$CR(c_j, c) = 1$ or	$FRC(c) = 1$ or $c \uparrow$
c_j ____F____ c	
inhibiting cat. rel.	decrease of the frequency of c:
(c_j inhibits c,	
c is inhibited by c_j):	
$CR(c_j, c) = -1$ or	$FRC(c) = -1$ or $c \downarrow$
c_j ____I____ c	
controlling cat. rel.	frequency of c changes:
(c_j controls c,	
c is controlled by c_j):	
$CR(c_j, c) \neq 0$	$FRC(c) \neq 0$
neutral cat. rel.	frequency of c remains the same:
(c is not controlled by c_j):	
$CR(c_j, c) = 0$	$FRC(c) = 0$ or $c =$

Basic assumption
Frequency courses of controlled behavior categories are a function of frequency courses of controlling behavior categories.

The totality of the category relations constitutes an interaction structure. Within any given interaction structure, the frequency dynamics, i.e., the frequency courses of the involved behavior categories, is of primary concern. The theory of behavior interaction poses certain restrictions on this frequency dynamics such that only certain frequency courses of controlled behavior categories are admissable given a particular interaction structure and specific frequency changes of the controlling behavior categories.

How the theory of which, so far, only the non-theoretical concepts have been introduced actually explains human behavior is easier to grasp if we trace the process of application of the theory which is summarized in Table 3.

TABLE 3. The process of application of the theory

Behavioral interactions between two or more individuals in the research setting during different time periods

(a) Observation and description of the behavior of the individuals involved in the interaction by means of a behavior category system

Interrelated sequences of behavior categories for each of the participants in the interaction

(b) Analysis of the sequences of categories with regard to:

    (1) Frequency of the categories within the different time periods

    (2) Frequency course of the categories across the different time periods

    (3) Mutual dependencies of the categories of different persons

Facilitating, inhibiting (generally: controlling) and neutral relations between the behavior categories of the participants of the interaction (i.e., interaction structure);
Frequency courses of the categories

(c) Identification of the antecedent conditions of the laws of the theory (for purposes of applying these laws)

Facilitating and/or inhibiting relations between behavior categories;
Frequency changes of controlling behavior categories

(d) Application of the special laws SL1-SL6 of the respective theory-element

Positive and/or negative influences on facilitated or inhibited (generally: controlled) categories

(e) Application of the special law SL7 of the respective theory-element

Expected frequency courses of controlled categories

(f) Comparison with actual frequency courses

Conclusions concerning the empirical corroboration of the theory within the intended domain of application, taking into account the special methodological and theoretical constraints and links assumed to hold for (a), (b), (c) and (e).

Note. The steps (a) to (f) are operations to be applied to the respective units listed directly above each operation; the respective units produced by an operation are listed directly below.

Up to point (c), everything is well known from other approaches to social interaction analyses (cf. Gottman and Roy, 1990; Wampold, 1992). But, so far, only a description of the interaction structure and the frequency dynamics within the interaction structure has been achieved. For explanation, of course, more is required: Laws are needed.

Figure 2 gives an example of a part of the interaction structure in family dyad FD 4 and the respective frequency dynamics. The symbols for the behavior categories used to describe the behavior interaction of the mother and the child are shortly characterized below the figure.

Figure 2. Segment of the interaction structure in family dyad FD 4 which is relevant for the explanation of the frequency course of behavior category c_2

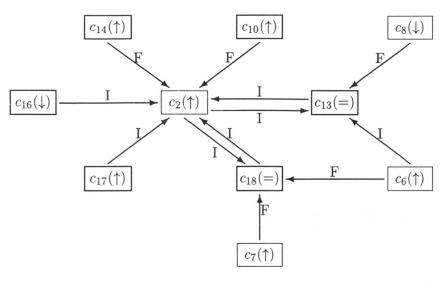

Categories for Parent Behavior (CS$_1$)

c_1 Punishing behavior
c_2 Socially rewarding behavior
c_3 Supporting behavior
c_4 Verbally rewarding behavior
c_5 Instructional behavior
c_6 Neutral talk (with the child)
c_7 Watching, listening
c_8 Neutral behavior without interaction
c_9 Supporting behavior in combination with neutral talk

Categories for Child Behavior (CS$_2$)

c_{10} Offensive and coercive behavior
c_{11} Damaging and soiling behavior
c_{12} Disturbing behavior
c_{13} Self-supporting behavior
c_{14} Play behavior
c_{15} Enjoying, showing affection
c_{16} Receptive and inactive behavior
c_{17} Neutral talk
c_{18} Self supporting behavior in combination with receptive behavior or neutral talk

Note: "F" and "I" stand for "facilitating" and "inhibiting" (category relations); "↑", "=", and "↓" stand for (the frequency of the category) "increases", "remains the same", and "decreases".

What is at stake in this case is the explanation of the increase of the frequency of c_2, that is, socially rewarding behavior of the mother. This behavior category is multiply controlled by behavior categories of the child and, mediated by child behavior categories, multiply controlled by categories of the mother herself.

The question is, what are the laws which are able to impose restrictions upon the frequency dynamics within a given interaction structure and which are necessary for the purpose of explaining the frequency course of controlled behavior categories? Six laws (SL 1-6) are part of the theory which express the influences which frequency changes of controlling behavior categories exert on a controlled category. They are summarized in Table 4. In the laws, two kinds of influences are differentiated: direct influences (DI) exerted on a controlled category c of a person by a controlling category c_j of the interaction partner, and indirect influences (IDI) exerted on a controlled category c of a person by a controlling category c_k of the person her- or himself mediated by a category c_j of the interaction partner which controls the controlled category c of the person and is, in its turn, controlled by the controlling category c_k of the person (for further details see Westmeyer, 1989).

TABLE 4. Illustration of the special laws of TE(BISG$_1$) and the theoretical concepts DI and IDI: Direct (DI) and indirect (IDI) influences on a controlled category c contingent upon category relations (CR) and frequency courses (FRC)

	$CR(c_k,c_j)$	$FRC(c_k)$	$CR(c_j,c)$	$FRC(c_j)$	$DI(c_j,c)$	$IDI(c_k,c_j,c)$ $c_j = c$	$c_k \neq c$
SL1			1 (F)	1 (↑)	1		
			1 (F)	−1 (↓)	−1		
SL2			−1 (I)	1 (↑)	−1		
			−1 (I)	−1 (↓)	1		
SL3	1 (F)	1 (↑)	1 (F)	0 (=)		2	1
	1 (F)	−1 (↓)	1 (F)	0 (=)		−2	−1
SL4	1 (F)	1 (↑)	−1 (I)	0 (=)			−1
	1 (F)	−1 (↓)	−1 (I)	0 (=)			1
SL5	−1 (I)	1 (↑)	1 (F)	0 (=)			−1
	−1 (I)	−1 (↓)	1 (F)	0 (=)			1
SL6	−1 (I)	1 (↑)	−1 (I)	0 (=)		2	1
	−1 (I)	−1 (↓)	−1 (I)	0 (=)		−2	−1
SL7	If SUMI$(c) < 0$, then FRC$(c) \neq 1$; if SUMI$(c) > 0$, then FRC$(c) \neq -1$.						

Def. SUMI$(c) := \sum_j \mathrm{DI}(c_j,c) + \sum_j \sum_k \mathrm{IDI}(c_k,c_j,c)$

The problem is that there are, in most cases, many positive and negative influences on one and the same controlled category. What is to be expected with regard to the frequency course of this category? What is needed to answer that question is something like striking the balance of all these influences, and this is done in a very simple way by introducing a seventh law which is shown in Table 4 as SL 7. *SUMI(c)* is the sum of all direct and

TABLE 5. Lists of contributing and counteracting causes of $FRC(c_2 = 1)$ in family dyad FD 4

F			Positive Influences	I			Negative Influences
$CR(c_{10}, c_2)$	$=$	1 &		$CR(c_{17}, c_2)$	$=$	-1 &	
$FRC(c_{10})$	$=$	1	1	$FRC(c_{17})$	$=$	1	-1
$CR(c_{14}, c_2)$	$=$	1 &		$CR(c_7, c_{18})$	$=$	1 &	
$FRC(c_{14})$	$=$	1	1	$CR(c_{18}, c_2)$	$=$	-1 &	
				$FRC(c_7)$	$=$	1 &	
$CR(c_{16}, c_2)$	$=$	-1 &		$FRC(c_{18})$	$=$	0	-1
$FRC(c_{16})$	$=$	-1	1				
$CR(c_6, c_{13})$	$=$	-1 &		$CR(c_6, c_{18})$	$=$	1 &	
$CR(c_{13}, c_2)$	$=$	-1 &		$CR(c_{18}, c_2)$	$=$	-1 &	
$FRC(c_6)$	$=$	1 &		$FRC(c_6)$	$=$	1 &	
$FRC(c_{13})$	$=$	0	1	$FRC(c_{18})$	$=$	0	-1
$CR(c_8, c_{13})$	$=$	1 &					
$CR(c_{13}, c_2)$	$=$	-1 &					
$FRC(c_8)$	$=$	-1 &					
$FRC(c_{13})$	$=$	0	1				
$CR(c_2, c_{13})$	$=$	-1 &					
$CR(c_{13}, c_2)$	$=$	-1 &					
$FRC(c_{13})$	$=$	0	2				
$CR(c_2, c_{18})$	$=$	-1 &					
$CR(c_{18}, c_2)$	$=$	-1 &					
$FRC(c_{18})$	$=$	0	2				
Sum of positive influences on c_2			9	Sum of negative influences on c_2			-3
			$SUMI(c_2) = 6$				

Note. In this table, "F" stands for "List of contributing causes", and "I" for "List of counteracting causes". "Positive influences" are positive values of the functions $DI(c_j, c)$ or $IDI(c_k, c_j, c)$, "negative influences" are negative values of these functions.

indirect influences exerted on the controlled category c. And if $SUMI(c)$ is smaller than 0, than the frequency of c will not increase (i.e., $FRC(c) \neq 1$), and if $SUMI(c)$ is greater than 0, the frequency of c will not decrease (i.e., $FRC(c) \neq -1$).

Now, we may consider the remaining phases of the application process as shown in Table 3 and exemplify them by looking at behavior category c_2 in family dyad 4 (cf. Figure 2). Table 5 lists all the contributing and counteracting causes and their weights which emerge with respect to category c_2 from the application of the theory. The classification of the frequency courses, given certain category relations, as contributing (list F) or counteracting (list I) causes refers to the respective terms F and I of the concept of causal explanation by Humphreys as explicated in Table 1.

If we strike the balance, we get a $SUMI(c_2)$ which is greater than 0 and which is compatible with the increase of the frequency of c_2 which has actually been observed.

On this way, an adequate explanation of the frequency course of c_2 has been achieved. But this is only one example. We did, for this domain of intended applications, several experimental single case studies of different mother-child dyads and found a rate of frequency courses, which are not in accord with the theory, of under four percent – not a bad result in the case of human behavior in a natural setting (cf. Westmeyer *et al.*, 1984). But the empirical corroboration of the theory is of minor importance in this context.

4. Discussion

The applicability of Humphreys' concept of causal explanation of specific events to the theory of behavior interaction notwithstanding, two problems remain to be settled: The first one addresses the question "What are the proper referents of the concepts of contributing cause and counteracting cause in the case of the theory of behavior interaction?". The second one concerns a general feature of Humphreys' concept.

With regard to the first problem, the contributing and counteracting causes could be identified with the events described in the *if*-parts of the special laws 1 to 6, i.e., with certain frequency courses of controlling behavior categories given certain relations to the controlled behavior category. In this case the direct and indirect influences on the controlled category mentioned in the *then*-parts of the special laws 1 to 6 are the consequences produced by the probabilistic causes.

As is shown in Table 6, a certain kind of influence on a category c is, under certain circumstances, i.e., if a certain event is to be explained, identical with the increase or decrease of the value of the chance of this event.

Another interpretation of the concepts of contributing and counteracting cause is the following: Direct and indirect influences on a controlled

TABLE 6. Interpretations of the concept of probabilistic cause

Probabilistic causes produce changes in the value of the chance of Y. (cf. Humphreys 1989, p. 287)

First option

(a) The probabilistic causes are the freqency courses of controlling behavior categories, given certain controlling category relations.

(b) The frequency courses of controlling behavior categories, given certain controlling category relations, produce positive and/or negative influences on the controlled behavior category the frequency course of which is the event to be explained, i.e. Y.

(c) The positive and negative influences on the controlled behavior category concerned are (express themselves in) the changes in the value of the chance of Y.

Second option

The probabilistic causes are the positive and negative influences on the controlled behavior category concerned.

category are the probabilistic causes. This interpretation is illustrated in Table 7.

If an increase of the frequency of the behavior category c is to be explained, a positive direct or indirect influence on c is a contributing cause, and a negative direct or indirect influence on c is a counteracting cause. Within the conceptual framework introduced by Humphreys, influences on c would produce changes in the value of the chance of the event to be explained, i.e., the respective frequency course of category c (cf. Table 1).

In the case of this interpretation, theoretical events, i.e., events described by theoretical concepts of the theory of behavior interaction, gain a causal status as is stressed by reorganizing Table 7 into Table 8.

TABLE 7. Causal status of the theoretical terms DI and IDI within the Theory of Behavior Interaction

Kind of influence	Event Y to be explained	
	$FRC(c) = 1$	$FRC(c) = -1$
Positive direct or indirect influence on c	Contributing cause	Counteracting cause
Negative direct or indirect influence on c	Counteracting cause	Contributing cause
Assumption: Direct and indirect influences on c *are* the probabilistic causes.		

I prefer the first interpretation given in Table 6, but the second one fits equally well into Humphreys' framework. In both cases, the requirement "For something to be a cause, it must invariantly produce its effect" is satisfied, since the special laws are strictly formulated deterministic assumptions.

The second and final problem I would like to address does not concern the theory of behavior interaction, but Humphreys' concept of causal explanation in general. Nevertheless it can be illustrated by a certain step which is an integrative part of the process of applying the theory of behavior interaction. I refer to step (e) of the process of application of the theory (cf. Table 3) in which the special law SL 7 of the respective theory-element is applied to the set of positive and/or negative influences on a controlled behavior category by striking the balance.

Nothing comparable is required in Humphreys' concept, it is compatible with Humphreys' concept, but not a part of it. His concept (cf. Table 1) does not explicitly presuppose that the contributing causes of Y are more powerful than the counteracting causes of Y. An assumption referring to the impact of the contributing causes on the event to be explained in comparison to the impact of the counteracting causes is missing in Humphreys' explication of causal explanations. Without such an additional assumption which states, for example, that the contributing causes of an event are more powerful than the counteracting ones, it is difficult to differentiate between adequate and inadequate explanations of specific events. Since the interpretation of the concepts of contributing and counteracting causes has anyway to be given in the context of the respective theory used for the purpose of explanation, the phrase "is more powerful than" can likewise be interpreted within this theory or its theoretical and methodological environment.

TABLE 8. Instances of contributing and counteracting causes in the case of the Theory of Behavior Interaction

Kind of cause	Event Y to be explained	
	FRC$(c) = 1$	FRC$(c) = -1$
Contributing cause	Positive direct or indirect influence on c	Negative direct or indirect influence on c
Counteracting cause	Negative direct or indirect influence on c	Positive direct or indirect influence on c

Assumption: Direct and indirect influences on c are the probabilistic causes.

References

Gottman, J.M. (1979) *Marital interaction: Experimental investigations*. New York: Academic Press.

Gottman, J.M. and Roy, A.K. (1990) *Sequential analysis: A guide for behavioral researchers*. Cambridge: Cambridge University Press.

Hahlweg, K., Schindler, L. and Revenstorf, D. (1982) *Partnerschaftsprobleme: Diagnose und Therapie*. Berlin: Springer-Verlag.

Humphreys, P.W. (1989) Scientific explanation: The causes, some of the causes, and nothing but the causes. *Minnesota Studies in the Philosophy of Science*, 13, pp. 283–306.

Ickes, W. (1981) Sex-role influences in dyadic interaction: A theoretical model. In C. Mayo and N. Henley (eds.), *Gender and nonverbal behavior*, New York: Springer, pp. 95–128.

Ickes, W. (1985) Sex-role influences on compatibility in relationships. W. Ickes (ed.), *Compatible and incompatible relationships*, New York:Springer, pp. 187–208.

Magnusson, D. (1980) Personality in an interactional paradigm of research. *Zeitschrift für Differentielle und Diagnostische Psychologie*, 1, pp. 17–34.

Nell, V. (1982) *Interaktives Verhalten in Situationen der Kontaktaufnahme*. München: Minerva.

Nell, V., Völkel, U., Winkelmann, K., Hannemann, J. and Westmeyer, H. (1988) Verhaltensinteraktion in Vater-Mutter-Kind-Triaden: Ein Theorie-Element und seine empirische Bewährung. *Zeitschrift für Sozialpsychologie*, 19, pp. 175–192.

Stegmüller, W. (1983) *Erklärung, Begründung, Kausalität* (2nd ed.). Berlin: Springer-Verlag.

Wampold, B.E. (1992) The intensive examination of social interactions. T.R. Kratochwill and J.R. Levin (eds.), *Single-case research design and analysis*, Hillsdale, NJ: Erlbaum, pp. 93–131.

Westmeyer, H. (1987) Zum Problem des empirischen Gehalts psychologischer Theorien. J. Brandtstädter (ed.), *Struktur und Erfahrung in der psychologischen Forschung*, Berlin: de Gruyter, pp. 35–70.

Westmeyer, H. (1989) The theory of behavior interaction: A structuralist construction of a theory and a reconstruction of its theoretical environment. H. Westmeyer (ed.), *Psychological theories from a structuralist point of view*. Berlin: Springer-Verlag, pp. 145–185.

Westmeyer, H. (1990) Theoriebezogene Validierung diagnostischer Verfahren: eine Monotheory-Multimethod Analyse. H. Feger (ed.), *Wissenschaft und Verantwortung*, Göttingen: Hogrefe, pp. 139–156.

Westmeyer, H. (1992) Sex-role influences in dyadic interaction: A structuralist reconstruction of W. Ickes' theory. H. Westmeyer (ed.), *The structuralist program in psychology: Foundations and applications*, Toronto: Hogrefe and Huber Publishers, pp. 249–281.

Westmeyer, H., Hannemann, J., Nell, V., Völkel, U. and Winkelmann, K. (1987) Eine Monotheory-Multimethod Analyse: Plädoyer für einen deduktivistischen Multiplismus. *Diagnostica*, 33, pp. 227–242.

Westmeyer, H., Winkelmann, K. and Hannemann, J. (1984) Eltern-Kind–Interaktion in natürlicher Umgebung: Darstellung einer Theorie und ihrer empirischen Bewährung. *Zeitschrift für personenzentrierte Psychologie und Psychotherapie*, 3, pp. 39–53.

Westmeyer, H., Winkelmann, K. and Hannemann, J. (1988) Intrasituationale dynamische Interaktion. *Zeitschrift für Differentielle und Diagnostische Psychologie*, 9, pp. 241–256.

SIMULATION AND STRUCTURALISM

KLAUS G. TROITZSCH
Social Science Informatics Institute
Koblenz–Landau University
Koblenz, Germany

1. Introduction

This paper aims at a logical reconstruction of some different approaches to
modeling processes of self-organization in the social sciences, and it uses
these approaches as examples which are to show that computer simulation
supports structuralist reconstruction of theories if it is done in a certain way.
In contrast especially to Troitzsch (1992) we do not concentrate on a specific
theory of attitude change although we shall take the axiomatizations carried
out there as but one example of a theory of self-organization.

1.1. SELF-ORGANIZATION

One of the central problems of social science is the problem of emergence
and of reducing collective phenomena to individual phenomena. While this
is not the place to retell the whole story of the doctrine of irreducibility
since Durkheim, we may rather return to the onset of social science in the
18th century when the problem of the "invisible hand" was discussed by
Adam Smith and others.

First, we have to discuss the term "self-organization". Following Haken
(1988, p. 11) we call a system self-organized

> "if it acquires a spatial, temporal or functional structure without specific
> interference from the outside. By 'specific' we mean that the structure
> or functioning is not impressed on the system, but that the system is
> acted upon from the outside in a nonspecific fashion."

Furthermore we refer to Hayek (1942, p. 288) who wrote even fifty years
ago about the social sciences that

183

R. Hegselmann et al. (eds.),
Modelling and Simulation in the Social Sciences from the Philosophy of Science Point of View, 183–207.
© 1996 *Kluwer Academic Publishers. Printed in the Netherlands.*

"the problems which they try to answer arise only in so far as the
conscious action of many men produce undesigned results, in so far as
regularities are observed which are not the result of anybody's design.
If social phenomena showed no order except in so far as they were
consciously designed, there could be no room for theoretical sciences of
society... It is only in so far as some sort of order arises as a result of
individual action but without being designed by any individual that a
problem is raised which demands a theoretical explanation."

During the last twenty years, the interdisciplinary research efforts of synergetics have invented methods that make new and powerful tools available
for the social scientist as well. It should be interesting to try a structuralist
reconstruction of these theoretical tools and to show that such a reconstruction is half way done once a computer simulation model is specified in
terms of the simulation technology applied in this paper.

Modern approaches to modeling processes of self-organization have in
common that they

- are carried out with the methods of mathematics and computer science,
- model complex stochastic systems,
- represent individuals and populations (or groups or other collectives
  of people) at the same time, as well as
- interactions between these two (or even more) levels, and, in many
  cases,
- interactions among the objects of each level.

We have to discriminate between two groups of approaches: In a first
group we have models in which individuals of one or more populations
do not interact directly, but change the state of the whole system (or of
their respective population) by their behavior and react on the change
of this collective with individual changes of their behavioral state. This
is the case in typical synergetic models, both in the social science and in
physics, chemistry, biology etc. where the most famous example of the latter
seems to be the slime mould case [Wurster, 1988]. The point is here that in
the beginning we have an "aggregate" of initially unconnected individuals,
which by itself turns into a system due to the effect that the individuals
are endowed with the ability to move in a potential which is built up as a
result of the sheer existence of these individuals. For a comparison of the
slime mould case and a theory of attitude change in human populations see
Troitzsch (1991).

On the other hand we have models in which individuals interact directly,
mostly in a network or, as it were, in a cellular automaton. Here too, the
population is modeled homogeneously in the beginning, then stochastically
influenced interactions change the state of the interacting individuals and
the relations between each pair of interacting individuals, and in the end

we find stable clusters, groups, subnets, or strata in the whole population [Hegselmann, 1994; Helbing, 1992b; Helbing, 1992a; Drogoul and Ferber, 1994].

1.2. MULTILEVEL MODELLING

Both approaches to modelling interaction between levels raise some difficult problems both to computer simulation and to a structuralist reconstruction. As for the case of computer simulation, MIMOSE is used as a specification and simulation tool which has been designed and developed by Michael Möhring and the author during the last seven years. Hence, MIMOSE is a possible solution to the problems raised by multilevel models.

The main goal of the MIMOSE project[1] was the development of a modelling language which satisfies the special demands of modelling in social science, especially the description of multilevel models in which several classes of objects are considered which change their states due to the states of other objects belonging to the same class or other classes. States may be quantitative or qualitative; state changes may be deterministic or stochastic, linear or nonlinear, and may include creation and deletion of new objects. Furthermore, the language concept should free modelers from programming and implementation details as well as support the development of structured, homogeneous simulation models, and hence improve the transparence of the "model programming process" and make model descriptions and even the corresponding simulation results easier to understand.

The language structure of MIMOSE is strongly influenced by the paradigms of functional programming languages. Because of its declarative, uniform language concept, functional programs are easier to understand, and more implementation independent than programs written in procedural programming languages.

This leads to the following main characteristics of MIMOSE models: Every MIMOSE model consists of a set of object types, from which concrete objects will be created during the model initialization. Objects, as formal representations of entities in reality (i.e. individuals, groups, or organizations), are structured by a set of attributes, which are formal representations of real properties (i.e. age or attitude). The values of all object attributes at a given time represent the state of this object. Each object attribute can take a state transition function, which evaluates the attribute

[1]The research project "Micro and multilevel modeling and simulation software development (MIMOSE)" was directed by Michael Möhring and the author and funded by the Deutsche Forschungsgemeinschaft under grant no. Tr 225/3–1 and –2 from 1988 to 1992.

value in each simulation step. The behaviour of an object is defined as its state change over time.

Regarding that the designers and developers had structuralist reconstruction in mind, it does not come as a surprise that the specification language for just these kinds of models in which an interaction between (and within) several levels occurs can easily be translated into a structuralist reconstruction — as is shown in several examples throughout this paper.

2. Models with indirect interaction

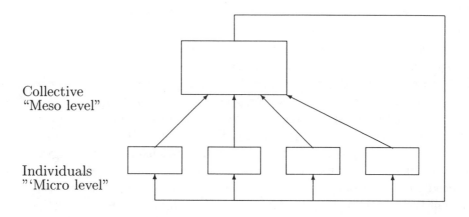

Collective
"Meso level"

Individuals
"'Micro level"

Figure 1.   Models with indirect interaction between individuals and the collective

Elements on the lowest level are endowed with the capability to react on a potential (or, more generally: on the state of a higher level) and at the same time to change this potential or higher level state (see figure 1, where arrows point from causes to effects: changes at the arrowhead are effected by the state of the arrow's origin). In the following example, the individuals have individual attitudes which change the majority or the "public opinion" on the collective level. On the other hand, they change their individual attitudes depending on "public opinion", trying to adapt to the major opinion trends. As has been shown elsewhere, the classical example of a self-organizing process in biology — the slime mould example — follows the same traits as ours [Troitzsch, 1991, pp. 112–113], and the same is true — after some modification with respect to the structure of the attitude space — for Weidlich's and Haag's (1983) model of opinion formation [Troitzsch, 1992] as well as for Glance's and Huberman's model of the "outbreak of cooperation" [Glance and Huberman, 1993].

2.1. A THEORY OF ATTITUDE CHANGE AND IT POTENTIAL MODELS

We shall now take this example to show what computer simulation and structuralist theory reconstruction have in common. Thus we take one version of the miniature theory of attitude change considered in Troitzsch (1992) and quote the major definitions of this paper.

We first define the theory element of **AC**:

Def TE (AC): TE(AC) := $\langle$ **K(AC), I(AC)**$\rangle$ where
  K(AC) := $\langle$ **M**$_p$**(AC), M(AC), M**$_{pp}$**(AC), Po[M**$_p$**(AC)], M**$_p$**(AC)**$\rangle$
  and
  I(AC) $\subseteq$ **M**$_{pp}$**(AC)** is such that members of **I(AC)** are data sets describing homogeneous populations whose individual members have attitudes which may be observed; there are exactly two continuous attitudes — say on a "left"-"right" continuum and on a "political satisfaction" continuum — which must (and can) be asked for in one or several consecutive surveys or even panels.

By this description of intended applications of **AC** we include

- singular surveys,
- repeated anonymous surveys encompassing the same population such that it is not possible to build time series of individual attitudes,
- panels of arbitrary length and intervals.

We continue with a definition of a potential model of **AC**.

Def M$_p$**(AC):** η is a potential model of **AC**, i.e. $\eta \in$ **M**$_p$**(AC)**, iff there exist $I, \mathcal{C}, A, T, \mathbb{R}, \ell, \Omega, a, \phi, \delta$ and ρ such that

1. $\eta = \langle I, \mathcal{C}, A, T, \mathbb{R}, \ell, \Omega, a, \phi, \delta, \rho \rangle$.

2. I is a non-empty, finite set [of persons].

3. $\mathcal{C}$ is a non-empty finite set of subsets of I [i.e. of populations or collectives], $\mathcal{C} \subseteq \mathrm{Po}(I)$.

4. $\langle A, \mathcal{A} \rangle$ is a measurable space [of attitudes] with $A = \mathbb{R}^2$.

5. T is a set [of instants].

6. $\ell : T \to \mathbb{R}$ is a bijective function [labeling (or coordinatizing) the instants ($\in T$) with real numbers ($\in \mathbb{R}$)].

7. $\langle \Omega, \mathcal{F}, P \rangle$ is a probability space with

   (a) Ω is a sample space,

   (b) $w : I \to \Omega$ is bijective,

   (c) $\mathcal{F} \subseteq \Omega$ is a family of events,

   (d) P is a probability measure defined on F.

8. a is a function with $\mathrm{Dom}(a) = I \times T$ and $\mathrm{Rge}(a) = A$, such that (with $i \in I$) $a(i, t)$ is the attitude of i at time t.

9. ϕ is a function with $\mathrm{Dom}(\phi) = C \times T$ and $\mathrm{Rge}(\phi) = \{f | f : A \to \mathbb{R}^+; \int_A f da = 1\}$, such that (with $C \in \mathcal{C}$) $\phi(C, t)$ is the function which describes the probability density of each $i \in C$ having attitude $a \in A$ at time $t \in T$, i.e. $\phi(C, t)(a)$ is the probability density of finding a $i \in C$ with attitude $a \in A$ at time $t \in T$.

10. δ is a function with $\mathrm{Dom}(\delta) = C \times T \times T$ and $\mathrm{Rge}(\delta) = \{d | d : A \to A\}$ such that (with $C \in \mathcal{C}$) $\delta(C, s, t)$ is the integral of the trend coefficient of the diffusion process from s to t, i.e. $\delta(C, s, t)(a) = \int_s^t \mu[a(., \tau), \tau] d\tau$. This means that $a(i, t)$ would be equal to $a(i, s) + \int_s^t \mu[a(., \tau), \tau] d\tau$ if there were no noise.

11. $(\rho_t)_{t \in T}$ is a stochastic process on the probability space $\langle \Omega, \mathcal{F}, P \rangle$ with values in the measurable space $\langle A, \mathcal{A} \rangle$ such that ρ is bivariate Gaussian white noise without cross-correlation which may be written $\boldsymbol{\rho}(i, t)$ further on [i.e. $\boldsymbol{\rho}(i, t)$ yields a pair of uncorrelated standard normal random numbers for each person $i \in I$ at any point of time $t \in T$].

2.2. EMPIRICAL APPLICATIONS

It should be clear that in empirical applications of our miniature theory, all terms mentioned up to item 9 of **Def $M_p(\mathbf{AC})$** are measurable without any theory concerning attitude *changes*. The measurement of a, of course, necessitates theories of attitude measurement, as the measurement of ϕ necessitates probability theory, but in both cases we do not have to know anything about the actual mechanisms behind the process of attitude change.

As an illustration for an empirical application we show the graphs of the density functions for several successive surveys taken from the "Politbarometer" carried out monthly by the Forschungsgruppe Wahlen e.V. for the German television system ZDF (cf. Fig. 2).

These empirical density functions indeed changed during autumn 1986 and during autumn 1987 — in 1986 in a somewhat systematic manner, winning a second maximum of density towards the Bundestag election held in early 1987, and in 1987 in a somewhat erratic manner, displaying a bimodal distribution of attitudes only in October after the "Barschel affair" —, and these changes can be detected without any knowledge of individual attitude changes, but the underlying process of individual attitude change calls for a theoretical explanation that will have to make use of the theoretical terms δ and *rho* which are the only **AC**-theoretical terms — as will be discussed in subsection 2.4.

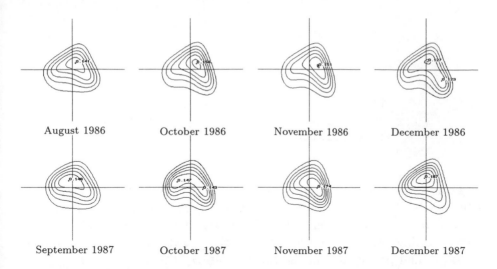

August 1986	October 1986	November 1986	December 1986
September 1987	October 1987	November 1987	December 1987

Figure 2. Empirical density functions for two series of monthly Politbarometer surveys

AC	MIMOSE
I	object type **person**
C	object type **pop**
A	type **list of real** of the attribute **att** of object type **person**
T	the set of simulation steps which is always implicit in MIMOSE
ℓ	no match, since time is always discrete in MIMOSE (and on any digital computer)
Ω	the set of seeds of MIMOSE's random number generators, each random attribute of each instantiation of each object type having its own seed
a	the attribute **att** of object type **person**
ϕ	the attribute **params** of object type **pop** — the range of **params** being the space of ϕ's function parameters
δ	the attribute **kparams** of object type **pop** — the range of **kparams** being the space of δ's function parameters
ρ	the attribute **r** of object type **person**

TABLE 1. Match between terms of **AC** and identifiers of the related MIMOSE simulation program

We are now in a position to compare the definition of the potential model of **AC** to a skeleton of a MIMOSE program which is complete but for some function applications:

```
pop := { params  : list of real
                 := ... persons.att ... ;
         kparams : list of real
```

```
                      := ... ;
            persons : list of person
         };
person := { r              : real
                           := ... ;
              att          : list of real
                           := ... \IX{population}.params_1 ... ;
            \IX{population} : pop
         };
```

From table 1 we see that MIMOSE models a function from $I \times T$ to an arbitrary set S as an attribute of the object type corresponding to I whose type is the type corresponding to S. If S is a set of functions, then the corresponding MIMOSE attribute is of type `list of real`, i.e. the space of real valued parameters of the functions $\in S$.

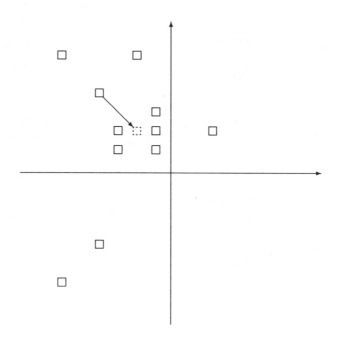

Figure 3.   Attitude space of **AC** with some individuals plotted. One individual is changing its attitude position to achieve a more densely populated position, thus changing the whole distribution on the attitude space.

In our example we suppose a two-dimensional continuous attitude space (for a graphical representation see figure 3). We restrict ourselves to a function ϕ whose domain consists of probability density functions of the type

$$
\begin{aligned}
f(\boldsymbol{a}; \boldsymbol{\theta}) \;=\; \exp\{ \quad & \theta_{00} && + \; \theta_{10}a_1 && + \; \theta_{20}a_1^2 && + \; \cdots && + \; \theta_{n0}a_1^n \} \\
& + \; \theta_{01}a_2 && + \; \theta_{11}a_1a_2 && + \; \cdots && && + \; \theta_{n-1,1}a_1^{n-1}a_2 \\
& + \; \cdots \\
& + \; \cdots && + \; \theta_{ij}a_1^i a_2^j && + \; \cdots \\
& + \; \theta_{0n}a_2^n \} \\
=\; & \exp\{-V(\boldsymbol{a})\}
\end{aligned}
\tag{2}
$$

that is, to probability density functions of the exponential family whose exponent is a polynomial up to fourth degree in $\boldsymbol{a}$. This is not an important restriction, however, because supposing a polynomial $V(\boldsymbol{a})$ up to fourth degree in the exponent of the probability density function only means taking a Taylor expansion of an arbitrary function $\tilde{V}(\boldsymbol{a})$ including only the lower powers up to the biquadratic terms. The last term of the Taylor expansion of $\tilde{V}(\boldsymbol{a})$ included into $V(\boldsymbol{a})$ must, of course, always be of even degree, since otherwise $\exp\{-V(\boldsymbol{a})\}$ will be no probability density function (and $V(\boldsymbol{a})$ will be no potential).

The θ's may be estimated by one of the two procedures (named `calcpdfparams` in the MIMOSE program text below) described by Troitzsch (1987, p. 169–170) and by Herlitzius (1990), respectively.

2.3. FULL MODELS OF AC

In this case, the attitude change function δ describes a diffusion process with

$$
\boldsymbol{\mu}(\boldsymbol{a}, t) \;=\; \frac{-\partial \gamma V(\boldsymbol{a})}{\partial \boldsymbol{a}}
\tag{3}
$$

$$
\boldsymbol{\Sigma}(\boldsymbol{a}, t) \;=\; \sigma_\rho^2 \boldsymbol{I}
\tag{4}
$$

with V the (negative) exponent of the probability density function of equations 1 and 2, and γ is a constant in which the dependence of δ on C becomes manifest. Perhaps we should have inserted an additional **AC**-theoretical term $\gamma : C \to \mathbb{R}$ into our definition of potential models of **AC**.

In this continuous case, $\boldsymbol{\rho}$ is a bivariate Gaussian white noise process with no cross correlation (i.e. with $D^2(\boldsymbol{\rho}) = \sigma_\rho^2 \boldsymbol{I}$ where σ is a constant in which a dependence of ρ on C unmentioned so far becomes manifest. Perhaps an additional **AC**-theoretical term $\sigma : C \to \mathbb{R}^+$ would have been appropriate.

Here, the definition of a model of **AC** reads:

Def M(AC): ζ is a model of **AC**, i.e. $\zeta \in$ **M(AC)**, iff there exist $I, C, A, T, \mathbb{R}, \ell, \Omega, a, \phi, \delta$ and ρ such that

1. $\zeta = \langle I, C, A, T, I\!R, \ell, \Omega, a, \phi, \delta, \rho \rangle.$

2. $\zeta \in \mathbf{M}_p(\mathbf{AC})$

3. V is an (auxiliary) function with $\mathrm{Dom}(V) = A \times T$ and $\mathrm{Rge}(V) = I\!R \backslash \{-\infty\}$ with

$$V(a, t) = -\ln \phi(C, t)(a)$$

and

$$\lim_{\Delta t \to 0} \frac{\delta(C, t, t + \Delta t)(a)}{\Delta t} = -\gamma_c \frac{\partial V(a, t)}{\partial a}$$

such that (informally)

$$a(i, t) = a(i, s) + \gamma_c \int_s^t -\frac{\partial V(a(i, \tau), \tau)}{\partial a} d\tau + \int_s^t \rho(i, \tau) d\tau$$

where γ_c is a constant function.

This leads to a complete MIMOSE program which reads

```
pop := { gamma    : real;
         sigma    : real;
         params   : list of real
                  := calcpdfparams(persons.att);
         kparams  : list of list of real
                  := derivative(params);
         persons  : list of person
       };
person := { r          : list of real
                       := normal;
            att        : list of real
                       := langevin(att_1,
                                   pop.gamma,
                                   pop.kparams,
                                   pop.sigma,
                                   r);

            \IX{population} : pop
       };
```

where the function named `langevin` performs the operations required by Definition $\mathbf{M}(\mathbf{AC}).2$.

Here, too, the parameters of the attitude change function δ can be measured by means of $\mathbf{AC}$ as was shown by Troitzsch (1990).

Figure 4 shows two realizations of the stochastic process on the collective level. Both realizations start from the same initial conditions on the individual level; they differ only in the seed of the random number number generator. Both realizations are approximately normally (at least: unimodally) distributed in the beginning. Their distributions develop into multimodality after a while, apparently becoming stationary in the end. There are, of course, also realizations that seem to remain unimodal forever.

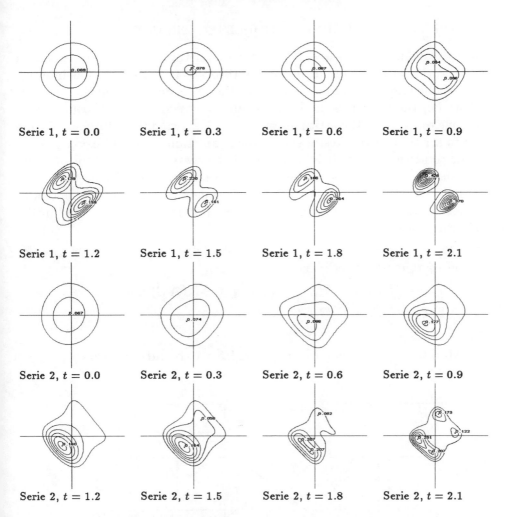

Figure 4. Two realizations of the stochastical process of the continuous case over eight time steps (development of probability density functions over time; contour lines for multiples of 0.025 if maxima are lower than 0.2, else for multiples of 0.050)

It should be noted here that the user interfaces provided by MIMOSE (which we shall not describe in this paper) have some facilities to hide attributes corresponding to theoretical terms from the user. A MIMOSE user using GEMM (the graphical editor for the MIMOSE modeling language, [Klee and Troitzsch, 1993]) will even be guided through the modeling process in a similar manner as the one adopted by structuralism: In step 1 of GEMM the user will have to define object types (base sets), while in step 2 attribute types (function ranges) are defined; step 3 allows the definition of function domains, while step 4 defines the function bodies which correspond to the axioms of the definition of the actual model.

2.4. THE PARTIAL POTENTIAL MODELS OF THEORY **AC**

We have defined the gradient function δ and the stochastic process ρ as **AC**-theoretical terms of our miniature theories. It is obvious that in every conceivable empirical application of our theory as described in the previous subsection the influences of δ and ρ cannot be separated. Of course, it is possible to measure the attitudes of all $i \in C$ at two different points of time and to arrive at estimates of the gradient function of the diffusion process of the continuous case. It is, however, first necessary to *know* (or *assume*) that the observed process is the diffusion process of equations 3 and 4.

This is why δ and ρ do not belong to the definition of a *partial* potential models. Thus we have to define the partial potential models of **AC** in the following manner:

Def $M_{pp}(AC)$: ξ is a partial potential model of **AC**, i.e. $\xi \in M_{pp}(AC)$, iff there exists η such that

1. $\eta = \langle I, C, A, T, \mathbb{R}, \ell, \Omega, a, \phi, \delta, \rho \rangle \in M_p(AC)$ and
2. $\xi = \langle I, C, A, T, \mathbb{R}, \ell, \Omega, a, \phi \rangle$

3. Models with direct interaction between individuals

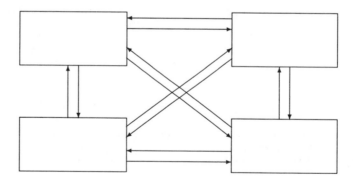

Figure 5.   Models with direct interaction between the individuals

A completely different approach to modeling processes of self-organization must be adopted when individuals are allowed to interact directly. Here, the individual is no longer isolated from the other individuals (and not only knows about their opinions by observation of the population as a whole, e.g. by listening to the radio or by reading newspapers) but changes its attitude due to the locally observable attitude of those individuals with whom it has personal contacts. Here, for example, we may suppose that the strength of *alter*'s influence on *ego* depends on how often *alter* and *ego* interacted in the past, and that the frequency of interactions depends on the similarity of attitudes. Thus again we have two coupled attributes, the frequency of pairwise interactions, and the individual attitude.

3.1. INTERACTIONS IN A THREE-PERSON GROUP: STARTING FROM A SIMULATION MODEL

The classical example of this type of model seems to be Kirk's and Coleman's model (1967) of pairwise interaction in a three person's group which goes back to Simon's (1957, pp. 99–114) famous formalization of Homans's (1950) theory of interaction in social groups. In this model, each of three individuals has a certain inclination to interact with one of the two others. This inclination (sympathy in the MIMOSE program below) is corroborated by any realized interaction. Moreover, sympathy towards all possible partners decreases in every time step. Whether an interaction occurs in a certain time step is decided by a randomly selected individual (the dominant individual) which then opens the interaction with respect to the individual it likes most. The third partner is left empty-handed in this round. Kirk and Coleman describe several variants of this model: first according to the probability that a certain individual becomes dominant in a certain round: this probability may

- be constant and equal for all three individuals, or
- depend on the frequency of interactions a certain individual has taken part in up to this round — either: the more frequent the participation up to now, the higher the probability to be dominant in the next round; or: the less frequent the participation up to now, the higher the probability to be dominant in the next round;

secondly according to the amount by which the inclination to interact with a certain partner is increased by an interaction:

- If the interaction occurred with the desired partner, the inclination to re-interact with him or her will be increased by more (or with higher probability, RA > RB in the MIMOSE program below) than in the case that the partner was not the desired partner.
- The increase in inclination to re-interact is not dependent on whether the interaction occurred with the desired or with the third partner (RA = RB in the MIMOSE program below) .
- The increase in inclination to re-interact with the same partner is higher (RA < RB in the MIMOSE program below) in the case when the interaction occurred with the third partner than in the case of the desired interaction.

The possible outcomes of this model are the following:

- In the long run, all individuals participate equally in the interactions (which, of course, does not seem to be very likely).
- In the long run, one individual is left out of the interactions while the two others form a stable pair.
- In the long run, one individual participates in nearly every interaction while the two others nearly never interact with each other.

Thus, in Kirk's and Coleman's model, too, we have the case of emergent structure in an initially homogeneous collective, and hence the case of self-organization in the sense of Haken (1988, p. 11) since the triad "acquires a ... functional structure without specific interference from the outside." The only inference from outside is the random effect which transforms a person's inclination to interact into a real interaction.

Returning to Hayek (1942, p. 288) we find that "regularities are observed which are not the result of anybody's design" since the consciousness of our model individual does not even refer to the concept of a "stable pair" or a "stable leader".

The simulation gives quite clear results as to which of the three possible outcomes may be expected from different parameter combinations of the model (see figure 6). Both the variation of the inclination increase alternatives and the dominance regulations play their roles: Constant and equal dominance probability ($\kappa = 0$) leads to an equal participation — at least if inclination towards non-desired partners is also increased ($RA - RB < 0.4$) —; increased dominance as a consequence of high participation ($\kappa > 0$) leads to a stable pair — at least if inclination towards desired partners is increased, too (the effect of the difference of RA and RB obviously depends on κ) —; and increased dominance as a consequence of low participation ($\kappa < 0$) leads to a stable leadership — at least if inclination towards desired partners is increased, too ($RA - RB > 0.4$): In the upper left corner of fig. 6 we find stable pairs, in the lower left corner we have a stable leader, and in the lower right we have equal participation

We now come to the presentation of a MIMOSE program[2] of Kirk's and Coleman's model (the variants are hidden in the functions `dominance` and `incsymp` which are of no interest for the definition of a potential model of the theory).

```
system :=
{ dominant : int := dominance(kappa, triple.qbusy);
  triple : list of person;
}
person :=
```

[2]The simulation whose results are presented in fig. 6 were actually achieved with a "C" program which runs considerably faster than the MIMOSE program.

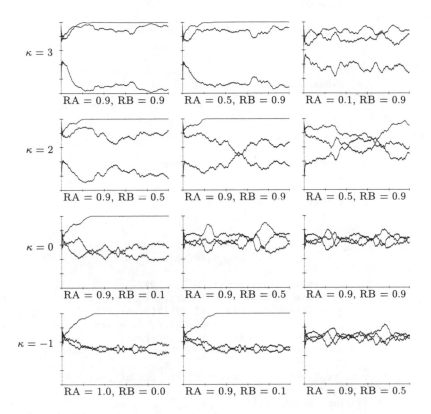

Figure 6. Simulation results for various parameter combinations of Kirk's and Coleman's model (abscissa is time or number of interactions, ordinate is the ratio of participation in interactions, hence the graphs show how strongly each of the three modelled persons took part in the interactions)

```
{ number : int;
  sys     : list of system;
  partners : list of person;
  qbusy : real := insert(plus, busy)/length(busy);
         /* counts the ratio of \IX{interaction}s */
  busy : list of int := append(tail(busy_1), (realpartner > 0));
         /* counts the number of \IX{interaction}s within the last */
 /* 500 rounds, elements of busy are either 0 or 1     */
  symp    : list of real
         := updsymp(symp_1, corr, SYMPDECR, realpartner);
  desiredpartner : int
                := dp(symp_1);
  realpartner : int
              := sys.dominant_1
                   if partners.desiredpartner_1[sys.dominant_1]
                                = number
                 else (desiredpartner_1
                   if sys.dominant_1 = number
```

```
                          else 0);
    corr : real
         := K if ((realpartner > 0) &&
                   (((desiredpartner_1 = realpartner) &&
                     (uniform(2, 0.0, 1.0) < RA)) ||
                    ((desiredpartner_1 != realpartner) &&
                     (uniform(2, 0.0, 1.0) < RB))))
            else 0.0;
}


\IX{probability} : real # real -> real
              := fct x, kappa to e
                    where e : real := exp(kappa*x)
                    end;


dominance : real # list  of real -> int
          := fct kappa, interact to d
                 where
                     d : int := 1 if r < head(i) else
                                2 if r < head(tail(i)) else
                                3;
                     r : real := uniform(3, 0.0, 1.0);
                     i : list of real := normdistrib(q);
                     q : list of real := apply2(\IX{probability}, p, kappa);
                     p : list of real := apply2(div, interact, sum);
                     sum : real := insert(plus, interact);
                 end;

updsymp : list of real # real # real # int -> list of real
        := fct syl, c, d, rp to
               decsymp(incsymp(syl, c, rp), d)
           end;

decsymp : list of real # real -> list of real
        := fct syl, d to
               apply2(times, syl, d)
           end;



incsymp : list of real # real # list of real
        := fct syl, c, rp  to
      apply2(plus, syl, c*ones)
      where ones : list of int :=
             [1, 0, 0] if rp = 1 else
     [0, 1, 0] if rp = 2 else
     [0, 0, 1]
           end;

dp : list of real  -> int
   := fct syl to d
        where
          d : int
             := 1 if r < head(s) else
                2 if r < head(tail(s)) else
                3;
```

```
    r : real := uniform(4, 0.0, 1.0);
    s : list of real := normdistrib(syl);
  end;
```

normdistrib(list) normalizes a list such that the sum of its elements becomes 1; apply2(function, list1, list2) yields a list of results of function(list1[i], list2[i]); uniform(selector, low, high) yields uniformly distributed random numbers between low and high; head(list) is the first element of a list, while tail(list) is the remainder of the list; identifiers in upper case letters are constant parameters.

3.2. SIMULATION-BASED STRUCTURALIST RECONSTRUCTION

From this simulation program a structuralist reconstruction of Kirk's and Coleman's theory should be possible. To achieve this goal we set up table 2 analogous to table 1. We remember that a variable MIMOSE attribute x of an object type y corresponds to a function from $Y \times T$ to X while a constant MIMOSE attribute z corresponds to a function from Y to Z. Furthermore we remember that we do not yet need the MIMOSE state change functions (not even their signatures, because a potential model corresponds to a MIMOSE program with everything cancelled between each ":=" and the next semicolon). We must only observe implicit applications of random functions like uniform which are allowed in MIMOSE.

Hence, we are ready to write down the definition of a potential model of **KC** — here we do not yet need the MIMOSE state change functions (not even their signatures, because a potential model corresponds to a MIMOSE program with everything cancelled between each ":=" and the next semicolon).

Def $M_p(KC)$: η is a potential model of **KC**, i.e. $\eta \in M_p(KC)$, iff there exist $\mathcal{P}, \mathcal{S}, A, T, \Omega, d, \sigma, \delta, \rho, \gamma$ such that

1. $\eta = \langle \mathcal{P}, \mathcal{S}, A, T, \Omega, d, \sigma, \delta, \rho, \gamma \rangle$.

2. $\mathcal{P}$ is a non-empty, finite set [of persons].

3. $\mathcal{S}$ is a non-empty finite set of subsets of $\mathcal{P}$ [i.e. of triads or, for the sake of generality, of m-ads, or systems], $\mathcal{S} \subseteq \mathrm{Po}(\mathcal{P})$.

4. $\langle A, \mathcal{A} \rangle$ is a measurable space [of attitudes] with $A = \mathbb{R}^{|\mathcal{P}|}$.

5. T is a set [of instants].

6. $\ell : T \to \mathbb{R}$ is a bijective function [labeling (or coordinatizing) the instants ($\in T$) with real numbers ($\in \mathbb{R}$)].

7. $\langle \Omega, \mathcal{F}, P \rangle$ is a probability space with

    (a) Ω is a sample space,

    (b) $w : I \to \Omega$ is bijective,

MIMOSE	KC
the set of simulation steps which is always implicit in MIMOSE	T
the set of seeds of MIMOSE's random number generators, each random attribute of each instantiation of each type having its own seed	Ω
the type `list of real` of the attribute `symp`	$A = \mathbb{R}^{|\mathcal{P}|}$
object type `system`	$\mathcal{S}$
attribute `dominant`	$d : \mathcal{S} \times T \to \{1, \ldots, |\mathcal{P}|\}$
attribute `triple`	—
object type `person`	$\mathcal{P}$
attribute `number`	$n : \mathcal{P} \to \{1, \ldots, |\mathcal{P}|\}$
attribute `sys`	—
attribute `partners`	—
attribute `symp`	$\sigma : \mathcal{P} \times T \to A$
attribute `desiredpartner`	$\delta : \mathcal{P} \times T \to \{1, \ldots, |\mathcal{P}|\}$
attribute `realpartner`	$\rho : \mathcal{P} \times T \to \{0, \ldots, |\mathcal{P}|\}$
attribute `corr`	$\gamma : \mathcal{P} \times T \to \{0, 1\}$

TABLE 2. Match between identifiers of the Kirk-Coleman MIMOSE simulation program and the terms of **KC**

    (c) $\mathcal{F} \subseteq \Omega$ is a family of events,

    (d) P is a probability measure defined on F.

8. d is a function with $\mathrm{Dom}(d) = \mathcal{S} \times T$ and $\mathrm{Rge}(d) = \{1, \ldots, |\mathcal{P}|\}$, such that (with $s \in \mathcal{S}$ and $t \in T$) $d(s,t)$ is the number of the dominant person in system s at time t.

9. σ is a function with $\mathrm{Dom}(\sigma) = \mathcal{P} \times T$ and $\mathrm{Rge}(\sigma) = A$, such that (with $p \in \mathcal{P}$ and $t \in T$) $\sigma(p, t)$ is the sympathy vector of p at time t, the elements of $\sigma(p, t)$ denoting p's sympathies towards his or her possible partners.

10. δ is a function with $\mathrm{Dom}(\delta) = \mathcal{P} \times T \times T$ and $\mathrm{Rge}(\delta) = \{1, \ldots, |\mathcal{P}|\}$, such that (with $p \in \mathcal{P}$ and $t \in T$) $\delta(p, t)$ is the number of the desired partner of person p at time t [which, of course, is the one corresponding to the highest element in p's sympathy vector $\sigma(p, t)$].

11. ρ is a function with $\mathrm{Dom}(\rho) = \mathcal{P} \times T \times T$ and $\mathrm{Rge}(\rho) = \{0, \ldots, |\mathcal{P}|\}$, such that (with $p \in \mathcal{P}$ and $t \in T$) $\rho(p, t) > 0$ is the number of the real partner of person p at time t, or $\rho(p, t) = 0$ means that person p has no partner at time t.

The definition of a model of **KC** is derived from the complete MIMOSE program in a straightforward manner, too:

Def M(KC): ζ is a model of **KC**, i.e. $\zeta \in$ **M(KC)**, iff there exist $\mathcal{P}, \mathcal{S}, A, T, \Omega, d, \sigma, \delta, \rho, \gamma$ such that

1. $\zeta = \langle \mathcal{P}, \mathcal{S}, A, T, \Omega, d, \sigma, \delta, \rho, \gamma \rangle$.
2. $\zeta \in$ **M$_p$(KC)**.
3. $d(s, t) = \ldots$
4. $\sigma(p, t) = \ldots$
5. $\delta(p, t) = \ldots$
6. $\rho(p, t) = \ldots$
7. $\gamma(p, t) = \ldots$

where the "$\ldots$" in items 3 to 7 stand for the MIMOSE state change functions of the program above which need not necessarily be written down here since the example serves only to make clear which steps must be taken to transform a MIMOSE simulation program into a structuralist reconstruction of the underlying theory.

We have to add one important point concerning the emergence of structure in a previously homogeneous group: Unlike the case of models with indirect interaction (where the density function may be used for detecting the change from unimodal to multimodal densities), here we have not introduced any term with which to describe an emerging structure within our group. It is, of course, possible to introduce (non-theoretical) terms which count the interaction within each possible pair in a group, or which express the frequency of interaction within each pair during the last, say, 500 steps (like **persons.qbusy**), but our theory (and our simulation program) is, as it were, blind against the emergent structure of, say, two close pairs without frequent interactions between those pairs. This deficiency may of course be overcome by linking our theory to cluster analysis which would then take an interaction frequency matrix and yield cluster membership information. We would then have an attribute containing the group membership of an individual as a new **KC**-theoretical term.

4. Summary

4.1. TRANSFORMATION FROM A MIMOSE PROGRAM INTO A STRUCTURALIST RECONSTRUCTION

We have seen in this paper that the formalization of a theory in a MIMOSE simulation program is equivalent to a structuralist reconstruction of this theory — which should not come as a surprise since the design of MIMOSE was partly guided by the thoughts of the founders of the structuralist program.

To perform the transformation from a simulation program into a structuralist reconstruction the following steps are necessary:

- Take the object types of the MIMOSE model as base sets of the definition of the potential model.
- Take T as a set of instants and — if any random functions are applied in the MIMOSE program — $\langle \Omega, \mathcal{F}, P \rangle$ as a probability space to be additional base sets of the definition of a potential model.
- Take any constant attribute of any MIMOSE object type as a function from this object type to the attribute type (for the sake of simplicity, we identify the "type" with the "set of its instances").
- Take any variable attribute of any MIMOSE object type as a function from the cross product of object type and the set of instants to the attribute type.
- After these first four steps the definition of the potential model is complete.
- Take any function application (state change function, between ":=" and the next semicolon) as an axiom of the definition of the model. This completes the definition of the model of the theory.

4.2. TRANSFORMATION FROM A STRUCTURALIST RECONSTRUCTION INTO A MIMOSE PROGRAM

The transformation of a structuralist theory reconstruction into an executable MIMOSE program is also straightforward:

- Take any base set from the definition of the potential model (except a set of instants and a probability space) and transform it into a MIMOSE object type. This is done by inserting

```
<base_set_name> := {
}
```

  anywhere into the MIMOSE program (but of course not between braces).
  If there is no set of instants the theory will not be about a dynamical process, and a simulation program is of no use. If the set of instants is continuous, a discretization will always be necessary for any kind of (digital) computer simulation. So first a redesign of the theory will be necessary.
- Take any function from the definition of the potential model and transform it into an attribute of the respective object type, considering the domain of the respective function. This is done by inserting

```
<function_name> : <function_range> ;
```

anywhere into the corresponding type definition (but of course only just after a semicolon).

- Take any axiom from the definition of the model of the theory and transform it into a MIMOSE state change function. This is most easily done by inserting

```
:= <axiom_name>(<terms in the right hand side of the axiom>)
```

between the corresponding `<function_range>` and the semicolon.

- Use MIMOSE's user interface to initialize all constants and variable attributes which must have taken values at simulation start time, and to fix the simulation parameters (time step size, break and stop condition), and run the program.

We may put this procedure to a test using an axiomatization of a simple psychological theory concerning stress by waiting times (**SbW**) [Holling and Suck, 1989] whose intended applications are experiments on subjects working at a computer and experiencing varying system response times, i.e. interruptions of their work. "Such interruptions often lead to anger, anxiety or other negative emotions, which may be summarized as stress." (1989, p. 187) Holling's and Suck's definitions are the following (1989, p. 190) (they are given here with some additional explanations in brackets and outside quotes):

Def M_p(SbW): x is a potential model of **SbW**, i.e. "$x \in M_p(\textbf{SbW})$, iff" there exist $H, R^+, \mathcal{L}, P, T, C, MS$ such that

1. "$x = \langle (R^+, R^+ \cap \mathcal{L}, P), T, C, M, S \rangle$",

2. H is a non-empty finite set [of human beings subject to the experiment of which only one element is of interest here — this seems to be why Holling and Suck omitted this part of the definition].

3. "$(R^+, R^+ \cap \mathcal{L}, P)$ probability space".

4. "T corresponding random variable as identity function: $T : \mathbb{R}^+ \to \mathbb{R}^+, T(t) = t$ for all t".

5. "C :" $H \times$ "$\mathbb{R}^+ \to \mathbb{R}^+$" [psychic costs in the guise of waiting times].

6. $MS : H \to \mathbb{R}^+$ "is the mean stress empirically measured during the interruption, e.g. mean systolic blood pressure".

Def M(SbW): x is a model of **SbW**, i.e. "$x \in M(\textbf{SbW})$ iff" there exist $H, R^+, \mathcal{L}, P, T, C, MS, S$ such that

1. "$x \in M_p(\textbf{SbW})$".

2. "C monotonically P-a.e. non decreasing" (a.e. = almost everywhere).

3. "$E(S) = MS$" [$E(.)$ is the expectation operator, $S(t)$ is "accumulated stress intensity" which is derived from one-time stress intensity $SI(t)$].

Although the definitions might not be fully clear to the reader, they are sufficient to derive a MIMOSE program from them. We have to define just one object type (with only one instance because experiments are performed on a single person at a time):

```
person := {
}
```

SbW	MIMOSE
H	object type **person**
T	the set of simulation steps which is always implicit in MIMOSE
R^+	the set of seeds of MIMOSE's random number generators, each random attribute of each instantiation of each object type having its own seed
C	attribute **psychicCosts**
SI	attribute **oneTimeStressIntensity**
S	attribue **accumulatedStressIntensity**
$E(S)$	no counterpart since S can be 'measured' within the model

TABLE 3. Match between terms of **SbW** and identifiers of the related MIMOSE simulation program

As above, we make up a table of correspondences (see Table 3) and write down the complete program (save for some details which are not fully specified by Holling and Suck, e.g. the function $C(t)$ which is only described as non-decreasing and which seems to have to contain an application of a random number generator):

```
person := {
 psychicCosts : real
            := nonDecreasing(psychicCosts_1);
 oneTimeStressIntensity : real
                    := someFunctionOf(psychicCosts);
 accumulatedStressIntensity : real
                        := accumulatedStressIntensity_1 +
                                oneTimeStressIntensity;
 history : list of real
        := append(history_1, accumulatedStressIntensity);
};
```

After an appropriate definition of the two functions **nonDecreasing()** and **someFunctionOf()** (which must be left to the authors of the article), only `psychicCosts_1` and `accumulatedStressIntensity_1` (presumably with zero for both) and `history_1` (with [] for an empty list) have to be initialized, and then the simulation may be run: `history` will afterwards contain the history of `person`'s stress by waiting times. It may be noted that simulation time steps have no real counterpart here: every simulation step stands for one task interruption; more elaborated versions of this simulation program (and of the theory) might define a dependence of the psychic costs not only on the length of each interruption but also on the time between interruptions.

4.3. STRUCTURALIST RECONSTRUCTION AND SIMULATION

We have shown in this paper that structuralist reconstruction of a theory and the procedure of building a simulation model are very much alike. This is not only true in the case where a tool like MIMOSE is used for simulation, but also for more classical approaches to simulation, though for example in the case of DYNAMO (i.e. the *Systems Dynamics* approach) only one object is possible (and — as it were — invisible); and if all-purpose programming languages like FORTRAN or even PASCAL are used for simulation, the composition of a simulation model — consisting of object types — may be hidden behind a bulk of lines of code necessary for running a model and obtaining graphical output having nothing to do with the simulation model proper. Thus, Schnell's recommendation "it is essential to use a language, which is widely available and as much self-documenting as possible. I believe, only PASCAL without any extensions fulfil all these points" (1992, p. 340) is questionable. Object oriented programming languages (like SmallTalk; MIMOSE is also close to object oriented) are far better appropriate to code just the core of a simulation model, hiding all the technical details before the eyes of the modeler. But even in those cases where a modeler does not use one of the more modern simulation tools, he or she is well advised to follow the same paths as recommended by a structuralist reconstruction: A simulation model designed this way will always win clarity and perspicuity.

5. Acknowledgement

I am indebted to my colleagues and students at the Social Science Informatics Institute at Koblenz–Landau University as well as the participants of the the conference on Modeling and Simulation in the Social Sciences from the Philosophy of Science Point of View, Bremen, Germany, February 17 to 19, 1994, for helpful criticism. This paper originates in the research project "Micro and multilevel modeling and simulation software development (MIMOSE)" directed by Michael Möhring and the author and funded by the Deutsche Forschungsgemeinschaft under grant no. Tr 225/3–1 and –2.

References

Drogoul, A., and Ferber, J. (1994) Multi-agent simulation as a tool for studying emergent processes in societies. Doran, J., and Gilbert, G.N., eds. *Simulating Societies: the Computer Simulation of Social Phenomena*, pp. 127–142. University of London College Press, London.

Glance, N.S., and Huberman, B.A. (1993) The outbreak of cooperation. *Journal of Mathematical Sociology*, 17 (4), pp. 281–302.

Haken, H. (1988) *Information and Self-Organization. A Macroscopic Approach to Complex Systems.* Springer Series in Synergetics, vol. 40. Springer, Berlin, Heidelberg, New York.

Hayek, F.A. (1942) Scientism and the study of society. *Economica*, 9, pp. 267–291; 10, pp. 34–63; 11, pp. 27–39.

Hegselmann, R. (1995) Experimentelle Moralphilosophie. Computersimulationen zu Klassen, Cliquen und Solidarität. Hegselmann, R., and Peitgen, H.O., eds., *Ordnung und Chaos in Natur und Gesellschaft.* Hölder–Pichler–Tempski, Wien, in press.

Helbing, D. (1992) A mathematical model for attitude formation by pair interactions. *Behavioral Science*, 37, pp. 190–214.

Helbing, D. (1992). A mathematical model for behavioral changes by pair interactions and its relation to game theory. *Angewandte Sozialforschung*, 17, pp. 179–194.

Herlitzius, L. (1990) Schätzung nicht-normaler Wahrscheinlichkeitsdichtefunktionen. Gladitz, J., and Troitzsch, K.G., eds., *Computer Aided Sociological Research. Proceedings of the Workshop "Computer Aided Sociological Research" (CASOR'89), Holzhau/DDR, October 2nd–6th, 1989.* Akademie-Verlag, Berlin, pp. 379–396.

Homans, G.C. (1950) *The Human Group.* Harpers, New York.

Holling, H., and Suck, R. (1989) Interruption of action and stress. Westmeyer, H., ed., *Psychological Theories from a Structuralist Point of View.* Springer, Berlin, Heidelberg, New York, pp. 187–202.

Kirk, J., and Coleman, J.S. (1967) Formalisierung und Simulation von Interaktionen in einer Drei-Personen-Gruppe. Mayntz, R., ed., *Formalisierte Modelle in der Soziologie*, volume 39 of *Soziologische Texte.* Luchterhand, Neuwied, pp. 169–190.

Klee, A., and Troitzsch, K.G. (1993) Chaotic behaviour in social systems: Modelling with GEMM. Troitzsch, K.G., ed., *Catastrophe, Chaos, and Self-Organization in Social Systems. Invited Papers of a Seminar Series on Catastrophic Phenomena in Soviet Society and Self-Organized Behaviour of Social Systems Held at the Institute of Sociology of the Academy of Sciences of the Ukrainian Republic, Kiev, September 4 to 11, 1992.* Universität Koblenz–Landau, Koblenz, pp. 81–104.

Schnell, R. (1992) Artificial intelligence, computer simulation and theory construction. Faulbaum, F., ed., *SoftStat '91. Advances in Statistical Software 3*, Gustav Fischer, Stuttgart, Jena, New York, pp. 335–342.

Simon, H.A. (1957) *Models of Man, Social and Rational. Mathematical Essays on Rational Human Behavior in a Social Setting.* Wiley, New York.

Troitzsch, K.G. (1987) *Bürgerperzeptionen und Legitimierung. Anwendung eines formalen Modells des Legitimations-/Legitimierungsprozesses auf Wählereinstellungen und Wählerverhalten im Kontext der Bundestagswahl 1980.* Lang, Frankfurt, Bern, New York.

Troitzsch, K.G. (1990) Self-organisation in social systems. Gladitz, J., and Troitzsch, K.G., eds., *Computer Aided Sociological Research. Proceedings of the Workshop "Computer Aided Sociological Research" (CASOR'89), Holzhau/DDR, October 2nd–6th, 1989.* Akademie-Verlag, Berlin, pp. 353–377.

Troitzsch, K.G. (1991) A comparison of some models of processes of self-organization.

Werner Ebeling, Manfred Peschel, and Wolfgang Weidlich, editors, *Models of Self-Organization in Complex Systems*. Akademie-Verlag, Berlin, pp. 106–116.

Troitzsch, K.G. (1992) Structuralist theory reconstruction and specification of simulation models in the social sciences. Westmeyer, H., ed., *The Structuralist Program in Psychology: Foundations and Applications*. Hogrefe & Huber, Seattle, Toronto, Bern, Göttingen, pp. 71–86.

Weidlich, W., and Haag, G. (1983) *Concepts and Models of a Quantitative Sociology. The Dynamics of Interacting Populations*. Springer Series in Synergetics, vol. 14. Springer, Berlin, Heidelberg, New York.

Wurster, B. (1988) Periodic cell communication. Mario Markus, Stefan C. Müller, and Grégoire Nicolis, editors, *From Chemical to Biological Organization*, Springer, Berlin, Heidelberg, New York, Paris, pp. 255–260.

CELLULAR AUTOMATA IN THE SOCIAL SCIENCES

Perspectives, Restrictions, and Artefacts

RAINER HEGSELMANN
Fachbereich 9 Philosophie
Universität Bremen
Bremen, Germany

Cellular automata (CA) based models are known for about fifty years. What made them attractive is that in those models simple basic or micro structures very often induce complex dynamics with surprising macro effects, which are fascinating and hardly understandable in analytical terms. CA became known to a more general audience as the *game of life*, which is a good example for the genre. In the most elementary form of that game, invented by John Conway in 1970, cells on a checkerboard have two states, "alive" and "dead". Time goes on step by step, i.e. in a discrete way. According to some rules the state of a cell in the next time step depends on its own present state and the states of all its surrounding cells in the present period. The rules are deterministic and the same for all cells on the board. A living cell will be alive in the next period if and only if it has 2 or 3 neighbors. Otherwise it will be dead (caused by overcrowding or by loneliness). A dead cell will start to live if and only if there are exactly 3 neighbors surrounding that cell. Otherwise it will remain dead. Neighbors are the 8 adjacent cells. By the dynamic of that game new and surprising patterns evolve, some stable, some changing and moving across the checkerboard. Because of the high complexity of the emergent structures, *game of life* and its derivatives have attracted the attention of an enormous number of people.

Originally CA was introduced by John von Neumann and Stanislaw Ulam at the end of the forties (cf. Neumann 1966) mainly to give a reductionist model of life and self-reproduction. Usually CA based models ask for a lot of iterated computations. As computers became more available, many new CA based models have been developed in science and engineering, and nowadays are used extensively. The applications range from crystal growth, soil erosion, diffusion and fluid dynamics to pattern growth and clashes of galaxies (Burks 1970, Demongeot *et al.* 1985, Wolfram 1986, Toffoli and

R. Hegselmann et al. (eds.),
Modelling and Simulation in the Social Sciences from the Philosophy of Science Point of View, 209–233.
© 1996 *Kluwer Academic Publishers. Printed in the Netherlands.*

Margolus 1987, Gutowitz 1991). Moreover, CA can be regarded as parallel processing computers and are therefore interesting for computer scientists. Some steps toward classifications of CA have been made [Wolfram, 1984], but the analytical understanding of CA has remained poor. Hence simulations play an important role. Because there are natural and easy to handle strategies to visualize low-dimensional CA visualization has become an important tool.

The following article will focus on CA used as a modelling framework to analyse social dynamics. I will firstly give a more detailed description of the basics of CA and similar modelling concepts and then relate them to fundamental features of social dynamics. In a second section I will describe a CA based model of a certain kind of social dynamics. A third section will focus on the danger of artifacts. Finally, I will discuss some other difficulties and restrictions of modelling social processes by CA.

1. CA, checkerboard models, and social dynamics

Basic features of a CA are:[1]

- There is a D-dimensional lattice.
- Time is advancing in *discrete* steps.
- There is a *finite* number of states. At each site of the lattice we have a cell, which is in one of the possible states. (To put it in another way: *Finite automata* are residing in all sites of the D-dimensional lattice.)
- The cells change their states according to *local* rules. Locality is both locality in space and time. So the state of a cell in a next period depends upon the states of neighboring cells in the last t periods. Usually only the last period affects the future. By definition each cell is part of its neighborhood.
- The transition rules usually used are *deterministic*; but *indeterministic* rules are allowed, too.
- The system is *homogenous* in the sense that the set of possible states is the same for each cell and the same transition rule applies to each cell.
- The updating procedure may consist of applying the transition rule *synchronously*. Another updating method may be to select cells *randomly*.

With regard to *spatial locality* there are different ways to define neighborhoods. Fig. 1 shows three different neighborhoods in 2-dimensional lattices. In Fig. 1 (a) we have the so called *von Neumann-neighborhood*. For this type of neighborhood, the neighbors of the dark center cell are the cells

---

[1]A short and easy to read introduction to CA is Casti (1992, vol.1, chapter 3).

Figure 1. Different neighborhood templates

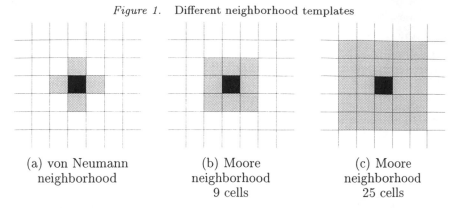

| (a) von Neumann neighborhood | (b) Moore neighborhood 9 cells | (c) Moore neighborhood 25 cells |

in the north, south, east, and west. Since the center cell is part of its neighborhood by definition, the center cell has 5 neighbors. In Fig. 1 (b and c), *Moore-neighborhoods* of different size are given. Moore neighborhoods include cells in diagonal directions as well and always form a square pattern. Obviously Moore-neighborhoods of all sizes (but restricted by the length of a given cellular world) are possible, but neighborhoods may have quite a different structure, being, for example, hexagonal rather than rectangular.

Analogous concepts of neighborhood can be defined for CA of higher dimensions.

Taking a more formal point of view transition rules are mappings: With Σ as the set of possible states of cells and N being the number of neighbors for a given neighboring template a *transition rule* is a *mapping*

$$\Theta : \Sigma^N \to \Sigma \qquad (1)$$

which assigns to all elements of the set of possible states of a neighborhood, i.e. the cross product Σ^N, a single element of the set of possible states of cells, i.e. Σ. Notice that according to this terminology it is *one* transition rule which governs the dynamics of a CA.[2] With Z as the number of elements of Σ we have Z^N possible neighborhood states. Therefore, the number of different transition rules totals up to $Z^{(Z^N)}$. This makes it obvious that there is an enormous number of different transition rules even for CA of the same dimension, same neighboring templates, and with the same set of possible states of cells. Consider a 2-dimensional CA with 2 possible cell states, based on a von Neumann-neighborhood, and with only the last period having influence on the next period. Even in that simple case we get $2^{(2^5)} = 2^{32} = 4\,294\,967\,296$ different transition rules.

---

[2]Sometimes people refer to the elements of Θ as transition rules. In that case we have more than one transition rule governing the dynamics of one CA.

Is CA not a gadget only, at the most applicable to problems in mathematics, natural sciences or engineering? Is CA too mechanistic a machinery to find any sensible application when dealing with social phenomena?

Looking from an abstract point of view, one recognizes that *CA and social dynamics have a lot in common* in their basic structure. Moreover, the *type* of problems which in other fields of science are successfully approached by using CA remain urgent and unresolved problems in the social sciences. Figure 2 gives an overview.

Figure 2. Similarities between CAs and social dynamics

	Cellular automata	Social dynamics
Basic units	Cells are the basic units or the atoms of a CA.	Individuals are basic units of a society.
Possible states	The cells are in states taken from a certain set of possible states.	Individuals make certain choices, adopt certain attitudes, and operate in certain emotional modes.
Interdependence	The state of a center cell affects the state of its neighbors et vice versa.	Individuals affect each other mutually.
Locality	Transition rules are local.	Individuals act and affect each other only locally, i.e. within a certain neighborhood, and the information about them are local as well.
Overlapping	Neighborhoods are overlapping.	Often interactions have an overlapping structure.
Applications and tasks	Fruitful applications in mathematics and natural sciences so far: modeling the emergence of order; macro effects explained by micro rules; modeling dynamic processes.	Important tasks for our understanding of social phenomena: understanding the emergence of order; understanding micro-macro-relations; understanding social dynamics.

Because of these fundamental similarities, it should not come as a surprise that within the social sciences we can find models which are CA in a literal sense or are at least based on the same spirit: Schelling (1969) analyzes segregation processes in a modelling framework which might be regarded as a 1-dimensional CA. Individuals are living on a line and belonging to two different classes (black and white, or more abstract: stars and zeros). Their neighborhood is a certain number of cells to the left and to the right. Cells get migration options. They use them to leave their given neighborhood in case that neighborhood does not have not the numerical ratio between the two classes which they want. For instance, individuals might wish to have a neighborhood in which their class is not a minority. In the case that requirement is not fulfilled, an individual will move to the nearest alternative place meeting that requirement. Schelling (1971) elaborates and analyses that model more in detail, additionally using a 2-dimensional checkerboard world as the underlying spatial (or logical) structure. A fundamental insight is that rules applied on a *micro level* by individuals might produce surprising *macro effects* that nobody intended. So total segregation may result from actions on an micro level which are not guided by the intention of living separated from another group at all.

In the same issue of the *Journal of Mathematical Sociology* which published the famous Schelling article one can find another, but less known article: *The Checkerboard Model of Social Interaction* [Sakoda, 1971]. There Sakoda develops a model in which members of two groups are living on a checkerboard. They have positive (valence $= +1$), neutral (valence $= 0$) or negative (valence $= -1$) attitudes to one another. The cells get the chance to move to empty sites in their 3×3 neighborhood. They use those chances to move to sites where they maximize the sum of valences summed up over the whole world and weighted by the inverse of distance to the power of some w. According to Sakoda the checkerboard model "grew out of an attempt to portray the interaction in a relocation center during World War II" [Sakoda, 1971, p. 120], where members of the Japanese minority in the US were evacuated after Japan's attack.[3] Sakoda stresses that the main purpose of those models is not a predictive one, but rather clarification of concepts and "insight into basic principles of behavior" [Sakoda, 1971, p. 121] which makes checkerboard modelling a promising approach for understanding social dynamics.

Checkerboard models share with CA basic features like the grid structure and the importance of neighborhood.[4] At the same time it is obvious that both Schelling's and Sakoda's checkerboard model essentially focus on migration, moving around looking for partners and attractive neighbor-

---

[3]In that context Sakoda refers to his unpublished dissertation [Sakoda, 1949] which is not accessible to me.

[4]In the case of Sakoda's models an individual is affected by a far distant individual much less than by an individual living close by.

Figure 3. Basic windows in a migration model

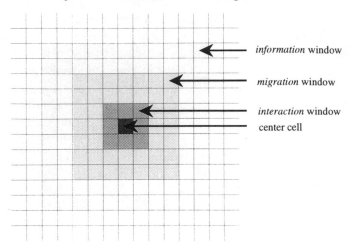

hoods. Whereas original CAs concentrate on cells changing their state at a given site, checkerboard models concentrate on changing a site as well. Therefore we can make a basic distinction between models which allow individuals to move, i.e. *migration models* and those which do not, i.e. *steady site models*. In the case of 2-dimensional models I will refer interchangeably to *both* kind of models as *CA based models* or *checkerboard models*. In the case of other dimensions I will classify steady site and migration models as *CA based models*.

For an easier description of *migration models* let us introduce additional terminology: all individuals have a certain neighborhood within which they are acting, influencing their neighbors and, conversely, are affected by them. I will refer to that part of their world as the *interaction window*. Usually the information individuals have is somehow local, too. That part of the world about which individuals have information will be called the *information window*. Often migrations are possible only within certain limits. I will refer to the area within which migration is possible as *the migration window*. Figure 3 shows how the different windows may be related to each other, but other arrangements in which, for instance, the migration window is bigger than the information window may be equally plausible. The windows may even differ in shape. Inside each window we may have different costs or probabilities for migrations or different degrees to which action of neighbors influence the situation of the individual in the center. It may depend upon distance how sound the available information is.

Though known since some decades it is only the last couple of years that CA- or checkerboard based models are used more frequently.[5] As to social psychology Nowak, Szamrej and Latané (1990) developed a 2-dimensional CA based model for the evolution of attitudes. The basic mechanism of that model is that individuals stick to a certain attitude or to switch to an opposing one depending on the number and strength and decreasing with the distance of all their neighbors having those attitudes. By that model one can understand how, for instance, minority opinions can survive after they got clustered which allows them to become local majority. As to economics, Keenan and O'Brien (1993) used a 1-dimensional CA to model and to analyze pricing in a spatial setting. [Axelrod, 1984, pp. 158ff.] took the first step in analyzing the dynamics of cooperation within a CA framework, although he did not refer to it as a CA. M. Nowak and May (1992; 1993) picked up the idea and studied the dynamics of cooperation within a 2-dimensional CA, using 2-person games as building blocks. Bruch (1994) and Kirchkamp (1995) followed the same line, but used different and more sophisticated strategies or learning rules. The same framework is used in Messick and Liebrand (1995), where the dynamics of three different decision principles are analyzed.[6] All models just mentioned are *steady site models*. Hegselmann (1994; 1994a) developed a checkerboard migration model in order to study the evolution of support networks among individuals who have greatly differing capabilities and have to find partners.

2. An example of CA based modelling: *N*-person prisoner's dilemmas

In this section I will describe a CA based model designed to analyze the dynamics of cooperation. This is not intended to present systematic results, since that will be done in another paper. Instead, I wish to demonstrate in more detail how interesting social dynamics can be approached by CA based modelling.

As an intuitive starting point consider spatially structured, overlapping dilemma situations. Examples for the intended situations are:

- Taking more or less of a certain resource, thereby affecting one's neighbors; e.g. water in a rural society.
- Listening to the radio in a loud or quiet mode in a group.

In a more abstract way we can describe the intended structure as follows:

---

[5]That is due to the fact that those models ask for great deal of iterated computation even if only very simple assumptions are used.

[6]For new overviews over research on social dilemmas cf. Schulz *et al.* (1994) and Liebrand *et al.* (1992).

Figure 4. Payoffs for cooperators and defectors

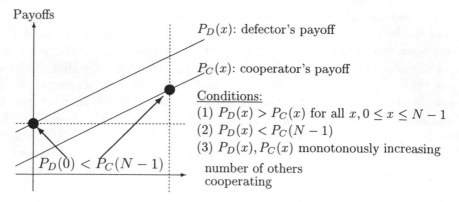

Payoffs

$P_D(x)$: defector's payoff

$P_C(x)$: cooperator's payoff

Conditions:
(1) $P_D(x) > P_C(x)$ for all $x, 0 \le x \le N - 1$
(2) $P_D(x) < P_C(N - 1)$
(3) $P_D(x), P_C(x)$ monotonously increasing

number of others cooperating

$P_D(0) < P_C(N - 1)$

(a) *Interdependence* The individuals have to make binary choices between a cooperative and a defective strategy. The payoffs of individuals depend on their own decision and the number of cooperative decisions made by their neighbors.

(b) *Problematic incentive structure* The individuals are facing a social dilemma. Following Schelling (1973) we can characterize the payoff structure by two functions. They characterize the payoff for a defective choice or a cooperative choice respectively, each of them depending on the number of cooperating *others*.

As a consequence of the conditions laid upon the functions each individual is better off by a defective choice independent of how many other may cooperate. So defection is a dominant strategy. But if all actors defect they get the payoff PD(0) and that is less then the payoff PC($N-1$), i.e. the payoff they obtain by everyone playing cooperative. So we have an N-person prisoner's dilemma based on binary choices.

(c) *Overlapping neighborhood structures* Each individual is acting within a certain spatial neighborhood of N players. Neighborhoods may be von Neumann neighborhoods or Moore neighborhoods of a certain size (figure 1). Due to the general geometry of CA we get an overlapping structure.

(d) *Different distances* In reality quite often actions of direct neighbors affect our situation much more than actions of those living further apart. So distance matters. The underlying 2-dimensional rectangular geometry allows for easy ways to model those effects. In a *first* step we introduce a simple measurement for distances, the *Euclidean distance*. The intuitive idea is that cells lying on the same circle around the center cell should have the same distance to the center, i.e. the *radius* is decisive.

Figure 5.   Measurement of distance

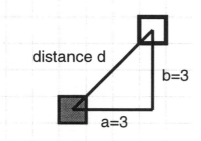

With a for the horizontal and b for the vertical distance between the grey center cell and the framed target cell we get the distance d by $d = \sqrt{a^2 + b^2}$. So, for the example in figure 5 we get $d \approx 4.24$.[7] The second step is to assign weights to players which somehow decrease with their distance from the center cell. A natural way to do that is to use the formula $\frac{1}{d^w}$ to assign weights. The advantage of using this formula is that the influence of distance can be controlled simply by controlling the parameter w. For $w = 0$ all members of the neighborhood have the same weight, i.e. distance doesn't matter. Assuming a Moore neighborhood of size 25 and $w = 1$ we get weights given by figure 6.

In a *last* step we base the payoff calculation for cooperative or defective choices on the sum of *weighted* cooperating others. Thus we get N-person dilemmas, where N is not an integer value any longer.

(e) Decision rules The individuals have to make choices between cooperation and defection. So we need rules for decision making. *One* way to approach the decision problem is as follows: All individuals have *two accounts*, one for payoffs resulting from cooperation and a second one for payoffs stemming from defective choices. Past payoffs may be discounted. Each actor can see the accounts of all neighbors within his neighborhood. When making decisions the individuals will adopt that strategy which was the most successful strategy in their neighborhood. "Most successful" may be understood either in the sense of highest average or in the sense of highest single success of a strategy in a given neighborhood.

To put it more precisely: For each player i we have two accounts. Account DEF_i is used to book the payoffs player i gets in periods in which

---

[7]Another measurement of distance is using $d^* = \max\{a, b\}$. This way we assign the same distance to cells which belong to the same square frame around the center cell.

Figure 6.   Weights of players for $w = 1$

0.35	0.45	0.5	0.45	0.35
0.45	0.7	1	0.7	0.45
0.5	1		1	0.5
0.45	0.7	1	0.7	0.45
0.35	0.45	0.5	0.45	0.35

he defects. The account $COOP_i$ is used to book payoffs from cooperative choices. Past payoffs are discounted using a discount parameter α with $0 \leq \alpha \leq 1$. The updating scheme for the two accounts is as follows:

$$DEF_i^{t+1} \quad \Leftarrow \quad DEF_i^t \cdot \alpha + def_i^t$$

$$COOP_i^{t+1} \quad \Leftarrow \quad COOP_i^t \cdot \alpha + coop_i^t \tag{2}$$

In that scheme def_i^t and $coop_i^t$ are the payoffs player i gets in period t by his defective or cooperative choice, respectively. Thus, for a cooperative choice of i in period t we have $def_i^t = 0$. If $\alpha = 0$, only payoffs of the last period count. For $\alpha = 1$ payoffs of all past periods have the same weight. With regard to single success and $\max_{N,i}[]$ as the maximum of a DEF or COOP account in the neighborhood N of i we get a principle:

$$\max_{N,i}[COOP_t] \geq \max_{N,i}[DEF_t]$$
$$\Rightarrow \quad \text{cooperate in } t + 1, \text{ otherwise defect!} \tag{3}$$

We get an alternative principle by substituting the 'conservative' biased "$\geq$" by "$>$". It should be noticed that there are no rules offering themselves as somehow natural principles. That is true even if we restrict ourself to rules, which might be called *success* oriented, thereby following or at least coming close to the maximizing tradition. With no natural and obvious decision principle at hand, one of the most interesting experiments with cellular worlds as described here consists of experiments with different decision principles.

It is not very difficult to develop a computer program for a CA, or rather for a *cellular society* based on the assumptions (a) to (e). Once that program is written it allows for a lot of interesting experiments of different types.[8]

In figures 7 to 9 three examples of experiments are shown. In all experiments the world starts with a primordial soup, i.e. cooperators and defectors are randomly distributed. The world is of size 50×50. So we have 2500 individuals. The shape of the world is a torus. Therefore all individuals have the same neighborhood. In all experiments the payoff functions given in eq. 4 are applied. x is the *relative frequency* of others cooperating.

$$P_D(x) \quad = \quad 0.1x + 1$$

$$\tag{4}$$

$$P_C(x) \quad = \quad 0.1x$$

All experiments start with half of the population being randomly selected cooperators. The individuals make their decisions asynchronously. A period is arbitrarily defined by 1250 decisions. The black cells in figures 7 to 9 are cooperators, white ones are defectors.

In all three experiments we find at least *two* important effects:

- an initial breakdown of cooperation;
- the evolution of clusters.

Just after the societies start running based on a *random* spatial distribution we get a dramatic loss of cooperation. Let us refer to this as the *initial breakdown effect* (IB-effect), which is quite a robust phenomenon. It will happen under a lot of circumstances, although varying in severity and shape and sometimes resulting in a total breakdown of cooperation. The IB-effect is a consequence of the fact that the random spatial distribution results in structures which allows for much of exploitation of cooperators. Many cooperators have in their neighborhood defectors which are better off than their best off cooperating neighbor. By observing that a lot of cooperators turn to defection. So, the frequency of cooperators will decrease. The payoffs for defectors will go down by that process, but the situation for many of the remaining cooperators will be even worse. So again cooperators turn to defection. If there is a single cooperator left in a neighborhood of defectors, then sooner or later (depending on α) this lonely cooperator will always switch to defection. But sometimes by a lucky coincidence we may have small islands of cooperation which can survive and even grow. So

---

[8]It has become progressively easier since comfortable programming languages and powerful debugging tools are at hand. Nowadays such equipment and the corresponding capabilities is also used in the social sciences.

clusters of cooperators emerge. Due to the decision rule described above, cooperators living deep inside such an island of cooperation will sooner or later (again depending on α) see positive COOP-accounts only, while the DEF-accounts around them go to zero. As for surviving clusters, those cooperators living at the border of areas of defection have in their neighborhood at least one cooperator, who is better off than the best off defector. Some clusters can grow until they get a certain shape and size.

More generally one might distinguish the primordial soup experiments (like the above ones) and invasion experiments. In *invasion experiments* small clusters of cooperators or defectors, respectively, invade a population of defectors or cooperators, respectively. Important research questions are: What are the conditions under which those clusters can grow? How is the process related to size and shape of clusters, payoff structures, size of neighborhoods, distance effects, influence of the past, and decision rules, that are in use? *Primordial soup experiments* start the world with different ratios of randomly distributed cooperators and defectors. Questions to be asked are for example: What's about the asymptotic behavior of such a world? What influences the shape of the IB-effect? How do the initial frequencies of cooperators and defectors affect the results? What are the decisive factors on which the possibility of surviving or growing of cluster depends? How are those factors related to the speed or upper limits of clustering? What happens there where growing clusters meet each other?[9]

Experiments conducted along these lines can give us both theoretical and practical insights. From a *theoretical* point of view we get a better understanding of some basic social dynamics. Experimental results can give fruitful hints to effects which afterwards can be understood or proved by analytical means. From a *practical* point of view we learn how to change behaviour. Suppose that due to some new knowledge common behaviour turns out to be a defective choice. Models like those described above demonstrate that in many cases a small number of persons randomly distributed will die out soon, whereas the same number coming as a cluster will not only survive but even grow. In such a case it is a good idea to change behaviour by something that might be called the *cluster approach*. Sometimes it may not be possible to install clusters as an island, free of 'enemies'. Primordial soup experiments can give insights how to overcome certain behaviour

---

[9]From a more abstract point of view one might consider both types of experiments as limiting cases of a more general experimental setting: Firstly, we have a certain area of a world where we can expect cells of a certain type at all. Secondly, there is a certain probability that cells within that area are of a certain type. Is that area the whole world and is the probability less than 1 we get primordial soup experiments. Is the area only a certain region of the world and is the probability equal to 1 then we have the invasion experiments.

Figure 7. Von Neumann neighborhood. Distance doesn't matter. $\alpha = 0$, so only the last period is relevant. Left: After 75 periods. Right: Frequencies of cooperators for the first 75 periods.

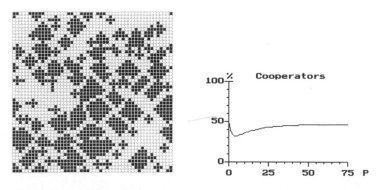

Figure 8. Moore neighborhood of size 3×3. Weights of players $1/d$. $\alpha = 0.7$, so more than the last period matters. Left: After 75 periods. Right: Frequencies of cooperators for the first 75 periods.

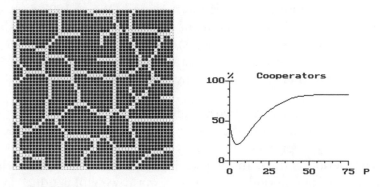

Figure 9. Moore neighborhood of size 7×7. Weights of players $1/d^2$. $\alpha = 0.5$. Left: After 75 periods. Right: Frequencies of cooperators for the first 75 periods.

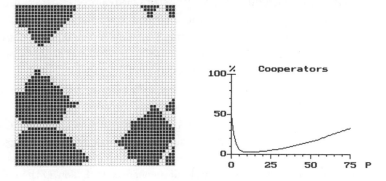

when change can not be started with closed islands of the intended, i.e. cooperative behaviour.[10]

3. The Danger of Artefacts

CA are fruitful modelling tools and are much more flexible than they appear at first glance. But they are *dangerous* instruments, too, because it is so easy to produce (often nice looking) artifacts. The easiest way to get artifacts is to use *synchronous updating*: We assume that there is something like a global clock and according to that all cells are updated simultaneously. To express the problem in terminology closer to the social sciences: We suppose a global clock and assume all individuals make their decisions synchronously according to the strokes of that clock.

In physics and the natural sciences it is well known that synchrony is a dangerous assumption (cf. Hogeweg 1980, Vichniac 1984, Ingerson and Buvel 1984). Since in almost all cases there is no global clock underlying human decision making one should be well aware of the fact that the danger of artifacts caused by synchronous updating is not confined to fields outside social phenomena. More recently Huberman and Glance 1993 pointed out that the results of a CA based model of the evolution of cooperation used by Nowak and May 1992 strongly depends on synchronous updating.

In what follows I will analyze in detail some strange effects closely related to the assumption of synchrony, by using the model described in the foregoing section.

Let's suppose we have the two payoff functions as given in eq. 4. Size of the world is 50 × 50 cells, shaped as a torus. Updating is always *synchronously*. The decision rule is given by eq. 3. We assume that $\alpha = 0$. So only the last period counts. In another interpretation this means that we have a memory of one period only. In that world we will conduct *invasion experiments* in which small clusters of cooperators are coming in a quadratic arrangement of different length. In figure 10 we see the evolution of a 2 × 2 cluster. Again black cells are cooperators. Grey cells are cooperators which have been defectors in the period before.

As figure 10 demonstrates a 2 × 2 cluster of cooperators will conquer a world of defectors. For clusters of size 4 × 4, 6 × 6, 8 × 8 etc. we would get the same asymptotic behavior, namely total cooperation.

Figure 11 shows how clusters of different *odd* length will end up (everything else being equal).

In general (but under the conditions described above) square clusters of an even length will turn their world to cooperation while clusters of an

---

[10]Since cooperation may consist for example in a conspiracy of silence, the behaviour one wants to change may be a cooperative one, too.

Figure 10. Evolution of a 2 × 2 cluster.

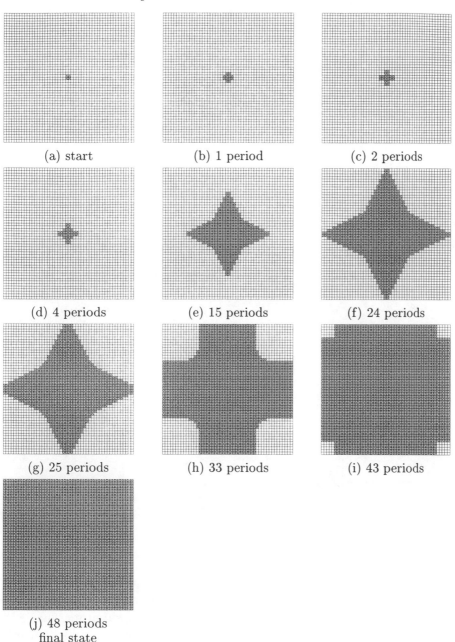

(a) start   (b) 1 period   (c) 2 periods

(d) 4 periods   (e) 15 periods   (f) 24 periods

(g) 25 periods   (h) 33 periods   (i) 43 periods

(j) 48 periods
final state

odd length will be less successful. They will conquer only certain regions. As a consequence smaller clusters can be more successful than larger ones, provided the former have an even length, e.g. 2, and the latter have an *odd*

Figure 11.   Final states for different clusters of an odd length

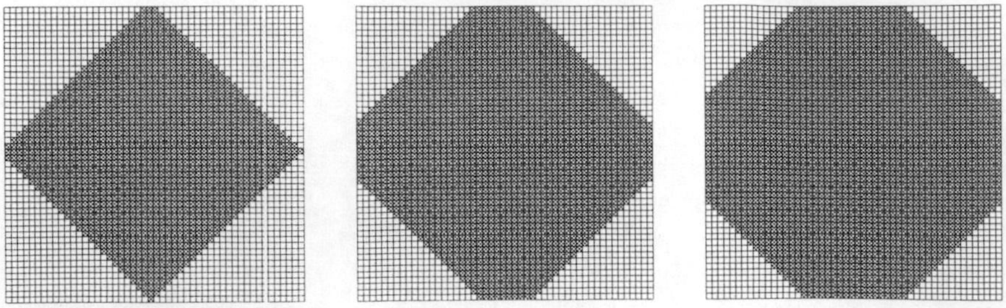

(a) start with size 3×3 (b) start with size   (c) start with size
                               11×11 19×19

length, e.g. 19.

Effects like that ask for explanations. To get some understanding one has firstly to focus on the underlying payoff structure given by eq. 4 and displayed in figure 12.

Figure 12.   Payoff structure based on eq. 4

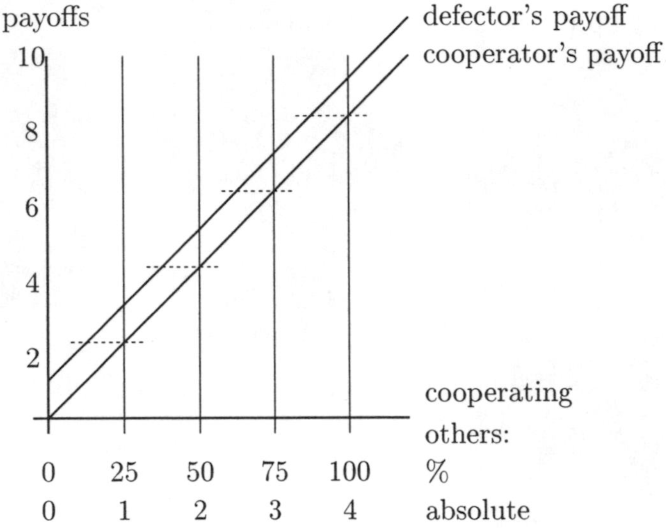

The payoff structure, the decision principle eq. 3, and the condition, that only the success of the last period counts ($\alpha = 0$), imply, that defectors living among defectors and cooperators living among cooperators will never change their strategies for the next period. For all other types of neighborhoods the following holds: An actor will cooperate in the next period whenever he has in his neighborhood a cooperator with more cooperating neighbors than the best off defector he can see; otherwise the actor will defect in the next period. With $N(D_{NC})$, $N(C_{NC})$ referring to the number of cooperating neighbors of a defector or a cooperator, respectively, we can state:

- Cooperators in a neighborhood of cooperators *only*, and defectors in a neighborhood of defectors *only* will stay with their strategy.
- For *mixed* neighborhoods it holds:
  If $\max_{N,i}[N(C_{NC})] > \max_{N,i}[N(D_{NC})]$, then i will cooperate in the next period; otherwise i will defect.

Now let us analyze the evolution of an 3×3 cluster.

Figure 13. Start of a 3×3 cluster

period 0

	A	B	C	D	E	F	G	H	I	J	K
I											
II											
III											
IV				0	1	1	1	0			
V				1	2	3	2	1			
VI				1	3	4	3	1			
VII				1	2	3	2	1			
VIII				0	1	1	1	0			
IX											
X											
XI											

period 1

	A	B	C	D	E	F	G	H	I	J	K
I											
II											
III			0	0	1	1	1	0	0		
IV			0	2	2	3	2	2	0		
V			1	2	4	4	4	2	1		
VI			1	3	4	4	4	3	1		
VII			1	2	4	4	4	2	1		
VIII			0	2	2	3	2	2	0		
IX			0	0	1	1	1	0	0		
X											
XI											

period 2

	A	B	C	D	E	F	G	H	I	J	K
I											
II		0	0	0	1	1	1	0	0	0	
III		0	0	1	2	3	2	1	0	0	
IV		0	1	2	3	4	3	2	1	0	
V		1	2	3	4	4	4	3	2	1	
VI		1	3	4	4	4	4	4	3	1	
VII		1	2	3	4	4	4	3	2	1	
VIII		0	1	2	3	4	3	2	1	0	
IX		0	0	1	2	3	2	1	0	0	
X		0	0	0	1	1	1	0	0	0	
XI											

period 3

	A	B	C	D	E	F	G	H	I	J	K
I	0	0	0	0	1	1	1	0	0	0	0
II	0	0	0	1	2	3	2	1	0	0	0
III	0	0	0	2	3	4	3	2	0	0	0
IV	0	1	2	2	4	4	4	2	2	1	0
V	1	2	3	4	4	4	4	4	3	2	1
VI	1	3	4	4	4	4	4	4	4	3	1
VII	1	2	3	4	4	4	4	4	3	2	1
VIII	0	1	2	2	4	4	4	2	2	1	0
IX	0	0	0	2	3	4	3	2	0	0	0
X	0	0	0	1	2	3	2	1	0	0	0
XI	0	0	0	0	1	1	1	0	0	0	0

Again, black cells are cooperators cooperating at least in the last two periods. Grey cells are former defectors which switched to cooperation in the present period. White cells are defectors. Arabic numbers within cells indicate the number of cooperating neighbors of that cell. The start of a 3×3

Figure 14.   3 × 3 cluster in a world of size 50 × 50

	A	B	C	D	E	F	G	H	I	J	K
I											
II		0	0	0	0	0	0	0	0	0	
III		2	1	0	0	0	0	0	1	2	
IV		3	2	2	1	0	1	2	2	3	
V		4	4	3	2	2	2	3	4	4	
VI		4	4	4	3	2	3	4	4	4	
VII		4	4	3	2	2	2	3	4	4	
VIII		3	2	2	1	0	1	2	3	3	
IX		2	1	0	0	0	0	0	1	2	
X		0	0	0	0	0	0	0	0	0	
XI											

period 23

	A	B	C	D	E	F	G	H	I	J	K
I											
II		1	0	0	0	0	0	0	0	1	
III		2	2	1	0	0	0	1	2	2	
IV		4	3	2	2	0	2	2	3	4	
V		4	4	4	2	2	2	4	4	4	
VI		4	4	4	4	2	4	4	4	4	
VII		4	4	4	2	2	2	4	4	4	
VIII		4	3	2	2	0	2	2	3	4	
IX		2	2	1	0	0	0	1	2	2	
X		1	0	0	0	0	0	0	0	1	
XI											

period 24

	A	B	C	D	E	F	G	H	I	J	K
I		0	0	0	0	0	0	0	0	0	
II		2	1	0	0	0	0	0	1	2	
III		3	2	2	0	0	0	2	2	3	
IV		4	4	4	2	2	2	4	4	4	
V		4	4	4	2	2	2	4	4	4	
VI		4	4	4	4	2	4	4	4	4	
VII		4	4	4	2	2	2	4	4	4	
VIII		4	4	2	2	0	2	2	4	4	
IX		3	2	2	0	0	0	2	2	3	
X		2	1	0	0	0	0	0	1	2	
XI		0	0	0	0	0	0	0	0	0	

period 25

	A	B	C	D	E	F	G	H	I	J	K
I		1	0	0	0	0	0	0	0	1	
II		2	2	0	0	0	0	0	2	2	
III		4	2	2	0	0	0	2	2	4	
IV		4	4	2	2	0	2	2	4	4	
V		4	4	4	2	2	2	4	4	4	
VI		4	4	4	4	2	4	4	4	4	
VII		4	4	4	2	2	2	4	4	4	
VIII		4	4	2	2	0	2	2	4	4	
IX		4	2	2	0	0	0	2	2	4	
X		2	2	0	0	0	0	0	2	2	
XI		1	0	0	0	0	0	0	0	1	

period 26

cluster of cooperators in figure 13 demonstrates something quite general: All border cells of all square clusters of any length $l \leq L-1$, where L is the length of the world, have at least two cooperating neighbors. All adjacent defectors for any length $l \leq L - 2$ have at most one cooperating neighbor. Therefore under such circumstances all cooperators have in their neighborhoods cooperators which are more successful than the best off defectors, if they see defectors at all. Therefore no cooperator will turn to defection. On the contrary the defectors V, E to VII, E, then V, G to VII, G and the corresponding groups in the north and south turn to cooperation in period 1. They realize that they themselves are the best off defectors in their neighborhood, being faced with cooperators which have two or three cooperating neighbors. So those cooperators are better off. In period 1 we get the same situation again: The new cooperators V, D to VII, D, then V, H to VII, H etc. see cooperators doing much better than their neighboring defectors while the neighboring defectors are attracted by the higher payoffs of their neighboring cooperators. Therefore in period 2 again in the east, west, north, and south strings of three cooperators will turn to cooperation thereby reproducing just that kind of situation which motivated them themselves to change their strategies. With period 3 the island of cooperation' starts to grow in an diagonal direction, too. In period 2 the cell IV,D observes cooperators with three cooperating neighbors while the best visible defector is the cell IV, D itself with only 2 cooperating neighbors. So that cell turns to cooperation in period 3 and so do the other three cells which live under the same circumstances.

Since our world is a torus of a finite length after a certain number of periods parts of the growing patterns will meet one another. In figure 13 the initial cluster is in the center of the 2-dimensional *projection* of that torus. In figure 14 we look at the same world some periods later. But to get a better view we readjust the view point in such a way that the growing patterns will meet one another in the center of our projection. Figure 14 zooms into that center and shows 4 decisive periods in sequence.

The figure shows that after 23 periods the cluster has grown up to a size that parts of the evolved pattern are going to meet one another. There is only the column F between them. The forefront of the cluster, i.e. V, E to VII, E and V, G to VII, G has a length of 3 cells, a direct consequence of the initial cluster size and length. For any length l of the initial cluster we will get a length l of the forefront. *Not all* of the defectors remaining in between the two forefronts turn to cooperation. The cells V, F and VII, F have 2 cooperating neighbors. Therefore they are better off than the best off cooperators they observe, which have 2 cooperating neighbors, too. But the cell VI, F will turn to cooperation and so would all defectors in a remaining column with exception of the uppermost and the bottommost defecting cell. At the same time we observe a diagonal growth of the cooperating cluster: defectors in a neighborhood like that of cell III, B or IV, D turn to cooperation in period 24. Thereby they create for the defectors II, A or III, C exactly that kind of neighborhood which made them changing to cooperation. That process continues until we get figure 11 (a) as the final structure. As shown in figure 15 the decisive point is the *diagonal* border between cooperators and defectors. Given that structure cooperators and defectors in such mixed neighborhoods, which constitute the border, have always two cooperating neighbors. Therefore the bordering defectors do not turn to cooperation. Since the bordering cooperators have cooperators in their neighborhood which have even four cooperating neighbors those cooperators don't change their strategy either.

Figure 15. Stable border

	A	B	C	D	E	F	G
I							
II		2	2	0	0	0	0
III		4	2	2	0	0	0
IV		4	4	2	2	0	0
V		4	4	4	2	2	0
VI		4	4	4	4	2	2
VII							

As a main result we get that the 3×3 cluster cannot conquer its world. The *strange thing* is that the 3×3 cluster would be able to conquer the world if we only had a world of size 51×51 instead of size 50×50. Figure 16 shows how that happens.

In that case parts of the growing patters are going to meet one another in period 23. But now there are *two* columns of defectors left, namely columns F and G. Contrary to the 50×50 world all remaining defectors in between the forefronts of cooperators have cooperating neighbors which are better off then the best off defectors. So they all turn to cooperation. By that the cells IV, E to IV, H and $VIII, E$ to $VIII, H$ get in period 24 into a situation which lets them turn to cooperation in period 25. Thereby that cells are bringing about a situation which makes the cells III, D to III, I and the cells IX, D to IX, I switching to cooperation and so on. So now after parts of the growing pattern have met each other again we do not get a stable diagonal border as in the case of a 50×50 world as shown in figure 15. Instead, we are faced with a process which will end up with the whole world being conquered by cooperators.

So, while (under our circumstances) a world of size 51×51 will be conquered by a 3×3 cluster, a world of size 50×50 can not be conquered by that cluster. Rethinking the explanations makes it clear that the decisive point is whether there are two columns or only one column of defectors left when the growing parts are going to meet one another. What makes the greatest difference therefore is whether we have a cluster of an even *or* odd length in a world of an even *or* odd length. We therefore have as a general effect under the given payoff structure, decision principle, and updating procedure:

- Cooperators coming as a square cluster of an odd [even] length will conquer worlds of an odd [even] length. But in worlds of an even [odd] length there will always be left an area of defection.

Figure 16.   3 × 3 cluster in a world of size 51 × 51

	A	B	C	D	E	F	G	H	I	J	K	L
I												
II		0	0	0	0	0	0	0	0	0	0	
III		2	1	0	0	0	0	0	0	1	2	
IV		3	2	2	1	0	0	1	2	2	3	
V		4	4	3	2	1	1	2	3	4	4	
VI		4	4	4	3	1	1	3	4	4	4	
VII		4	4	3	2	1	1	2	3	4	4	
VIII		3	2	2	1	0	0	1	2	2	3	
IX		2	1	0	0	0	0	0	0	1	2	
X		0	0	0	0	0	0	0	0	0	0	
XI												

period 23

	A	B	C	D	E	F	G	H	I	J	K	L
I												
II		1	0	0	0	0	0	0	0	0	1	
III		2	2	1	0	0	0	0	1	2	2	
IV		4	3	2	2	1	1	2	2	3	4	
V		4	4	4	3	3	3	3	4	4	4	
VI		4	4	4	4	4	4	4	4	4	4	
VII		4	4	4	3	3	3	3	4	4	4	
VIII		4	3	2	2	1	1	2	2	3	4	
IX		2	2	1	0	0	0	0	1	2	2	
X		1	0	0	0	0	0	0	0	0	1	
XI												

period 24

	A	B	C	D	E	F	G	H	I	J	K	L
I		0	0	0	0	0	0	0	0	0	0	
II		2	1	0	0	0	0	0	0	1	2	
III		3	2	2	1	1	1	1	2	2	3	
IV		4	4	3	3	3	3	3	3	4	4	
V		4	4	4	4	4	4	4	4	4	4	
VI		4	4	4	4	4	4	4	4	4	4	
VII		4	4	4	4	4	4	4	4	4	4	
VIII		4	4	3	3	3	3	3	3	4	4	
IX		3	2	2	1	1	1	1	2	2	3	
X		2	1	0	0	0	0	0	0	1	2	
XI		0	0	0	0	0	0	0	0	0	0	

period 25

	A	B	C	D	E	F	G	H	I	J	K	L
I		1	0	0	0	0	0	0	0	0	1	
II		2	2	1	1	1	1	1	1	2	2	
III		4	3	3	3	3	3	3	3	3	4	
IV		4	4	4	4	4	4	4	4	4	4	
V		4	4	4	4	4	4	4	4	4	4	
VI		4	4	4	4	4	4	4	4	4	4	
VII		4	4	4	4	4	4	4	4	4	4	
VIII		4	4	4	4	4	4	4	4	4	4	
IX		4	3	3	3	3	3	3	3	3	4	
X		2	2	1	1	1	1	1	1	2	2	
XI		1	0	0	0	0	0	0	0	0	1	

period 26

This effect — it might be called the *even/odd-effect* — may be interesting for its own. But is it a good candidate for a deep insight about our social world? The main problem here is that there is at least one assumption to which the model reacts extremely sensitive and which is totally wrong from an empirical point of view: In the model just described all individuals make their decisions *synchronously*. So we assume something like a global clock, giving a universal rhythm for decision making. If we give up that assumption and update cells *asynchronously* the even/odd-effect disappears at once. Clusters of about the same size will have about the same success in conquering a world of defection. The same will turn out, if we do not abandon synchronous updating, but add a sufficiently high probability of making mistakes when making decisions (noise).

The main lesson to be learnt here is that CA based models may react quite sensitively to updating procedures. So, choosing the 'right' updating procedure is not a trivial task. Therefore, whenever modelling within the CA framework one should make use of the following rules:

- Regard synchronous updating as a prime suspect which has to be proven innocent before being applied.

There are some other simple rules which one should have in mind also:

- Find out whether your results are stable under different sizes of your world.
- Make sure that your results aren't affected by an even or odd length of your world.
- Analyze the influence of different shapes and boundary conditions for your world.
- Analyze the influence of different neighborhood definitions.
- Find out how introducing some noise affects the results.

Following those rules will make it fairly likely that one does not fall victim to artifacts.

4. Other restrictions and flexibility

Besides artifacts like that ones analysed in the foregoing section there are other restrictions, too. The most severe restrictions are those on possible interaction structures, caused by the *geometry* of neighborhoods. This geometry suggests as a natural interpretation spatially based or anchored social interactions for which real world neighborhoods may be paradigm cases. However, one should notice that from a more abstract point of view what the geometry of CAs does is imposing on interactions certain *logical structures and restrictions*. It is possible, but not necessarily so, that the logical structure has a spatial interpretation.

Important and characteristic consequences of the imposed logical structure are:

- As long as we do not define the whole world as a cells neighborhood individuals have interactions only within a *subset* of the total population.
- I'm a neighbor of all my neighbors, but not all neighbors of my neighbors are neighbors of mine, too. More formally, the neighborhood relation is *reflexive*, but *not transitive*.
- If we have a cluster of cells of a certain type, for instance an island of cooperation, than such a cluster has at least some bordering cells which have at least some neighboring cells of another type. So there are *bordering phenomena*.

Those consequences seem to be quite "natural" or appropriate under a lot of circumstances. They look far less natural if stated in terms of what is thereby excluded:

- With increasing distance between two cells the probability of having interactions with one another is not simply decreasing. What is more, it is logically excluded to have interactions with cells outside the neighborhood.
- Suppose we have a von Neumann neighborhood and we allow for moving. If an actor moves a distance of one cell to the north, south, east or west, then he will have lost all his former neighbors. If we allow for diagonal moves, too, the minimal loss of neighbors is 50 %. In small Moore-neighborhoods we get similar results. So, close range moves may imply dramatic changes in neighborhood.
- It is logically excluded that all members of a cluster of a certain type of cells have the same number of interactions with cells of just that type (given that the cluster is a subset of the whole population). So it is logically excluded to have something like clusters without bordering phenomena.

Therefore, being aware of features like that one has to find out whether it is reasonable to model certain dynamics within a CA framework or not.

In general, CA based models allow for much more flexibility than it might seem to be at first sight. As to social dynamics the CA framework does not exclude to have individuals which are quite different from each other. The individuals may differ in the decision making principle they use. Related to that, they may differ in how often they evaluate their decisions and they may have quite different degrees of misperceptions. As far as payoffs are involved, they may change in the course of events or even depend on the site where an actor is living. What I called *migration models* allows us to model dynamics driven by looking for attractive partners and neighborhoods. So a lot of prima facie restrictions are remediable by some refinements.

But despite of that, there are conceptually based restrictions for reasonable applications of a CA framework and there is even a danger of falling victim to very strange artifacts. At the same time there is strong evidence, provided by successful applications, that CA based models contribute to a much better understanding of social dynamics and micro-macro relations. One should notice that reasonably constructed CA based models enable us to study social self organization. CA models may address the question, to what extent the "invisible hand" works fine.

Acknowledgements

This article was written during an academic year I spent as a fellow of the Netherlands Institute For Advanced Study. I would like to thank for hospitality and the stimulating intellectual environment. The article was significantly improved by criticisms of Wim Liebrand and Andrzej Nowak. Therefore they are to a certain extent responsible for what I have written here. I hope they do not suffer too much from that burden. Männi thängs tu Mark Geller vor korrekting mei inglisch.

References

Axelrod, R. (1984) *The evolution of cooperation*, New York (in German 1987. *Die Evolution der Kooperation*, Oldenbourg, München).

Bruch, E. (1994) *The evolution of cooperation in neighbourhood structures*, manuscript, Bonn University

Burks, A.W. (1970) *Essays on cellular automata*, University of Illinois Press, Urbana IL.

Casti, J.L. (1992) *Reality rules. Picturing the world in mathematics* (Vol.I: The fundamentals; Vol.II: The frontier), Wiley, New York.

Demongeot, J., Goles, E., Tchuente, M. (eds.) (1985) *Dynamical systems and cellular automata*, Academic Press, London.

Friedman, J.W. (1986) *Game theory with applications to economics*, 2nd ed. 1991, Oxford UP.

Gutowitz, H. (ed.) (1991) *Cellular automata — theory and experiment*, MIT Press, Cambridge MA.

Hegselmann, R. (1994) Zur Selbstorganisation von Solidarnetzwerken unter Ungleichen — Ein Simulationsmodell. K. Homann, Hrsg., *Wirtschaftsethische Perspektiven I — Theorie, Ordnungsfragen, Internationale Institutuionen*, Duncker & Humblot, Berlin, pp. 105–129.

Hegselmann, R. (1994a) Solidarität in einer egoistischen Welt — Eine Simulation, J. Nida-Rümelin, Hrsg., *Praktische Rationalität — Grundlagen und ethische Anwendungen des rational choice-Paradigmas*, de Gruyter, Berlin, pp. 349–390.

Hegselmann, R. (1995) Modelling social dynamics by cellular automata, K.G. Troitzsch *et al.*, eds. *Social Science Microsimulation: A Challenge to Computer Science*, Springer, Berlin (to appear).

Hogeweg, P. (1980) Locally synchronised developmental systems, conceptual advantages of the discrete event formalism, *Internat. J. Gen. Systems* 6, pp. 57–73.

Huberman, B.A. and Glance, N.S. (1993) Evolutionary games and computer simulations, *Proc. Natl. Acad. Sci. USA*, 90, pp. 7716–7718.

Ingerson, T.E., and Buvel, R.L. (1984) Structure in asynchronous cellular automata, *Physica D* 10D, pp. 59–68.

Keenan, C.D., and O'Brien, M.J. (1993) Competition, collusion, and chaos, *Journal of Economic Dynamics and Control* 17, pp. 327–353.

Kirchkamp, O. (1995) Spatial evolution of automata in the prisoners' dilemma, K.G. Troitzsch *et al.*, eds. *Social Science Microsimulation: A Challenge to Computer Science*, Springer, Berlin (to appear).

Messick, D.M. and Liebrand, W.B.G. (1995) Individual heuristics and the dynamics of cooperation in large groups, *Psychological Review* 102, pp. 131–145.

Liebrand, W.B.G., Messick, D.M., and Wilke, H.A.M. (eds.) (1992) *Social dilemmas — Theoretical issues and resarch findings*, Pergamon, Oxford.

Neumann, J.v. (1966) *Theory of self-reproducing automata* (edited and completed by Arthur W.Burks), University of Illinois Press, Urbana.

Nowak, A., Szamrej, J., and Latané, B. (1990) From private attitude to public opinion — Dynamic theory of social impact, *Psychological Review* 97, pp. 362–376.

Nowak, M.A., and May, R.M. (1992) Evolutionary games and spatial chaos, *Nature* 359, pp. 826–829

Nowak, M.A., and May, R.M. (1993) The spatial dilemmas of evolution, *International Journal of Bifurcation and Chaos* 3, pp. 35–78.

Sakoda, J.M. (1949) *Minidoka — An analysis of changing patterns of social interaction,* PhD thesis, University of California, Berkeley.

Sakoda, J.M. (1971) The checkerboard model of social interaction, *Journal of Mathematical Sociology* 1, pp. 119–132.

Schelling, T.C. (1969) Models of segregation, *American Economic Review* 59, pp. 488–493.

Schelling, T.C. (1971) Dynamic models of segregation, *Journal of Mathematical sociology* 1, pp. 143–186.

Schelling, T.C. (1973) Hockey helmets, concealed weapons, and daylight saving — A study in binary choices with externalities, *Journal of Conflict Resolution* 17, pp. 381–428.

Schulz, U., Albers, W., and Mueller, U., eds. (1994) *Social dilemmas and cooperation,* Springer, Berlin.

Taylor, M. (1987) *The possibility of cooperation* (revised edition of: *Anarchy and cooperation,* 1976), Wiley, London.

Toffoli, T., and Margolus, N. (1987) *Cellular automata machines — A new enviroment for modeling,* MIT Press, Cambridge MA.

Vichniac, G.Y. (1984) Simulating physics with cellular automata: *Physica D* 10D, pp. 96–116.

Wolfram, St. (1984) Universality and complexity in cellular automata: *Physica D* 10, pp. 1–35.

Wolfram, St., ed. (1986) *Theory and applications of cellular Automata,* World Scientific, Singapore.

COMPUTER SIMULATIONS OF SUSTAINABLE
COOPERATION IN SOCIAL DILEMMAS

WIM B.G. LIEBRAND
Institute of Social Science Information Technology
University of Groningen
Groningen, The Netherlands

and

DAVID M. MESSICK
Kellogg Graduate School of Management
Northwestern University
Evanston IL, USA

1. Introduction

The understanding of the dynamics underlying cooperative and competitive behavior in large groups constitutes a dazzlingly complex challenge to social and behavioral scientists. The complexity arises from at least two mechanisms, each of which we barely understand at the moment. The first is the mere existence of self-sacrificial behavior. The second is the dynamical relation between micro level behavior and collective outcomes. We decided to accept the challenge and used computer simulations of very simple societies in order to get some feeling for the processes involved. This chapter provides an account of our experiences. It turned out that the computer simulations provided unexpected and surprising results, which, after a while, we were able to generalize into a theoretical model.

Why do people continue to cooperate in situations in which cooperation goes against their immediate self-interest? This problem has been the topic of many studies in several disciplines. Evolutionary biologists are trying to identify processes within evolutionary theories to explain the problem of cooperation or altruism. Thus far, kin selection, reciprocity and group selection have been identified as possible mechanisms. For human decision making, however, these processes seem to have limited explanatory power

R. Hegselmann et al. (eds.),
Modelling and Simulation in the Social Sciences from the Philosophy of Science Point of View, 235–247.
© 1996 *Kluwer Academic Publishers. Printed in the Netherlands.*

because they are heavily based on cooperation between related persons or assume either extensive knowledge of a person's past behavior, or a functional segregation between groups.

Psychologists have acknowledged time after time that people are irrational decision makers. However, people are quite persistently irrational, thus the psychologists modified some of their decision making models, by introducing the concept of bounded rationality. The new models still heavily rely on a rational analysis of the problem, the decision however, will not only be based on this rational analysis but also will take individual values and generalized norms into account.

The present study takes an alternative approach by assuming that decisions generally are not based on a rational analysis of all options, nor on a (sub)conscious strategy to gain evolutionary strength. Would cooperation be possible if we just assume that decision making is based on very simple decision making heuristics? Heuristics which have served us in the past and which are used as a rule of thumb in situations to complex for us to analyze completely. Our basic research question thus is, is cooperation sustained in a world populated with simple organisms seeking personal positive reinforcement?

The second major theme in this research deals with the level of analysis problem. In order to understand why cooperation in groups can be sustained, we need to understand the intricate relationship between behavior at the micro level and its consequences at the macro level. Researchers thus far focussed in their level of analysis either on very small groups, or on large populations, and had to assume that the effects of behavior on the individual level would simply add up and shape the characteristics of the collective. However, as Schelling (1971) pointed out almost two decades ago, behavior which might seem to be unpredictable at the individual level, can have either desirable or undesirable consequences at the collective level. Examples being racial segregation or the separation of swimmers and surfers on the beach. Vice versa, the same line of reasoning holds: a description of aggregate behavior then provides inadequate insight into the underlying individual choice processes.

To address both the problem of cooperation as well as the level of analysis problem we will use computer simulations of decision making within the social dilemma framework.

Social dilemmas are interpersonal and intergroup situations that are characterized by an unfortunate pattern of incentives (for reviews see Dawes 1980; Liebrand *et al.* 1992; Messick and Brewer 1983). The pattern is unfortunate because it tends to create highly unsatisfying outcomes for the participants. Outcomes that are unsatisfying because all of the actors involved realize that each of them individually would have been much better off if they had behaved differently, that is if they had cooperated. In our study we use the well-known PRISONER'S DILEMMA (PD), displayed in Figure 1, as a prototypical social dilemma. In basic two–person–two–alternative PD, each of the two players has a choice between two options, C (cooperate) or D (not cooperate or defect). The entries in the outcome matrix are assumed to represent positive payoffs to the players.

Figure 1.   The Basic Prisoner's Dilemma used in the simulations

Neighbor

		Cooperate	Defect
Actor	C	② 2	⓪ 3
	D	③ 0	① 1

The dilemma is obvious. Whatever the other does, it is better to choose D because in both cases the D-chooser will get one unit more. However, a mutual D choice results in the second worst outcome cell for both players whereas each could have got their second best by choosing C, the dominated option. Hence it is in each person's best interest to choose D, however, a socially optimal allocation of resources is only reached by choosing C.

2. Reconstruction of Decision Making

In this study we investigate the global consequences of decision making in basic Prisoner's Dilemmas. That is we assume that all interactions between pairs of actors in a population are described by PD contingencies, and that all actors use the same simple but self rewarding heuristic for making decisions. Would we then in the long run still observe some cooperation between actors?

Figure 2.   The Simulation Environment. (The cellular automata within the rectangle constitute a neighborhood.)

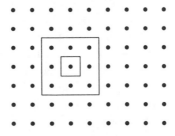

To address this research question a cellular automata approach generating the artificial environment shown in Figure 2 was used. It consists of a

population of size $n \times n$. All edges and corners are connected to each other, so we are dealing with a torus consisting of n^2 actors, or cellular automata. To start with all actors randomly got assigned a probability of 0.5 to choose C at the next play, while all actors in a given population use the same decision heuristic. This heuristic is one out of the three decision rules described below in more detail. In addition all payoff values of the basic Prisoner's Dilemma (Figure 1) are distributed randomly among the actors, causing the expected payoff of any actor or neighborhood to be 1.5.

Actual decision making takes place in the following framework. First, the subject is randomly chosen from the population. Next, again randomly, one of the eight immediately surrounding neighbors is selected. Subject and neighbor then play the PDG. Immediately after the play, both players will adjust their outcomes resulting from the last play. Only the initiator of the game, that is our subject, will evaluate and if necessary change the choice that will be made on the next move. A more detailed description of this evaluation is presented in the section on choice heuristics. All random sampling is done with replacement, which of course does not guarantee that all the actors have been engaged in exactly the same number of games. In the long run, however, any irregularities in the number of games played either as actor or as neighbor, roughly will be the same for all actors in the population.

3. The Choice Heuristics

The choice heuristics used in this study are known to be forceful decision making strategies for the PD. They are all contingent heuristics because they will react to changes in the environment.

The first one, of course, is *Tit-for-Tat* (TFT), in which a subject's next choice will be identical to the choice of the neighbor in the present game. TFT has been shown to be a decision heuristic which in some conditions is capable to outperform others (Axelrod 1984). In our case, however, we have to point out that the reciprocal effect of TFT is not necessarily directed to the neighbor who interacted lastly with the subject. Here we are dealing with generalized reciprocity because the subject evaluates outcomes immediately after playing with the neighbor, and then determines the choice to make in the next play with a randomly selected neighbor. From a psychological point of view we see this generalized reciprocity as realistic because we often will not be able to focus our reaction to a changing environment solely to the cause of that change.

The remaining two heuristics, *Win-Stay Lose-Change* (WSLC) and *Win-Cooperate Lose-Defect* (WCLD), are slightly more elaborated than TFT's mimic strategy. Both heuristics have two components, an evaluation and an action component. In the evaluation part of both heuristics the subject determines whether the outcome of the last play is perceived as a Win or a Loss. Ordinarily, game theoretical studies use the absolute payoffs in the matrix for this evaluation process. Again in an attempt to model social reality more closely, in this study we use a social comparison process for the evaluation component. The comparison is based on the payoffs the subject and the eight neighbors got for the last time they played. A subject will code any payoff which is at least as high as the average of the eight neighbors' payoffs as a Win. All other payoffs will be treated as a Loss. WSLC and WCLD do not differ with respect to this evaluation component, they can differ in their action component, that is in the way they respond to 'winning' or 'losing'. Both heuristics however, do not accept exploitation by the other: if the neighbor defects while the subject cooperates, then the subject will retaliate in the next move and defect as well.

Win-Stay Lose-Change (WSLC) is based on well known reinforcement principles. It prescribes to stay with the behavior that yielded positive rewards, in case of negative experiences the behavior has to change. The heuristic in some research areas known as the Pavlov strategy (Wu and Axelrod 1995) may also be seen as a kind off approach-avoidance strategy.

Win-Cooperate Lose-Defect (WCLD) differs from WSLC most in terms of the behavior following a win or a loss. Winning will put WCLD players in a good mood and in that state the subject will cooperate on the next play. On the other hand the feeling that they have lost will dictate a D choice to to the subject for the next play. Also WCLD is supported by prior research. Amog others, Krebs and Miller (1985) showed that positive moods does have generalized beneficial effects. These effects under the WCLD rule resemble the generalized reciprocity in TFT. Under the WCLD rule we will cooperate with our neighbors, only when we have the feeling that we can afford that to do, otherwise we will defect. In a way WCLD can be seen as a social-affective generalization of TFT. In a two-person PDG, WCLD and TFT will prescribe exactly the same responses to C or D behavior from the other person. Only in n-person games we get differences between the two heuristics if we define them as we did a moment ago.

4. Results of Simulation

The complexity of the situation under study is sufficiently high to preclude momentarily an analytical description of the global consequences of each of the three decision heuristics. We therefore used computer simulations to gradually get a better understanding of the dynamics involved. The simulations described in this paper are based on the Warsaw Simulation System (Gasik 1990). All major results were cross-checked in an independent simulation program written by the first author.

First we will present the results of our studies using a population of size 9. Please remember that in this case the size of the neighborhood is identical to the population size. The most important result for groups of size 9, in which all actors use the same decision heuristic, is that they soon will turn into a stable population consisting of only cooperators or defectors. For the imitative TFT strategy, a homogeneous population of cooperators will emerge with probability 0.5. The probability that the population turns into a homogeneous set of defectors is also 0.5. Given our starting configuration in which each actor has a probability of 0.5 to be a cooperator, this result is to be expected. TFT does not benefit either cooperation or defection, it just imitates the neighbor's strategy.

For WSLC it is impossible to turn into a homogenous world of cooperators. Under this choice rule the last defector in a world of all cooperative neighbors will always win and persist in using the D-choice next time around. The reason lies in the evaluation component of WSLC: a defector playing a cooperator will get 3 payoff points, the cooperator zero. Hence the payoff for the defector is always higher than the average mean of the eight neighbors, because that includes at least one neighbor having zero points. However, a homogeneous population of all defectors is possible under WSLC, and that is what we observed in these simulations with a groupsize of nine actors.

For WCLD there also is only one homogenous and stable end state, the all-C state. In this state all members receive two payoff points, that will be coded as a win causing the subject to play C again on the next play. An all-D state will not be reached under WCLD. Suppose all players were D and had only one payoff point for their last play. Their outcome would then be at least as high as the average of the neighbors and subsequently be coded as a win. Under the present heuristic that would imply a change from D to C.

5. Manipulation of Groupsize

The simulations with groups of size 9 showed that each choice heuristic would lead to a homogeneous population. In addition, the preceding analysis showed why these convergences occurred.

The question now is whether the probability of entering a homogeneous state depends upon the size of the population. We therefore manipulated groupsize and run simulations for groups of size 9, 16 and 25. The results are shown in Table 1.

The first thing to note here is that convergences do occur for all three heuristics in small groups. For TFT we extended the group size manipula-

TABLE 1. Mean number of generations to convergence into homogeneous all-C or all-D state. Each mean is based on 50 runs, one generation consists of N two- person plays, where N equals population size.

Population Size

CHOICE HEURISTIC	9	16	25
TFT	5	16	19
WCLD	15	102	614
WSLC	65	1645	*)

*) The simulation system allowed a maximum of 32767 generations. Only 8 out of 40 groups reached convergence within this limit.

tion to include several population sizes up to a population of 400. In these simulations we found that the time for convergence under TFT is a perfect linear function of group size.

More interesting however, are the results for the two other choice heuristics. Even for small groups (Table 1) the results for WCLD and WSLC show that the time for convergence clearly is not a linear function of group size. We conducted thereupon numerous simulations in which the population size systematically varied to up to 196. Never did we observe a homogenous all-C, or all-D state in populations of 50 ore more. Hence,what we observe here is a striking effect of group size, pointing out that small and large groups are qualitatively different.

The simulations in which we manipulated group size showed two things. First we saw that in principle homogeneous states for all three choice rules are possible. Second, however, we observed that in larger groups the probability that these absorbing states will be reached practically is zero. This leads to the question whether, in larger groups, stable levels of cooperation would be one of the global consequences from decision making under the WCLD and WSLC rule.

In our simulations we indeed did observe stable levels of cooperation which emerged already after a few generations. A prototypical result is shown in Figure 3. For WSLC cooperation rates settle at about 40 percent, for WCLD the level of cooperation is 55 percent. The results in Figure 3 are based on a population size of 100, similar results were obtained in population sizes of 50 up to 10,000 using very broad time scales.

Figure 3.   Mean prevalence of cooperation for WSLC and WCLD as a function of generations. Each mean is based on 50 simulations

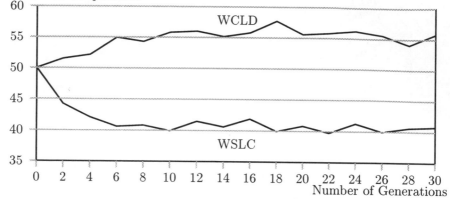

6. Theory

The simulations thus far have shown that there exist qualitative differences between small groups and large groups for WSLC and WCLD. Both choice rules support a stable level of cooperation in groups that have exceeded a critical size. In these cases the global consequences of individual decision making, clearly cannot be explained by using simple linear mechanisms. After analyzing several simulations, however, we were able to formulate an analytical description of the mechanisms that lead to sustainable cooperation. In this section we will present that analysis. In addition, we present a generalization of these processes to the aggregate or population level.

TABLE 2. Unconditional probabilities of choice pairs and conditional probabilities of cooperation for WSLC and WCLD. The overall prevalence of cooperation is a, S = Subject, N = Neighbor.

Choice Pair (S,N)	C,C	C,D	D,C	D,D
Probability of pair	a^2	$a(1-a)$	$(1-a)a$	$(1-a)^2$
P(C—S,N) for WSLC	w_1	0	0	w_2
P(C—S,N) for WCLD	w_1	0	1	w_3

Assume that the overall level of cooperation in the population is a, and that actors make their decisions independently. This will allow us to specify the unconditional probabilities for each of the four outcome cells (Table 2, row 2). Next we turn to the conditional probabilities of getting a cooperative choice under WSLC and WCLD respectively.

For WSLC we have described earlier that the off-diagonal outcome cells, with the payoff values 3 and 0, will not lead to a C-choice of the subject on the next play. The corresponding conditional probabilities therefore are 0. For the CC outcome cell we can expect another C-choice by the subject IFF the last cooperative choice is rewarded (coded as a win). That is if the mean of the neighbors' payoffs is less or equal to two. We denote that conditional probability as w_1. Similarly, after a DD-outcome the subject will switch to a C-choice IFF defection has been punished (coded as a loss), that is if the average others' payoff is more than one unit. The conditional probability of a cooperative choice under WSLC after a DD-outcome is w_2. We are now able to express the overall level of cooperation in the population for WSLC as a sum of the products of conditional and unconditional probabilities for the diagonal cells:

$$a = w_1 a^2 + w_2 (1 - a)^2 \tag{1}$$

Solving Equation (1) for w_1 yields:

$$w_1 = a - 1 - ((1 - a)/a)^2 w_2 \tag{2}$$

We are now able to plot the relationship between the conditional prlities w_1 and w_2, for different values of a. Figure 4 shows these equi-cooperation contours for WSLC.

The visualization in Figure 4 reveals two important features. The first surprising discovery is an implication of expression (2) namely that WSLC can not generate more than 50% cooperation in the population. The second feature is that the equilibrium equation depends more on w_2 than on w_1. In other words, the overall cooperation level is more sensitive to changes in the probability that defection will be punished, than to changes in the probability that cooperation is rewarded.

Next we turn to a similar set of operations for WCLD. For the off-diagonal cells the conditional probability of cooperation after a DC-outcome is 1 because that outcome is coded as a win; after a CD-outcome the subject will defect on the next trial. Following the CC-outcome the subject will cooperate again with probability w_1, the probability that this outcome is coded as a win. Conversely, after DD-outcome a cooperative choice is possible IFF DD is coded as a win, that is if the mean of the average neighbor is less or equal to 1. This conditional probability, the complement of w_2, is denoted w_3. Again, cross multiplying and summing conditional and unconditional probabilities yields the equilibrium equation for WCLD:

$$a = a^2 w_1 + a(1 - a) + (1 - a)^2 w_3 \tag{3}$$

Solving expression (3) for w_1 shows the negative linear relationship between both conditional probabilities in terms of parameter a:

Figure 4.   Equal cooperation contours for WSLC. W_1 is the probability following rewarded mutual cooperation, and W_2 is the probability following punished defection.

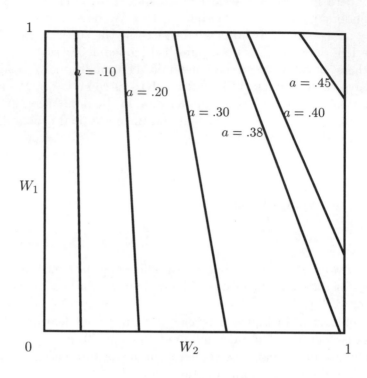

$$w_1 = 1 - ((1 - a)/a)^2 w_3 \qquad (4)$$

The corresponding equi-cooperation contours for WCLD are shown in Figure 5. It is immediately obvious that for WCLD the overall level of cooperation can range between zero up to a homogeneous state of all cooperators. As explained before, under WCLD all-D is not possible. Furthermore, compared to WSLC we now observe a symmetry in the effects on overall cooperation levels, if either conditional probability changes from 0 to 1.

7. Summary and Conclusions

Cooperation is sustainable in large groups where people interact under the highly competitive PD structure while using a competitive type of social comparison for the evaluation of their outcomes. This is seen as the most important conclusion of the study described here.

The two choice heuristics in which action was contingent upon an explicit check whether own outcomes were at least as high as that of the average direct neighbor, unexpectedly produced stable levels of cooperation in large

Figure 5.   Equal cooperation contours for WCLD. W_1 is the probability following rewarded mutual cooperation, and W_2 is the probability following punished defection.

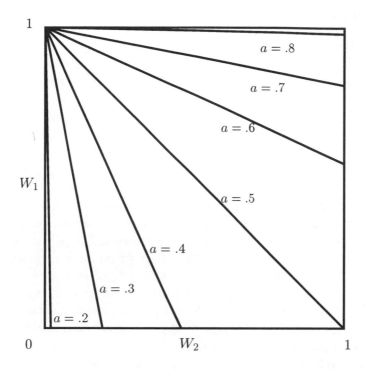

groups. Thus far it has been observed by several scholars (Axelrod 1984; Taylor 1976; Friedman 1971) that conditional cooperation should be possible in case of iterated games where the players do not discount future outcomes too heavily. However, as has been pointed out by Bendor and Mookherjee (1987), these scholars all doubt that conditional cooperation will be feasible in very large populations (p. 131). In addition, Dawes (1980) observes that "all experimenters who have made explicit or implicit comparisons of dilemma games with varying numbers of players have concluded that subjects cooperate less in larger groups than in smaller ones".

In contrast, the research described here shows that there exists qualitative differences between small and large groups for WSLC and WCLD. After a critical group size cooperation stabilizes for these choice heuristics. This finding demonstrates that for example Taylor's (1982) argument is inaccurate. According to Taylor (cited in Bendor and Mookherjee 1987, p. 131)

"voluntary cooperation is less likely to occur in large groups than in small ones, since a conditional cooperator must be able to monitor the behavior of others in the group so as to reassure himself that they are doing their parts and not taking advantage of him. Clearly, as the size of the group increases, this mutual monitoring becomes increasingly difficult and the "tacit contract" of conditional cooperation becomes increasingly fragile."

In our opinion mutual monitoring of all actors in very large groups is not only difficult, it clearly is impossible. Fortunately, this condition is not a necessary condition for securing cooperation in iterated games. The competitive monitoring of some persons to which we are socially related, is not only a more realistic option, but our theoretical approach shows that it also can be a sufficient condition for maintaining some level of cooperation.

The theoretical model allows us to predict changes in cooperation levels for two decision rules as a function of different payoff distributions in social neighborhoods. The model therefore adds to our knowledge of decision making under conflict, but its generality of course is limited. We assume, among others, prisoner' s dilemma contingencies, symmetrical interaction patterns a fixed neighborhood and a homogeneous choice rule. Currently we are not in a position to assess the effect of relaxing or changing our basic assumptions. Each modification in that respect will very likely complicate the analysis substantially. However, based on some modifications we have implemented thus far, we do know that observing stable levels of cooperation is not limited to the present set of basic assumptions.

The second major theme underlying this research concerns the dynamical relationship between micro level behavior and aggregate level outcomes. In this study we have seen that stable levels of cooperation 'emerged', a phenomenon which cannot be fully understood if the level of analysis had been restricted to either the micro- or the macro-level. Recently more researchers have been making the same point. In using methods from statistical thermodynamics Glance and Huberman (1994) show that cooperation spontaneously can arise and stabilize in small groups in which all members initially were defecting. Nowak and May (1992) have shown that highly simplistic choice rules can generate "chaotically changing spatial patterns of extreme richness and beauty", patterns in which cooperators and defectors persist indefinitely.

We feel that computer simulations offer us the tools to analyze the dynamical relationship between individual behavior and the characteristics of the collective system. When we started out this line of research we were overwhelmed by the complexity and reactivity of our artificial world of decision makers. Gradually we gained more control and are now able to outline some basic mechanisms. The theoretical model developed here, however, still is rudimentary. The next step will be to introduce more "richness" in our reconstruction of decision making under uncertainty.

8. Acknowledgement

The research described in this paper was performed while the second author was a visitor at the Institute for Social Science Information Technology (SWI) at the University of Groningen, the Netherlands. It was completed while the first author was a Fellow at the Netherlands Institute for Advanced Studies (NIAS) in Wassenaar, the Netherlands. The work could not have been done without the generous support of both Institutes. A more extensive description of this research is published in Messick and Liebrand (1995).

References

Axelrod, R. (1984). *The evolution of cooperation*. New York: Basic Books.

Bendor, J., and Mookherjee, D. (1987) Institutional structure and the logic of ongoing collective action. *American Political Science Review*, 81, pp. 129–154.

Dawes, R. (1980) Social Dilemmas. *Annual Review of Psychology*, 31, pp. 169–193.

Friedman, J. (1971) A Non-Cooperative Equilibrium for Supergames. *Review of Economic Studies*, 38, pp. 1–12.

Gasik, S. (1990) *Definition of the Warsaw Simulation Language*. Unpublished manuscript.

Glance, N.S., and Huberman, B.A. (1994) The Dynamics of Social Dilemmas. *Scientific American*, March, pp. 58–63.

Krebs, D.L., and Miller, D.T. (1985) Altruism and aggression. In G. Lindzey and E. Aronson (eds.), *The Handbook of Social Psychology*, 1, Hillsdale, NJ: Erlbaum.

Liebrand, W.B.G., Messick, D.M., and Wilke, H.A.M. (1992) *Social Dilemmas; theoretical issues and research findings*. Oxford: Pergamon Press.

Messick, D.M., and Brewer, M.B. (1983) Solving social dilemmas: A review. In Wheeler, P., and Shaver, P. (eds.) *Review of Personality and Social Psychology*, 4, Beverly Hills, CA: Sage, pp. 11–44.

Messick, D.M., and Liebrand, W.B.G. (1995) Individual heuristics and the dynamics of cooperation in large groups. *Psychological Review*, in press.

Nowak, M.A., and May, R.M. (1992) Evolutionary games and spatial chaos. *Nature*, 359, pp. 826–829.

Schelling, T.C. (1971) Dynamic models of segregation. *Journal of Mathematical Sociology*, 1, pp. 143–186.

Taylor, M. (1976) *Anarchy and Cooperation*. New York: Wiley.

Taylor, M. (1982) *Community, Anarchy, and Liberty*. Cambridge, Cambridge University Press.

Wu, J., and Axelrod, R. (1995) How to cope with noise in the iterated Prisoner's Dilemma. *Journal of Conflict Resolution*, pp. 1–39, pp. 183–189.

MODELING SOCIAL CHANGE WITH CELLULAR AUTOMATA

ANDRZEJ NOWAK
Institute for Social Studies
University of Warsaw
Warszawa, Poland

and

MACIEJ LEWENSTEIN
Institute for Theoretical Physics
Polish Academy of Sciences
Warszawa, Poland

In this paper we will first discuss computer simulations of social processes as models of qualitative understanding. In the second part of the paper we will present the cellular automata model of dynamic social impact (Nowak *et al.* 1990) and its applications in the areas of the formation of public opinion and social change as an example of a model of qualitative understanding.

1. Computer Simulations in the Social Sciences — Models of Qualitative Understanding

1.1. THE CONTROVERSY CONCERNING COMPUTER SIMULATIONS OF SOCIAL PROCESSES

As computer simulations are becoming increasingly popular tools in the social sciences their popularity generates a mixed reaction among social scientists. Advocates of computer simulations often write about advantages of the simulation approach. The arguments of the opponents of computer simulations are mainly formulated verbally during discussions following presentations of simulation models. Proponents of simulations view them as a powerful tool that will transform social sciences by bringing precision and rigor, into social theories (see for example Meadows *et al.* 1982; Whicker and Sigelman 1991; Séror 1994; Doran and Gilbert 1994; Hegsel-

R. Hegselmann et al. (eds.),
Modelling and Simulation in the Social Sciences from the Philosophy of Science Point of View, 249–285.
© 1996 *Kluwer Academic Publishers. Printed in the Netherlands.*

mann 1994, in press, and his contribution to this volume). By utilizing the speed and power of modern computers they allow to examine the consequences of complex rules and to study the dynamics of large systems. This makes them the tool of choice for dealing with complexities inherent in the domain of the social sciences. In more general terms computer simulations bring social sciences more in line with the recent developments in mathematical and natural sciences.

Opponents of computer simulations in social sciences express concern that the most valuable aspects of humanistic and social sciences cannot be captured by a formal tool. According to this view computer simulations can never fully describe the richness of psychological and social processes. Humans, especially in a social context, are too complex to be captured by a simulation program. In the social reality the outcome of most processes depends on many factors. In the social sciences the probability of making a false assumption is higher than in the natural sciences. Thus taking into account the multitude of factors that effect every social process, it is not possible that all the assumptions of a simulations are correct. Furthermore, many of factors decisive for prediction are idiosyncratic and practically impossible to be accounted for in computer simulations. The decision of an actor, for example, may depend both on his or her childhood experiences and on the emotions of the moment.

An additional criticism concerns the reproductive nature of knowledge generated by computer simulations. A computer simulations will produce the output which is totally dependent on the input to the simulation. In this respect no knowledge is generated by running computer simulations because all the knowledge had to be there initially. According to opponents of computer simulations, simulations generally trivialize our understanding of social processes by offering a mirage of precise description. In the best case this diverts attention from those theories that aim at deep understanding of the nature of social processes. In the worst case it can destroy the richness and the humanistic heritage of the social sciences.

The degree to which each of those conflicting approaches is judged to be true depends both on our understanding of the goal and nature of the social sciences and our understanding of the aims and methods of computer simulations. The goals, computer simulations are to fulfill, are usually different for computer simulations used in applications of science as contrasted to the uses of simulations in theory development.

1.2. QUANTITATIVE MODELS VS. MODELS OF QUALITATIVE UNDERSTANDING

In engineering computer simulations are widely used to solve such problems as aircraft wing designs, bridge constructions, or designing power plants.

Such applications take existing theories, describe them in some computer language, and apply them to solve practical problems. In order to get reliable quantitative answers usually a multitude of factors has to be taken into account. Usually as much detail as possible, concerning the modeled phenomena, is introduced into the simulation to make prediction more accurate. Applied simulations usually do not benefit the theory, nor they are supposed to do so. The main reasons to use this kind of simulations are: first to deal with problems of high complexity, where easy analytical solutions do not exist or would be too time consuming without the use of computer, and second, to save time and money by testing solutions with the help of a computer that would require costly testing otherwise.

A quantitative precise and reliable answer is usually expected from practically applied quantitative simulations. The cost of arriving at a wrong solution, for example by utilizing false assumption, is high. This kind of simulation needs to be based on a well tested theory, since otherwise wrong acceptance of simulation results could lead to a disaster such as an aircraft crash. In most areas of the social sciences the theories are not advanced enough to be a basis for quantitative simulation. Obtaining reliable quantitative answers from computer simulations in the social area is made even more difficult by the fact that in the social sciences the relationships between variables are usually in the form of a weak stochastic dependency.

In natural sciences enormous progress was recently achieved due to widespread use of computer simulations. Computer simulations have been one of the main tools that have allowed for the development of the theory of non-linear dynamical systems(Arnold 1978, 1983, Eckmann and Ruelle 1985; Glansdorff and Prigogine 1971; Glass and Mackey 1988; Haken 1978, 1983; Hao-Bai-Lin 1987; Moon 1987; Rasband 1990; Peitgen and Richter 1986; Ruelle 1989; Schuster 1984; Zaslavsky and Sagdeev 1988), probably the most significant event in the natural sciences in the eighties. Computer simulations also are one of the main tools of modern biology (cf. Langton *et al.* 1992; Sigmund 1993). The use of computer simulations have also underlied the creation of a new interdisciplinary area of science — complex systems. (Haken 1985; Pines 1987; Stein 1988; Weisbuch 1992; for a critical view see Horgan 1995).

The nature of those simulations is usually very different from simulations used to solve applied problems. In the development of the theory, simulations are often used to build and test models of qualitative understanding. In this approach, instead of trying to model the phenomenon in its natural complexity, one tries to isolate the most important characteristics of the modeled phenomenon, and to build the simplest possible model that can reproduce these characteristics. The phenomenon of interest, then, is explained only qualitatively through analogy to the model. The model itself, however, is very precise and specified in terms of analytical equations or rules of computer simulation. This approach allows one to apply hard models to relatively soft data sets without worrying that one's investigation is less than scientific.

Such models are widely used in physics to model such phenomena as magnetism, hydrodynamics (Ruelle and Takens 1971), laser physics (Haken 1982), biology of ecosystems (May 1981), and modeling neural systems (Amit 1989). These models are built to gain qualitative understanding of phenomena of interest. The simplest model of ferromagnet — the Ising model is a good example. This model, built to explain magnetic phenomena, assumes that each particle of a magnetic material may have only two orientations of its magnetic moment, which are usually represented as arrows pointing up or down. This model has very little in common with physical reality, where any particle may have any orientation in a three-dimensional space but nevertheless serves as a paradigm to understand phase transitions and other collective phenomena. Relatively simple systems of differential equations such as the Lorenz model (1963) similarly serve as paradigms for our understanding of multiple phenomena in meteorology, hydrodynamics, nonlinear optics and so forth.

The application of qualitative methods in physics is well represented in statistical physics. Statistical physics deals with large systems consisting of many interacting elements, and as the name implies, uses statistical methods. Within statistical physics, one can describe collective phenomena, the emergence of patterns, self-organizations of systems, and so on. One of the main insights of statistical physics concerns the universality of qualitative mechanisms and phenomena. Although almost any system can be described by a multitude of variables, usually only a few of those variables are critical for the qualitative properties of the system. First of all, the qualitative behavior of large systems often does not depend a great deal on details involving the behavior of individual elements. This observation is of major importance for researchers. To achieve qualitative understanding, it is often possible to build simple models that capture only essential properties of the interactions in the system, which nevertheless allow for the proper description of aggregate behavior.

Second, research done in statistical physics has shown that similar rules apply to external variables influencing the system, so called control parameters. Although usually there are often many external variables influencing the behavior of any real system, some are clearly more important than others. Many variables lead only to some quantitative effect on the system's behavior. Other variables, however, may cause qualitative changes in the system's behavior. For example, a system might tend toward stability for some values of such a variable, but begin to display a breakdown in stability above a certain value of the same variable. Usually for a given system, a relatively small set of variables that influence its qualitative behavior exists. From this notion, it is clear that description of the effect of some variables is much more enlightening about the system than is description of the effect of the other variables. If we are interested only in qualitative predictions of behavior, it is usually sufficient to know the values of only a few variables describing the system. The remaining variables may be simply ignored. We may, however, want to account for the influences of those variables if we need more precise quantitative predictions.

In summary, the results of analytical reasoning and computer simulations done in natural sciences, point out that systems composed of very simple elements may behave in a very similar way to a system in which the elements are characterized in a rich way. It also shows that of the almost innumerable variables influencing a system, it may be possible to select only a few that are most important, namely, those that qualitatively change the dynamics of the system. The main challenge is how to find those critical variables. Two strategies seem possible here. The first one is simply to start from forming a more comprehensive model, such which would be acceptable from the perspective of our theoretical understanding of the phenomena. The next step is then to reduce the complexity of the model by eliminating those assumptions that are proven by subsequent research not to be critical for the behavior of the system. This strategy is likely to lead to gradual simplification of the model, but is not likely to produce a radical reduction in complexity. As an example, if the initial model is in the form of partial differential equations, this strategy may lead to reduction of the number of equations, but is unlikely to result in formulating the model in much simpler terms, such as, for example, cellular automata.

In the second strategy one tries to build the simplest possible model that would have qualitative properties of the phenomena to be modeled. The goal is to try to capture just the essence of the modeled phenomena disregarding the details. There is no simple algorithm indicating how to achieve it. Often the construction of such a model is simply an act of an insight. Sometimes just by playing with models a researcher observes that the behavior of the model resembles some real phenomenon and then the models is reconstructed to model this phenomenon explicitly. Usually the knowledge of results from previous work with similar models may be very useful. A lot of work has been done within mathematics and physics that concentrated on features of models that are critical for dynamics. Since the behavior of systems usually does not depend on the details concerning individual elements, analogies with existing models in other disciplines of sciences may prove to be very useful in constructing a model in a particular discipline. Models of cooperative behavior designed in biology (May and

Nowak 1992, 1993) provide rich insights into our understanding of features critical for the emergence of patterns of cooperation in human societies.

1.3. SIMULATIONS IN THE SOCIAL SCIENCES — MODELS OF QUALITATIVE UNDERSTANDING

One of the main reservations concerning computer simulations of social processes is related to inherent complexity of the subject matter of social sciences. According to this reservation, computer simulations of social processes simply cannot be right if one takes into account the complexity of social phenomena. Human thought and behavior, which depends both on numerous external factors and the internal workings of often elusive psychological mechanisms, is among the most complex phenomena to be explained by science. The interdependencies of different individuals increase such complexity in a multiplicative manner, making the social aspect of human behavior probably the most complex phenomena science can investigate. Even if general laws in this area can be formulated, usually they admit to important exceptions. In this view, the complexity of the causal mechanisms producing social and psychological phenomena requires models of high complexity to explain such phenomena (cf. Bandura 1982; Meehl 1978).

People and social groups, of course, are clearly different than fluids, lasers, and weather systems. The discoveries concerning the behavior of complex systems done in natural sciences are, however, of major importance for building models of social phenomena. The possibility that only a few of the variables are decisive about qualitative properties of the system's behavior provides a means by which complex social and psychological phenomena may be described in relatively simple ways. Although realistic modeling of any social group would require computer models of enormous size, an adequate model of qualitative understanding of phenomena of interest concerning this group may be constructed in a relatively simple way. Computer simulations are the tool of choice in this approach in two respects. First, they are critical for discovering which variables must be specified when building a model, second they allow to test the consequences of the model.

When a general model or a class of computer simulation models exist for a social phenomena, it is possible to vary systematically all the assumptions of the model and to observe the effects of changing both the general assumptions of the model and values of specific variables. In such a procedure one would usually observe that dropping some assumptions or substituting them with other assumptions does not have much impact on the system's behavior. Some other assumptions are, however, critical for the behavior of the system. Even slight changes of their values lead to dramatic changes in the system dynamics. The researcher may then focus, in the proper model, on the most important assumptions or factors. In other words computer simulations may greatly simplify the process of model building by eliminating the unnecessary variables and assumptions of the model.

As an example, when we were building the model of dynamic social impact of the emergence of public opinion (Nowak *et al.* 1990) we were greatly concerned with how many theoretical assumptions we had to make. It seemed that it was impossible to set the right values of all the parameters needed for the simulation and to make the right choices concerning the mechanism of change of individual characteristics. Even if the chance that each of the assumptions needed for the model was correct equaled to 90 % (an overestimate as compared to our subjective judgments), the chance that the model i.e. all of the assumptions was right, is equal to about .07 for 25 variables and assumptions. While running the simulations we discovered that variations of most of the factors did not lead to significant differences in simulation runs. A simulation program SITSIM was built to allow systematic variation of values of variables and simulation assumption (Nowak and Latané 1994). Later analytical considerations (Lewenstein *et al.* 1992) and computer simulations (Nowak and Latané 1994; Latané and Nowak in preparation) have shown that of all the factors only a few are of critical importance for the qualitative behavior of the model.

The fact the often only few variables really matter for the qualitative nature of simulation results has important implications for estimations of correctness of simulations and thus for judgments of value of simulations as tools of social sciences. Assuming that the probability of each individual assumption of a simulation model equals to .9, the probability that a model utilizing 100 assumption will be right is less than 3 in a million, model based on 25 assumptions has about .07 probability of being correct and a model based on just 5 critical assumptions is correct with a probability roughly .6 . When we take into consideration, that it is easier to make correct assumptions when one can concentrate on a smaller number of factors, the estimate is even more in favor of models of qualitative understanding.

1.3.1. *Emergence*

Serious doubts regarding computer simulations of social processes concern their usefulness. Since computer simulations require a theorist to formulate all the rules and assumptions before simulations are run, what else can we learn from actually running simulations? The same problem formulated in more positive way translates into the question: what are the main advantages computer simulations offer in the social sciences? While many specific gains may be associated with the simulation approach in the social sciences, such as imposing precision on the theory, testing theories for contradictions and so on, in our opinion the main advantages of computer simulations in the social domain are connected to insights they can offer regarding emergence and dynamics. Social scientists commonly stress that properties and behaviors of social groups and societies cannot be reduced to the averaged properties and behaviors of the individuals comprising the group or society. Rather, social groups and societies exhibit emergent properties (e.g., Durkheim 1938), that are not present at the level of individuals.

Within the framework of the dynamical systems approach in the natural sciences, it has been demonstrated that many systems studied both theoretically and experimentally in hydrodynamics (Ruelle and Takens 1971), meteorology (Lorenz 1963), laser physics (Haken 1982), and biology (Glass and Mackey 1988; Başar 1990; Amit 1989; Othmer 1986 etc.) have emergent properties. Emergence is especially characteristic of systems consisting of elements that interact in a non-linear fashion. Even if the elements of the system and their interactions are relatively simple, nonlinearities may lead to amazingly complex dynamic behavior, such as self–organization and pattern formation (see for example Haken 1978, 1982, 1983; Kelso 1984, 1988; Kelso *et al.* 1991 and refs. therein). It follows that a system composed of relatively simple interconnected elements may exhibit much greater complexity than each of the elements separately. This gain in complexity with movement from micro to macro levels of description is clearly relevant to the notion of emergence, which is often considered to be a key feature of social phenomena. The emergence of both order and chaos, for example, has been documented in systems such as neural networks (Amit 1989) and cellular automata (Wolfram 1986), where the elements are essentially binary. In such systems, emergent properties are usually exhibited at the global or macroscopic level. In other words, regularities, irregularities, and patterns can typically be detected in terms of some macroscopic rather than microscopic variables.

Emergence in systems consisting of many interacting elements is often created in process of self-organization in which order (e.g., coordination, pattern formation, growth in complexity) may arise from low level interactions without any supervision from higher order structures. The notion of self-organization can resolve the seeming paradox of complexity in the social sciences of the existence of the highest order agent responsible for imposing order on the otherwise disorganized phenomena. The success of computer simulations in the natural sciences is due in large part to their ability to provide simple explanations for complex phenomena that had previously resisted theoretical understanding. Indeed, the major accomplishment of the nonlinear dynamical systems approach, which was propelled to large degree by computer simulations, is the discovery of how complexity can arise from simplicity. Therefore often complexity may be regarded as a reverse of simplicity. By this, it provides a hope that simple easily understood, yet precise, explanations may exist for otherwise extremely complex phenomena in the domain of social sciences.

By the very definition, emergent properties cannot be trivially derived from the properties of individual elements, and are therefore difficult to predict. Computer simulations, however, allow to model individuals and their interactions and to observe to consequences of such interactions on the group level. It is important to stress the importance of visualization in this approach. Since computer simulations of this kind are of exploratory value, the researcher often does not know exactly what kind of phenomena are of interest and thus should be measured. If the results of a simulation are visualized, often naked–eye inspection may reveal the emergence of new properties on the macro level. The quality of visualization is of critical importance. The more properties of elements and the system are made visible by appropriate use of color, shape, spatial arrangement, the more apparent to the naked eye observation are the emergent properties (Brown 1995; Grave *et al.* 1994). Often just visual inspection may be sufficient to study emergent phenomena such as patterns being formed during the system's evolution. In such cases visualization serves not only the role of a heuristic for discovery, but also as a scientific proof. In other cases, however, once visual inspection discovers phenomena of interest, appropriate more precise measures may by utilized to quantitatively characterize the phenomena of interest.

1.3.2. *Dynamics*

For many years, the social sciences have been concerned with the dynamics of social processes. The importance of temporal characteristics of human thought and behavior was well recognized in social psychology where such notions as group dynamics and dynamics of attitude change are central to the theory (Shaw 1976; Brown 1988). James (1890) talked about the continuous and ever-changing stream of consciousness as the most salient characteristic of human thought. Although most researchers would agree on the importance of dynamics, there is far less consensus concerning how best to characterize dynamics theoretically and empirically (Vallacher and Nowak 1995, Nowak and Lewenstein 1994).

For the most part, dynamics have been analyzed within the framework of cause – effect assumptions. This approach is based on manipulating one or more external factors at Time 1 and observing the effects at Time 2. Despite the success of this general paradigm, however, it encounters two fundamental limitations as a general meta-theoretical model of social processes. First of all, only two points in time and their temporal order are considered (i.e. cause has to precede in time the effect). Everything that happens between those two points escapes consideration. The second is the assumptions that the dynamics simply reflects changes in some state due to some causal factor.

In natural sciences dynamics refers to either continuous change (when changes are modeled by differential equations) or to time series of discrete changes. In fact, often the temporal trajectory of changes of a system is the most revealing source of the information concerning the internal workings of a system. The same applies to the social sciences. The importance of temporal trajectory of changes is well recognized in both demography and in economical sciences. The time dimension also seems critical to understanding most of the social processes. The same social change, for example, might be considered an evolution if it occurs within the course of a century and a revolutions if it occurs within a course of a few months.

Second, traditional cause-effect approach is not well-suited to capture the insight that social processes are to a large degree internally caused, in that these processes display patterns of change even in the absence of external influences and sometimes in opposition to such influences. Such intrinsic dynamics (Vallacher and Nowak 1994) were what from beginning of social sciences was considered to be the essence of human thought and behavior. Thus, according to James (1890) thought never stood still, Cooley (1964) discussed humans' penchant for action in the absence of motives and reward contingencies, and Lewin (1936) argued that overt behavior and thinking are constant struggles to resolve conflicting motivational forces, including those operating from within the person. On the level of a social group, intrinsic dynamics is revealed for example in the phenomenon of attitudes polarization, which term refers to the fact that attitudes in an interacting group become more extreme in the course of time without any external influences (Myers and Lamm 1976; Moscovici and Zavalloni 1969).

Recent developments in the mathematical theory of dynamical systems (Arnold 1978, 1983; Eckmann and Ruelle 1985; Glansdorff and Prigogine 1971; Glass and Mackey 1988; Haken 1978, 1983; Moon 1987; Rasband 1990; Peitgen and Richter 1986; Ruelle 1989; Schuster 1984; Zaslavsky and Sagdeev 1988), however, have provided a new perspective for modeling the dynamics of social phenomena (Vallacher and Nowak 1994; Vallacher and Nowak 1995; Nowak and Lewenstein 1994). One of the factors decisive for the rapid progress in the theory of non-linear dynamical systems was the widespread use of computer simulations, which allows the precise study of the dynamical consequences of models, which cannot be solved by analytical methods. Increasing numbers of social scientists have adapted this new paradigm with a growing number of researchers reconceptualizing their theories and looking for explanations of various phenomena in terms of dynamical concepts and tools (see, for instance, Troitzsch in this volume; Hegselmann in this volume; Axelrod 1984; Nowak *et al.* 1990; Lewenstein *et al.* 1992; Newtson 1994; Başar 1990; Skarda and Freeman 1987; Vallacher and Nowak 1994, 1995, Hoyert 1992; Kohonen 1988; Baron *et al.* 1994; Eiser 1994; Hegselmann 1994; Messick and Liebrand 1995; Newtson 1994; Latané and Nowak 1994; Newtson 1994; Ostrom *et al.* 1994; Weidlich 1991; Weidlich and Haag 1983; Tesser and Achee 1994; Vallacher *et al.* 1994; Vallacher and Nowak 1995).

The dynamical systems approach is based essentially on two assumptions. First, a dynamical system exhibits intrinsic dynamics. The state of the system at a given time determines to a certain degree the state of the system at the next time, in accordance to some rule. It is important to note, however, that the rules of dynamics in general do not have to be purely deterministic and they might even involve elements of randomness. Second, a dynamical system is characterized by some extrinsic factors, which might drastically change the course of intrinsically generated dynamics. These factors are frequently called control parameters since they usually can be externally controlled. Control parameters may change in time, but this is not always the case. In contrast to variable describing the state of the system, the evolution of control parameters does not follow the rules of intrinsic dynamics. The setting of control parameters, however, determines the course of intrinsic dynamics. Even a small change of one or more control parameters can dramatically effect the system's intrinsic dynamics.

Social groups and societies may be viewed as a dynamical systems. Social groups and societies are neither static nor passive, and they display intrinsic dynamics. Even in the absence of external influences and without new information input, social processes continue to evolve and never come to rest. At the same time, it is obvious that social groups and societies react to changes in external conditions. Such reactions, of course, sometimes consist only of resistance, with little change in the state of the system. But at other times, the social system may show an exaggerated response to external factors. In other words, changes in external conditions can lead to unexpected and surprising dynamical effects.

Computer simulations are practically the most powerful tool that allows to study the dynamical consequences of social theories. In the social sciences many empirically observed relationships involve interaction effects between variables. In practice systems, that involve interactions cannot be modeled by linear equations. Since, usually only linear systems of equations are solvable by analytical means, computer simulations are often the only way to examine dynamical properties of systems. It follows that if the social sciences want to put more emphasis on dynamics of social phenomena, computer simulations are usually not only the most convenient tool but often they are the only tool that can be used to study dynamics.

In the following section of the paper we will discuss the model of dynamic social impact (Nowak *et al.* 1990) as an example of a model of qualitative understanding. Research done with this model was focused on two basic problems: how does public opinion emerge from interactions between individuals, and what is the role of individual interactions in a process of social change.

2. Dynamic Social Impact

In the social sciences, there are many theoretical approaches available for describing human interactions (Thibaut and Kelley 1959; Latané 1981; Kando 1977; Krech *et al.* 1962; Shaw and Constanzo 1982; Losada and Markovitch 1990). Individuals may affect states and behaviors of other individuals in two ways. First of all, interdependency may exist between the individuals (Thibaut and Kelley 1959; Conte and Castelfranchi 1994). If such a relationship exists between two individuals, the choices of one individual have direct consequences for the other individual. Relationships of this kind are described by game theoretical models. The best known example concerns the prisoner's dilemma. Social dilemmas (Liebrand *et al.* 1992) provide examples of a similar type of interdependency on a larger scale: the level of society. The prototypical example of social dilemma is provided by smog in Los Angeles. Each individual's decision to use a car for a given trip makes it easier to travel for him or her but contributes to smog effecting other individuals. Computer simulations of those models have provided many important insights into consequences of various types of interdependence, most notably solving a mystery of how cooperation may emerge between egoistic individuals (Axelrod 1984). Papers by Hegselmann and Liebrand are representative of this approach in this volume, in which individual attitudes may be viewed from a functional point of view, as interwoven into the interdependencies of interests between individuals.

Individuals, however, may also exert direct influence on other people. Persuading, giving orders, providing information or just being models other imitate, all belong to this class. In the process of decision making, people interact by influencing others and being influenced, consulting others and being consulted. These interactions can be absolutely crucial to individual decisions and the emerging social processes. In fact, to a great extent interactions among the individuals comprising the social system govern the intrinsic dynamics of the system (Abelson 1964; Nowak *et al.* 1990). In the following sections of this paper we will concentrate on computer simulations of social influence processes and how they explain social change. We will use cellular automata as a vehicle to model processes of social influence and their dynamical consequences, especially social transitions.

2.1. THE MODEL OF DYNAMIC SOCIAL IMPACT

2.1.1. *Theoretical Assumptions*

An important area where the operation of processes of social influence is visible is the emergence of public opinion (Nowak *et al.* 1990). Formation and change of public opinion has been one of the main concerns of social sciences. (Moscovici 1963; Noelle-Neumann 1984; Converse 1964; Crespi 1988; Iyengar and McGuire 1992). From a theoretical point of view, the understanding of how public opinion emerges is important because it gives insight into the differences between micro (individual) and macro (social) levels of analysis (Nowak *et al.* 1990). From a practical point of view, emergence and change are relevant to important social issues, such as voting behavior, consumer preferences, and public decision making. Public opinion is also critical for the course of social transitions.

Our models are constructed as follows. In our simulations social group is assumed to consists of a set of individuals. Each individual is assumed to have an opinion on a particular issue. In the simplest case, it may be one of two possible "for" or "against" opinions, or a preference for one of two alternatives, such as choosing between two candidates in elections. In other cases, there may be more possible attitudes or opinions. People in our models differ in their respective strength, that is, in their abilities to change or support each other's opinions. Individual differences in strength are very important for the behavior of the models. It is obvious that in all real social groups individuals differ in strength. The importance of leaders for the processes taking place in groups is well recognized by the social sciences.

People interact most often and are mostly influenced by those who are close to them, such as family members, friends, and co-workers. People are also much more likely to interact with neighbors, that is, those who live close to them in physical space (Latané *et al.* 1994; Bradfort and Kent 1977; Hillier and Hanson 1990; Hillier and Penn 1991). In our simulations we assign to each individual a specific location in a social space (see Nowak et al. 1994). In most of our simulations social space was conceptualized as a two dimensional matrix of N rows and M columns of points representing locations of individuals (see also Dewdney 1985). Below we will discuss the importance of the geometry of social space for the outcome of social interaction processes (see also Nowak et al. 1994). Our choice of a 2-dimensional lattice represents quite well the physical distribution of people on flat surfaces. The results of studies conducted in Boca Raton, Warsaw, and Shanghai, have shown that the probability of social interactions is decreasing as a square of physical distance (Latané *et al.* 1994; Kapuściarek and Nowak 1993).

To model social interactions, we assume that individuals communicate with others to assess the popularity of each of the possible opinions. Opinions of others located close to the subject and of those who are most influential are most highly weighted. Individual's own opinion is also taken into account in this scenario. In the course of the simulation, individuals adopt the opinions that they find prevailing in the process of interacting with others. This simple model of social interactions is not only intuitive but also agrees with a number of empirical studies. The theory of social impact (Latané 1981), built as a generalization of empirical results, states that in diverse situations where a group of people is exerting impact on an individual or on another group, the strength of this impact can be specified as a universal function of peoples' strength, immediacy, and number. Our models can incorporate these features of social interaction and lead to similar conclusions (Nowak *et al.* 1990, Nowak *et al.* 1994, Hegselmann and Nowak 1994; Szamrej *et al.* 1992, Nowak *et al.* 1993).

2.1.2. *Construction of the Models*

The computer models discussed here belong to a class of models called *cellular automata* (Wolfram 1986) that are widely used in physics (for example, to model the flow of fluids), and in various domains of biology, including neuroscience (Amit 1989; Başar 1990) and population dynamics (May 1981). Probably the best known example of cellular automata is the Game of Life (Gardner 1970). In this volume Hegselmann discusses in detail cellular automate as models of social processes. Cellular automata are dynamical systems, that is, they evolve in time. For simplicity, we may assume that time proceeds in discrete steps, t, $t + 1$, $t + 2$ etc. We consider individuals to be a set of formal automata located in some space. In the example discussed above example, individuals are located at the vortices of the 2–dimensional square lattice, so that their coordinates are (i, j) where i, j are integers. The state of the individual located at (i, j) at time t (that is his or her opinion) may be represented quantitatively by some set of numbers. In the case of "yes" or "no" choice, it can be conveniently represented as a number $s_{ij}(t) = \pm 1$, with 1 for "yes" and -1 for "no". The individuals influence and are influenced by other individuals. Again, in the simplest case we may assume that the individual located at (i, j) interacts most strongly with his or her nearest neighbors, that is, individuals located at $(i + 1, j)$, $(i - 1, j)$, $(i, j + 1)$, and $(i, j - 1)$.

Each individual at (i, j) adjusts his or her opinion in the following time steps according to an *updating rule*. In particular, the updating rule can have a form

$$s_{ij}(t+1) = \text{sign}(\sum_{i',j'} J_{ij}^{i'j'} s_{i'j'}(t)),$$

where $J_{ij}^{i'j'}$ measures the impact of a individual located at (i',j') on the individual located at (i,j). Obviously, according to the theory of social impact (Latané 1981), $J_{ij}^{i'j'}$ decays rapidly as the distance between (i',j') and (i,j) grows. Moreover, $J_{ij}^{i'j'}$ increases with the strength of (i',j') – the individual who is the source of the impact. We shall denote the strengths parameters for individuals by f_{ij}.

In our models we have used a slightly more generalized version of the updating rule described above. Generally,

$$s_{ij}(t+1) = s_{ij}(t)\text{sign}(I_s - I_p),$$

where I_s is the *supportive impact*, while I_p is the *persuasive impact*. Quite generally, one can assume that the supportive impact is a function of $\sum'_{i',j'} J_{ij}^{i'j'}$, where $\sum'$ limits the sum to supporters of $s_{ij}(t)$, i.e. to those (i',j') for which $s_{i'j'}(t) = s_{ij}(t)$. Similarly, I_p is assumed to be a function of $\sum''_{i'j'} J_{ij}^{i'j'}$, where $\sum''$ is restricted to (i',j') for which $s_{i'j'}(t) = -s_{ij}(t)$.

One of the most important aspects of the models discussed here is that individuals differ in strength and thus can be characterized by strength parameters f_{ij}. The coefficients $J_{ij}^{i'j'}$ that measure the impact of the (i',j')–th individual on the (i,j)–th individual are thus proportional to $f_{i'j'}$. In the model with nearest neighbor interactions only, we simply set $J_{ij}^{i'j'} = f_{i'j'}$ for $(i',j') \neq (i,j)$, and $J_{ij}^{ij} = \beta$. The parameter β in the latter formula determines a *self–supportiveness*, that is, the impact of (i,j)–th individual on herself or himself. The updating rule may be then termed a *weighted majority rule*. The individual located at (i,j) at the time $t+1$ would vote "yes," provided the weighted sum of opinions of his or her nearest neighbors including him or herself at time t was positive. Mathematically speaking, that means $s_{ij}(t+1) = 1$ provided

$$\beta f_{ij} s_{ij}(t) + f_{i+1j} s_{i+1j}(t) + f_{i-1j} s_{i-1j}(t) + f_{ij+1} s_{ij+1}(t) + f_{ij-1} s_{ij-1} \geq 0,$$

and $s_{ij}(t+1) = -1$ otherwise.

Another important aspect of our models is that typically the updating rules do not include the nearest neighbors only. Updating rules that are restricted to nearest neighbors are commonly used in physics. On the contrary, our models include the influence of every individual in the group. The influence of those individuals on a given person, however, decreases with increases in their distance from that person. We have also studied

models where each individual interacts with just a few other randomly selected from those that live not further than a specified distance (Nowak *et
al.* 1993; Szamrej *et al.* 1992).

2.2. INTRINSIC DYNAMICS AND EMERGENT PROPERTIES

We started the simulations from a random distribution of opinions. This
may be interpreted as representing the situation where initially each individual comes to his or her opinion unaware of the opinions of others. This
opinion may be the result of a number of factors not accounted for in our
model, such as vested interests, previous experiences, or simple reasoning
about the issue. In Fig. 1, the majority of the population choose the "no"
option represented by the light color. The minority of individuals, represented by the dark color, are for a "yes" choice. The height of the bars
indicates the strengths of the individuals. The strength is also distributed
randomly among individuals and in this model it does not change in the
course of the simulation.

Figure 1.   Initial random distribution of opinions in a group of 400 individuals. Color
represents attitude, height represents strength.

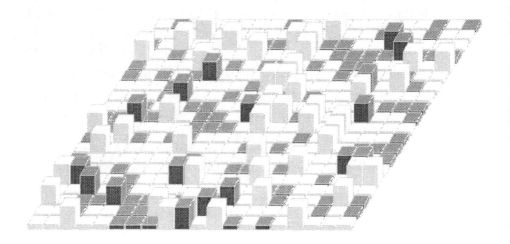

As individuals interact in the course of the simulation, those who find
the opposite opinion prevailing, change opinions. Finally, after some simulation steps an equilibrium is reached in which no one changes opinion
(see Fig. 2). In comparison to the initial distribution of opinions, the final
distribution is different in two respects. First of all, opinions are no longer
randomly distributed. Individuals holding minority opinion are grouped in
clusters or "bubbles." Clustering is reminiscent of a wide class of a real

Figure 2. Final distribution of attitudes in the group shown in Figure 1. After discussion minority opinions have polarized and clustered.

word phenomena such as the spread of accents, fashions, beliefs, and political preferences. In fact, it is difficult to find an example of opinion, fashion, or custom that is not clustered, if it was acquired through social interaction. Clustering of attitudes was empirically demonstrated in a survey study of attitudes at an MIT housing project, where people living within the same neighborhood tended to develop similar attitudes on issues concerning local policy (Festinger *et al.* 1950). The second phenomenon visible in the comparison of Fig. 1 through Fig. 5 is that the number of people holding the minority view has declined. Such a phenomenon is often referred to as *polarization* of opinions (Myers and Lamm 1976; Moscovici and Zavalloni 1969).

The formation of clusters or bubbles is a very universal phenomenon of a wide class of computer models of social change. They will form in any model that incorporates local interactions in which an individual is likely to adopt the prevailing opinion in his or her local environment. In particular, cluster formation occurs in models that incorporate elements of randomness and noise, provided the noise is not too large. It is interesting that bubbles will form even if all the individuals prefer to be in a global majority. Individuals try to establish the relative proportion of attitudes in the society by generalizing from their interaction with others. Since people are most likely to interact with those who are close to them, the result of their sampling is strongly biased in a clustered society. In fact, most individuals, even from the minority group, would come to the conclusion that their opinion is consistent with a global majority.

Figure 3. The simplest stable configuration of a minority: leader surrounded by followers.

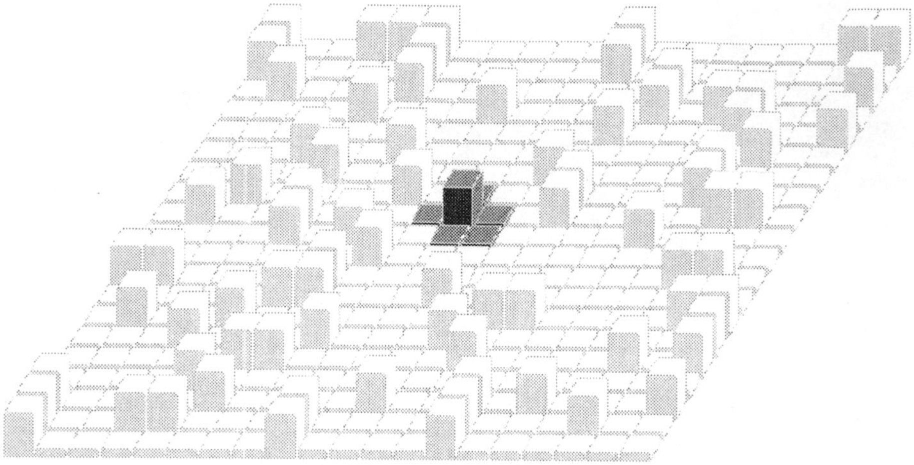

Figure 4. Several strong minority figures form a "stronghold".

2.3. FEATURES CRITICAL FOR DYNAMICS

We have analyzed a large class of models belonging to the category discussed above (Lewenstein *et al.* 1992; Latané and Nowak in preparation; Nowak *et al.* 1993, Nowak *et al.* in press). This analysis led us to the conclusion that the qualitative phenomena exhibited by the models, such as polarization and clustering, are caused by the features of the models critical for dynamics. These are:

Figure 5. A "wall" consists of several strong minority members surrounding a group of a weaker minority members.

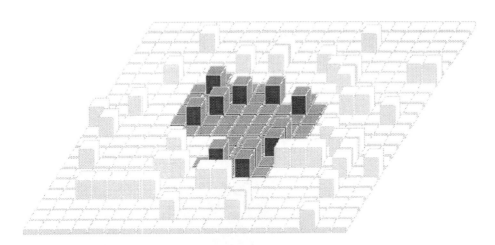

2.3.1. *Nonlinearity of Attitude Change*

In the traditional approach of the social sciences, any attitude held by an individual was considered to be a continuous variable. Change in attitude, moreover, was assumed to be proportional to the influence (or social forces) exerted on this individual (Abelson 1964). One should stress, however, that "social forces" very often act in a nonlinear manner. That means that sometimes a small change in social forces can bring about a drastic change of attitudes. In consequence, people frequently adopt only extreme attitudes (either very positive or very negative).

This phenomenon has been illustrated empirically by Latané and Nowak (1994). In this research, the authors have measured the distribution of attitudes (measured on a Likert scale from 1 to 5) concerning diverse issues. The subjects were also asked to indicate the personal importance of these issues. It turns out that the distribution of attitudes on issues that were judged as unimportant is approximately normal. On the other hand, the distribution of attitude concerning subjectively important issues has a well developed U shape. People tend to adopt extreme attitudes on such issues. This implies that for important issues, binary representation of attitudes is adequate. Such a representation implies necessarily nonlinearity of attitude change.

2.3.2. *Individual Differences.*

Our computer simulations indicate that the time course of polarization and clustering strongly depends on the assumption that individuals differ in their respective strength parameters. The introduction of individual dif-

Figure 6.   Hierarchical geometry. Top: initial configuration of attitudes, bottom: final distribution

ferences into the model is probably the most important factor that makes our model different from those studied elsewhere (Wolfram 1986; Landau and Lifshitz 1964). Mathematics and physics suggests, in fact, that the dynamics of systems consisting of "equal individuals" in the presence of infinitesimally small noise should inevitably lead to uniformity of opinions. In our case, individual differences provide a major source of minority group survival. We were not able to prove mathematically that in general such survival is eternal in the presence of some randomness. The results of our computer simulations, however, clearly indicate that the life time of minority clusters can be practically infinite.

The question "How can minority opinion survive?" may be answered by noting the formation of clusters. Those inside clusters are surrounded by others who share the same opinion. Only those located on borders are exposed to the pressure of individuals holding majority opinion. The survival of the clusters depends thus on what happens to individuals located on the borders of clusters. The borders of minority clusters are mostly convex. Members of minority located on the borders are on the average surrounded by the prevailing number of individuals holding the majority opinion. If there had not been individual differences, most individuals on the borders of minority clusters would not be able to hold their opinions. The minority clusters would gradually decrease in size and eventually vanish. The existence of individual differences in strength stops this process. An especially influential minority member located on the border may counter the prevailing number of majority members. The findings of our simulations correspond to the observations in social sciences that point to the importance of leaders in social movements.

A minority can survive in simple configurations, shown in Fig. 6. In the simplest configuration, a single strong individual is surrounded by several of weaker minority members (Fig. 3). The weaker members would not be able to maintain their opinion without the leader. It would also be difficult for the leader to survive without the followers since they also provide him or her with some support and, in addition, isolate him or her from the majority. A second configuration "stronghold" (Fig. 4) consists of a group of stronger minority members located close to each other and supporting each other. The whole group is usually surrounded by a weaker members. Note that in both of the discussed configurations, strong minority members are located inside the cluster. In the third scenario (Fig. 5), a "wall" of strong individuals is located on the borders of the cluster; it shields the weaker individuals located inside the cluster. The wall does not need to be compact, since weaker individuals who are located between stronger minority members ones are not outnumbered by majority neighbors. We should note that in our simulations we quite often observe much more complex configurations composed of combinations of simpler ones. For example, several strongholds may be located on the wall of a large minority cluster. Without other sources of influence, all the configurations described above are stable, that is, they would never change.

Usually individuals are subjected to a number of factors influencing their opinions in addition to the influence of the immediate social context. Such factors include selective exposure to media, recall of particular memories, personal experiences, and so on. The cumulative effects of such factors may be represented as a random factor or "noise" that adds up to the effect of social influence. If all the other factors have relatively small effect, as compared to the effect of social influence — or in other words, the noise is small — the picture described above will not change significantly. Even if from time to time some weak minority members change their opinion, typically their own group will be able to convert them back.

A slightly larger value of the noise, however, might cause one of the leaders in the minority group to change his or her opinion. If this leader is a part of the wall shielding weaker minority members, the cluster will start to decay rapidly. In the presence of larger noise, therefore, the clusters in principle become unstable. Typically, however, such decay will terminate in some new equilibrium as the shrinking border of the cluster encounters the resistance of the remaining strong individuals. For large clusters, such a scenario may take place many times. If we plot the size of the cluster as a function of time, the plot will consist of several periods of rapid decay separated by relatively long intervals of apparent equilibria that can be almost infinitely long. We call this type of decay staircase dynamics (Lewenstein *et al.* 1992). It should be stressed that this is a generic form of decay in systems consisting of many agents where the agents are different in strength.

2.3.3. *Geometry of Social Space.*

Figure 7.   One-dimensional geometry; top: initial configuration, bottom: final distribution

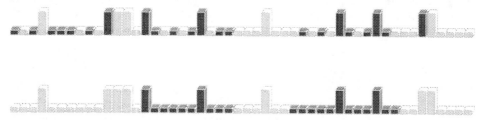

Another crucial feature of our model, yet the one that is perhaps the most difficult to relate to empirical data, concerns the geometry of social space. Our computer simulations indicate very clearly that the geometry of social space determines simultaneously the character of cluster formation, the shapes of the resultant clusters, and their likelihood of survival. We illustrate this statement in the next figures , in which we compare the results of the simulation for a group of individuals experiencing (a) hierarchical (Fig. 6) and (b) one dimensional strip geometry (Fig. 7). In the case of hierarchical geometry, people are divided into groups, subgroups etc. The distance between the individuals within the same hierarchy group are relatively small, but it increases rapidly between the individuals occupying the same group at the higher hierarchy level. This feature of the geometry causes the formation of clusters mostly at the same hierarchy group. Hierarchical geometry may represent an idealization of the distances within an organization. In the case of the one dimensional geometry, the interactions of individuals are particularly weak, since they are mainly mediated through the "left" and "right" neighbors. For this reason, in one dimension two strong individuals on the borders of the cluster are sufficient for its survival. This geometry may approximate spatial arrangement of people living along a road or a river.

2.4. THE APPROACH OF STATISTICAL MECHANICS

Computer simulation models are constructed on the micro level of social reality and are based on assumptions of concerning decisions of individuals in their social context. Corresponding rules on the macro level can be formulated with the tools of statistical mechanics. Those tools allow to predict many of the results of our computer simulation (Lewenstein *et al.* 1992). The tools that we use are closely related to the synergetic approach (Weidlich 1991, Weidlich and Haag 1983, Haken 1978, 1983). These tools are particularly useful when both the number of interacting individuals and the range of interactions increase toward infinity. A strong advantage in combining computer simulations with the analytical solutions is the possibility to cross-check the results. There always is a risk that a computer simulation program may contain a bug. The risk of arriving at erroneous conclusions may be reduced if more than one computer simulation program is written and their results are cross-checked. In fact, we have by now used 7 different computer simulation programs written in FORTRAN, PASCAL, Object Oriented Pascal, C, and Warsaw Social Simulation Language to simulate emergent properties of social influence. All the computer simulations may still be based on some common assumptions of which the researchers are not aware. Computer simulations also always dealing with groups of some finite size. This limitation which may lead to treating results true only for some group sizes as general laws. Research using computer simulations has shown, that group size may be a very important factor for the results of simulations. Although this risk may be minimized by using large groups in simulation runs (some simulations of our model were done using group size roughly equal to roughly 250 thousand individuals) still the danger is there. It is also worth to note that analytical derivations of solutions for class of problems require to use complex methods of statistical physics, and there is also a chance of an error when deriving a formula or assuming a simplification. Cross checking analytically derived results with the outcomes of compute simulations allows to discover such errors.

Let us again denote individuals' opinions by $s_i = \pm 1$, for $i = 1, \ldots, N$. Each individual is characterized by a strength parameter f_i which is a random number that remains stable over time. We assume some probability distribution of the strength parameters $p(f_i)$ that is the same for all i. We also assume that "distances" of a given individual to all others are equal N, whereas the distance of an individual to him or herself is set to be β, where β is some number. The supportive impact can be then defined as

$$I_s = \frac{1}{2N} \sum_{j \neq i}^{N} f_j(1 + s_i s_j) + \beta f_i,$$

whereas the persuasive impact can be defined as

$$I_p = \frac{1}{2N} \sum_{j \neq i}^{N} f_j(1 - s_i s_j),$$

so that the dynamics becomes

$$s_i(t + 1) = \text{sign}(\beta f_i s_i(t) + m(t)),$$

where the *weighted majority-minority difference* $m(t)$ is defined

$$m(t) = \frac{1}{N} \sum_{j \neq i}^{N} f_j s_j.$$

Even for such simple dynamics, the dynamical order parameters are quite complex (Lewenstein *et al.* 1992). Nevertheless, in the limit of large N one is able to derive a mean field equation for the quantity $\bar{m}(t)$ which is the mean value of $m(t)$ averaged over the distributions of f_i's and over the initial distribution of attitudes. Such a mean field equation has a form of a discrete map,

$$\bar{m}(t) = f(\bar{m}(t)),$$

where the function $f(\bar{m})$ depends functionally on the distribution of f_i's and on the distribution of initial opinions. We have shown that in general, the map introduced above may have several stationary points, corresponding to increasing values of the absolute value of $\bar{m}(t)$. Each of these stationary points corresponds to subsequent "conversion" of stronger minority subgroups. In the absence of noise, each of the stationary points is stable. The introduction of small noise causes the system to jump between the stationary points. The jumps occur typically in the direction of increasing absolute value of $\bar{m}(t)$, that is, of increasing polarization. The jumps are separated by long periods of apparent quasi–stability. This is a simple example of the "staircase dynamics" discussed above. It is worth stressing that the mean field approach can be generalized to other geometries such as hierarchical geometry, random connection matrix, as well as noise. For the case of hierarchical geometry, the mean field theory predicts clustering within the hierarchy groups and staircase decay of clusters from lower to higher hierarchy levels. To trigger the decay, however, one needs much larger values of noise because the clusters in the hierarchical geometry are much more stable. Random connection networks behave similarly to fully connected networks discussed above. Interestingly, however, due to the randomness of interactions between individuals, such networks inevitably incorporate self–induced noise.

2.5. HOW CAN MINORITY OPINION GROW: MODELING SOCIAL CHANGE.

For a social change to occur usually some external conditions have to change. In the case of recent social, political and economical transitions in Eastern and Central Europe the changes of the situation in Soviet Union and the degree of control exerted by this country was clearly one of the decisive factors. In some other cases such external factors may include a change in economy, technical inventions, military interventions and so on. The change of global factors, however, almost never directly changes individual opinions and attitudes. The effects of all the factors are mediated through the process of social interactions. Before adopting an opinion individuals usually consult with others, also an opinion, that has been recently been formed usually is a subject to intense discussions. In the following section of the paper we will consider how global change is produced by joint effects of global factors and social influence, in other words how does intrinsic dynamics combine with external influence in producing a social change.

In our class of models, if social influence is the only force causing attitude change, minorities cannot grow. On the other hand, the fact, that sometimes minorities grow and become majorities is one of the main factors causing social change. The recent social transitions in Eastern and Central Europe provide examples of global change in attitudes, such that initial minorities grow and eventually become majorities. Clearly, something has to be added to our model if is to account for social change. In our previous discussion, we have assumed that all the available opinions or attitudes are equally attractive. In reality some attitude positions usually are more functional for the individual, better reflect his or her values, or are simply advocated by propaganda. Some of these preferences are specific for every individual since they reflect individual experiences, etc. Sources of other preferences are external with respect to individuals. External influences, such as provided by the media, the fiscal system, and mechanisms of political control usually change in the time of transition, so a new set of opinions becomes more attractive.

We can account for one attitude position being more attractive than the other by introducing a factor of preference, a *bias*, into the rules of opinion change. In the presence of bias, each of the individuals would more easily adopt one position than the other, taking into account not only the results of their social interactions, but also externally given preferences.

The effects of social interactions in the presence of bias are presented in Fig. 8 to 13. These figures correspond to different stages of a transition. For the beginning of simulation (Fig. 8), we have chosen a state in which 10 percent of randomly located individuals have the favorable "new" opinion represented by the light color. The dark color represents the opinion of the majority, which corresponds to old attitudes.

Figure 8.   Dynamics of attitudes in the presence of "bias". (a) Initial distribution of "new" opinions.

To model the asymmetry of opinions a constant -"bias" - to the impact of the minority opinion was added. Figures 8 through 11 show the growth of the minority of the followers of the new as a result of social interactions in the presence of bias. As indicated in Figure 9, clusters are starting to form around those individuals who can be interpreted as seeds of the transition. These clusters of new continue to grow in the course of time and eventually connect to each other. The growth of the new, especially when the bias

Figure 9. Dynamics of attitudes in the presence of "bias". (b) Bias is in favor of the "new". Opinions after several rounds of discussion. Notice that the "new" enters through "bubbles" centered around the initial innovators.

Figure 10. Dynamics of attitudes in the presence of "bias". (c) Bubbles connect as the "new" is gaining more adherents.

is not too strong follows the staircase dynamics for the reasons describe above. Eventually the new equilibrium is reached. Interestingly, despite the fact the bias is still present there are clusters of old well – entrenched in the final state of the transition.

Judging by the numbers, in the situation portrayed in the figure 11

Figure 11. Dynamics of attitudes in the presence of "bias". (d) Opinions after the "new" has prevailed. Notice that the "old" can survive in well–entrenched clusters.

Figure 12. Dynamics of attitudes in the presence of "bias". (e) When the bias is withdrawn "old" regains popularity

the social transition has been complete. The "new" has completely pre-vailed and its proponents would be able to overwhelmingly win any elections or a referendum. The situation is, however, more complex. The minority owes it prevalence to the continuing presence of bias. The reason is that the strongest members of the original majority had most chance to resist change. Although they are low in numbers their average strength is high

Figure 13.   Dynamics of attitudes in the presence of "bias". (f) Bias has been reversed and now it is in favor of the "old". It takes only a few simulations steps for the "old" to prevail. Note that in this process the "new" has formed clusters in which it can sustain the pressure of the opposing group.

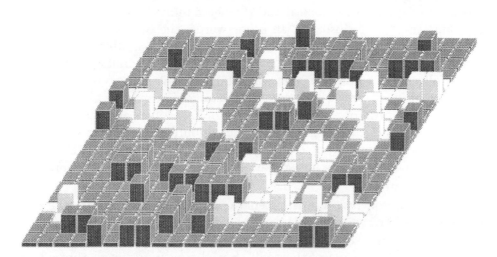

and in fact they form strongholds of powerful interconnected individuals. From those strongholds they can initiate an attack on the "new", whenever the situation is more favorable to them. The figure 12 shows the new equilibrium achieved when the bias was withdrawn corresponding to a situation when the "new' has lost some of it's initial appeal, and both attitudes became symmetrical. The "old" has regained much of its ground and the proportion of people having the "new" and the "old" attitude is close to being balanced. It may well happen that the bias reverses in favor of the "old". For example people may have found that the new is less favorable to them than they assessed, or mass media may reverse the direction of their influence. The effect of the reversal of bias is shown in the picture 13. After a short time proponents of the "old" win support of the overwhelming majority of the population. Although, the "new" is small in number , during the changes it had an opportunity to cluster. It forms strongholds in which it can resist the pressure of the "old" and wait till the bias changes in its favor. It took adherents of "new" 40 simulations steps to achieve prevalence. The victory of the "old" happens in just 5 steps. This phenomenon may be compared to the memory of a society. It looks as if the society has remembered its previous states. The tendency of a society to return to its previous states is a general one and will happen in any model in which the strongest individuals are most likely to resist the pressure to change. This mechanism may underlie a regularity observed in almost all the countries in Europe that underwent transition into a market economy. In all those countries in the first free elections the opposition had overwhelmingly won. To everybody's surprise, however, parties of communist provenience have won in the next elections.

The effect of societies having a tendency to come back to its previous states would be reversed if the strongest individuals were most likely to adopt the "new" positions. In other words, the bias would have to be positively correlated with strength. In such a situation, after the strongest individuals become advocates of the "new", other individuals in the group could follow, due to mechanisms of the intrinsic dynamics, even if the bias favoring the "new" was withdrawn. Such a scenario would be likely if adopting the "new" position was in the interest of individuals and social pressure was keeping them changing. Such a description fits the change of economic orientation during recent transformations in Eastern Europe. Starting a private business although unpopular according to old values in general may be highly profitable.

The major conclusion of the scenario presented above is that the transition occurs through the local centers of change: growing clusters of new within the sea of "old". During a transition two realities coexist: the reality of the "new" and that of the "old". Social transitions occur through the change of proportions of the two realities in favor of the "new", rather than through gradual change of all individuals from the old position to the new position. There may be, under some circumstances, a tendency for societies to return to their previous opinions.

Our analysis has some practical implications concerning facilitation of social change. Our results suggest that an effective social intervention should concentrate on changing spatially coherent areas of social network rather than dispersed individuals. Such an intervention should be aimed at well defined places, potential centers of change. Those centers will have a chance to develop into clusters and the change can diffuse from those centers to other areas. The same efforts spread spatially could run into the risk that those individuals who have changed will not find enough support for their new attitudes in their social networks, and under social pressure will revert to the old attitudes. Also important is the role of the leaders and the place in the social network the leaders are located. For example, by forming a wall a small number of leaders can protect a large number of followers.

2.6. GENERALITY OF THE MODEL.

As discussed in the beginning of the paper the behavior of the models often does not depend on the precise assumptions concerning individual elements, the same models and their qualitative results may be thus interpreted as modeling different phenomena. The model of dynamic social impact was originally constructed to model the dynamics of opinions and attitudes. Its extension allows, however to model economic transitions (Nowak *et al.* 1994, Nowak *et al.* in press). Economic and voting data in Poland suggest that in fact the "new" reality forms spatially coherent clusters. All the economic growth occurs through the expansion of existing centers of growth, regions far from the clusters of economic growth still decline despite the overall growth of the economy. The clusters of "new" go beyond economic measures and are also visible in voting for pro-reformist parties.

The connection between the changes of attitudes and economic change may seem far reaching, there is, however a clear connection between the two. Let us consider individuals, who try to start a private enterprise. First, of all, an individual must believe that starting an enterprise can be profitable. Clearly processes of social influence play an important role there. Second the individual must also know what to do to start a new enterprise. Information also to large degree disseminates through personal contacts. The spread of innovations also occurs through growing clusters and "bubbles" (Sperber 1990). To be sure, economic factors are also of major importance in this process. The survival of any enterprise to a large degree depends on local economic factors. Our models are not unique in predicting clustering effects during social transitions. Clustering effects have been demonstrated also in models stemming from the game theory (Axelrod 1984; Hegselmann this volume) in which it was demonstrated that cooperating individuals tend to cluster. Clustering effects have also been extensively discussed in economics (Myrdal 1958, Friedmann 1966, 1973). Our point is that different processes of both social and economic nature may have some common underlying mechanisms and show similar qualitative properties.

3. Conclusions and Generalizations

In first part of the paper we argued that computer simulations can be used as models of qualitative understanding of social processes. Their main functions would be to look into the emergent properties and into the dynamics of the social processes.

We have concentrated on the processes of social influence and specifically how the result in the intrinsic and extrinsic mechanism of social change. Cellular automata models are built to model social change produced by social influence. It is important to stress, that the description of individuals is greatly simplified in this approach. In fact, the description of individuals is trivial in such models and one should not claim that such description adequately represents a person in a social context. This model is not a model of an individual in a social system, but a model of aggregate behavior. The model displays intrinsic dynamics. Social change is due to intrinsic dynamics when it occurs without external influence. Clustering and polarization of opinions are the main emergent properties of intrinsic dynamics of social change. The decay in minority opinion follows a staircase dynamics. The main factors critical for the behavior of the model are: nonlinearity of attitude change, the existence of individual differences and the non-linearity of attitude change. Several factors such as the existence of randomness or the assumption about symmetry of the attitudes were important, and most of the assumptions (such as whether the simulation space has borders or is torus shaped) did not matter for the qualitative picture of simulation runs, although they could matter to some degree for quantitative results.

Social change may occur when initial minority of opinion grows and eventually prevails. Social change may be analyzed as a modification of the intrinsic dynamics by external factors. Usually in the process of social transitions same factors make the "new" more attractive than the "old". The transition occurs through the growing clusters of "new" within the sea of "old". Clustering is the necessary feature of the rapid evolution toward the "new". This is due to the fact that the fate of individuals both in the social end in the economic sense, depends to a large degree on their local environment.

We would like to stress that there are other models that attempt to describe phenomena similar to those we have considered. Some models, for example, give explicit attention to the movement of individuals in social space, and to individual interests. Thus, Schelling (1971, 1969) has shown that clustering may be achieved by the simple fact that individuals move if they are in local minority. Similar ideas have been developed by Hegselmann (this volume) and May and Nowak (1992) to describe social dynamics from the point of view of game theory. These ideas have much in common with the development of social cooperation formulated within the "prisoners dilemma paradigm" by Axelrod (1984). Although our models thus far incorporate only some of the features stressed by such theories, the mechanisms discovered in our models and the qualitative results we have obtained are observed in other models as well.

4. Acknowledgement

We acknowledge the support of a grant from ISS. UW. for this research. This manuscript was partially written while A.N. was a fellow at NIAS.

References

Abelson, R.P. (1964) Mathematical models of the distribution of attitudes under controversy. N. Fredricksen and H. Gullicksen (eds.), *Contributions to mathematical psy-*

chology, New York: Holt, Rinehart and Winston, pp. 142–160.

Amit, D.J. (1989) *Modeling Brain Function: The World of Attractor Neural Networks.* Cambridge: Cambridge University Press.

Anderson, R.M., and May, R.M. (1991) *Infectious Diseases of Humans: Dynamics and Control.* Oxford: Oxford University Press.

Arnold, V. I. (1983) *Geometrical Methods in the Theory of Ordinary Differential Equations.* New York: Springer.

Arnold, V. I. (1978) *Mathematical Methods of Classical Mechanics.* New York: Springer.

Axelrod, R. (1984) *The evolution of cooperation.* New York: Basic Books.

Bandura, A. (1982) The psychology of chance encounters and life paths. *American Psychologist,* 37, pp. 747–755.

Baron, R.M., Amazeen, P., and Beek, P.J. (1994) Local and global dynamics of social relations. Vallacher, R., and Nowak, A. (eds.) *Dynamical systems in social psychology.* San Diego, CA: Academic Press

Başar, E. (1990) *Chaos in Brain Function.* Berlin: Springer.

Batten, D., Casti, J.L., and Johansson, B. (eds.) (1987) *Economic Evolution and Structural Adjustment.* Lecture Notes in Economics and Mathematical Systems, 293, Berlin: Springer.

Bradfort, M.G., and Kent, W.A. (1977) *Human Geography: Theories and their applications.* Oxford: Oxford University Press.

Brown, J.R. (1995) *Visualization: using computer graphics to explore data and present information.* New York: Wiley

Brown, R. (1988) *Group processes: Dynamics within and between groups.* Oxford: Basil Blackwell.

Conte, R., and Castelfranchi, C. (1994) Mind is not enough: the precognitive basis for social interaction. N. Gilbert, and J.E. Doran (eds.), *Simulating societies: The computer simulations of social processes.* London: University College London Press, pp. 267–286.

Converse, P. (1964) The nature of belief systems in mass public. D.E. Apter (ed.), *Ideology and discontent.* New York: The Free Press.

Cooley, C.H. (1964) *Human nature and social order.* New York: Schonen Books.

Crespi, I. (1988) *Pre-election polling.* New York: Russel Sage Foundation.

Dewdney, A.K. (1985) Computer recreations. *Scientific American,* p. 4.

Doran, J.E., and Gilbert, N. (1994) Simulating societies: an introduction. N. Gilbert, and J.E. Doran (eds.), *Simulating societies: The computer simulations of social processes.* London: University College London Press, pp. 1–18.

Durkheim, E. (1938) *The rules of sociological method.* Chicago: University of Chicago Press.

Eckmann, J.P., and Ruelle, D. (1985) Ergodic theory of chaos and strange attractors. *Review of Modern Physics,* 57, pp. 617– 656.

Eiser, J.R. (1994) Toward a dynamic conception of attitude consistency and change. Vallacher, R., and Nowak, A. (eds.) *Dynamical systems in social psychology.* San Diego, CA: Academic Press

Festinger, L., Schachter, S., and Back, K. (1950) *Social Pressures in informal groups.* Stanford, CA: Stanford University Press.

Friedmann, J. (1966) *Regional development policy: A case study of Venezuala,* Cambridge, MIT Press.

Friedmann, J. (1973) *Urbanization, planning, and national development,* Beverly Hills: Sage.

Gardner, M. (1970) Mathematical Games. *Scientific American,* p. 9.

Glansdorff, P., and Prigogine, I. (1971) *Thermodynamic Theory of Structure, Stability and Fluctuations.* New York: J. Wiley.

Glass, L., and Mackey, M.C. (1988) *From Clocks to Chaos: The Rhythms of Life.* Princeton, NJ: Princeton University Press.

Grave, M., LeLous, M., Hevitt, W.T., (eds.) (1994) *Visualization in scientific computing.*

Berlin: Springer.

Haken, H. (1978) *Synergetics*. Berlin: Springer.

Haken, H. (ed.) (1982) *Order and chaos in physics, chemistry, and Biology*. Berlin: Springer.

Haken, H. (1983) *Advanced Synergetics*. Berlin: Springer.

Haken, H. (ed.) (1985) *Complex systems: operational approaches in neurobiology, physics and computer science*. Berlin: Springer.

Hao Bai-Lin (ed.) (1987) *Directions in Chaos*. World Scientific, Singapore.

Hillier, B., and Hanson, J. (1990) *The social logic of space*. New York: Cambridge University Press.

Hillier, B., and Penn, A. (1991) Visible colleges: Structure and randomness in the place of discovery. *Science in context*, 4, pp. 23–49.

Hegselmann, R. (in press) Modeling social phenomena by cellular automata. W.B.G. Liebrand, A.Nowak and R. Hegselmann (eds.), *Computer modeling of network dynamics*, Sage, in press

Hegselmann, R. (1994) Zur Selbstorganisation von Solidarnetzwerken unter Ungleichen – Ein Simulationsmodell. K. Homann (ed.), *Wirtschaftsethische Perspektiven I (Theorie, Ordnungsfragen, Internationale Institutionen)*, Berlin: Duncker and Humblot.

Hegselmann, R., and Nowak A. (1994) *The bargaining model of social interaction*. Unpublished.

Horgan, J. (1995) From complexity to perplexity. *Scientific American*, 272, pp. 104–109.

Hoyert, M.S. (1992) Order and chaos in fixed interval schedules of reinforcement. *Journal of the experimental analysis of behavior*, 57, pp. 339–363.

Iyengar, S., and McGuire, W.J. (eds.) (1992) *Explorations in political psychology*. Durham: Duke University Press.

James, W. (1890) *The Principles of Psychology*. New York.

Kando, T.M. (1977) *Social Interaction*. Sant Luis: C. W. Mosby Company.

Kapuściarek, I., and Nowak A. (1993) *Physical distance and social interaction among the resident of Warsaw*. Technical report.

Kelso, J.A.S. (1981) On the oscillatory basis of movement. *Bulletin of the Psychonomic Society*, 18, p. 63.

Kelso, J.A.S. (1984) Phase transitions and critical behavior in human bimanual coordination. *American Journal of Physiology: Regulatory, Integrative and Comparative Physiology*, 15, pp. R1000-R1004.

Kelso, J.A.S. (1988) Order in time: How cooperation of hands informs the design of the brain. H. Haken (ed.), *Neural and Synergetic Computers*, Berlin: Springer, pp. 305–310.

Kelso, J.A.S., Ding, M., and Schöner, G. (1991) Dynamic pattern formation – a primer. A. B. Baskin (ed.), *Principles of organization in organisms*. New York: Addison-Wesley, pp. 397–439.

Kohonen, T. (1988) *Self-organization and associative memory*. New York: Springer.

Krech, D., Crutchfield, R.S., and Ballachey, E. (1962) *Individual in society*. New York: McGraw-Hill.

Landau, L.D., and Lifshitz, E.M. (1964) *Statistical Physics*. Oxford: Pergamon Press.

Langton, C.G., Taylor, C., Farmer D.J., and Rasmumssen, S. (eds.) (1992) *Artificial Life II*. Redwood City: Addison Wesley.

Latané, B. (1981) The psychology of social impact. *American Psychologist*, 36, pp. 343–356.

Latané, B., Liu, J.H., Nowak, A., Bonavento, M., and Zheng, L. (1994) Distance matters: physical space and social influence. *Personality and Social Psychology Bulletin*, in press.

Latané, B., and Nowak, A. (1994) Attitudes as catastrophes: From dimensions to categories with increasing importance. R. Vallacher and A. Nowak (eds.), *Dynamical systems in social psychology*. San Diego, CA: Academic Press.

Latané, B., Nowak, A., and Liu, J.H. (1994) Measuring emergent social phenomena:

dynamism, polarization and clustering as order parameters of dynamic social systems. *Behavioral Science*, 39, pp. 1–24,

Latané, B., and Nowak, A. (in preparation) *The causes of polarization and clustering in social groups.*

Liebrand, W.B.G., Messick, D.M., Wilke, H.A.M. (1992) *Social Dilemmas*, Oxford: Pergamon Press.

Lewenstein, M., Nowak, A., and Latané, B. (1992) Statistical mechanics of social impact. *Physical Review A*, 45, pp. 703–716.

Lewin, K. (1936) *A dynamic theory of personality.* New York: McGraw-Hill.

Lorenz, E. (1963) Deterministic nonperiodic flow. *Journal of Atmospheric Science*, 20, pp. 282–293.

Losada, M., and Markovitch, S. (1990) Group Analyser: A system for dynamic analysis of group interaction. *Proceedings of the 23rd Hawaii International Conference on System Sciences, IV, Emerging technologies*, pp. 101–110.

May, R.M. (ed.) (1981) *Theoretical ecology: Principles and applications.* Oxford: Blackwell Scienific Publications.

May, R.M., and Nowak, M.A. (1992) Evolutionary games and spatial chaos. *Nature*, 359, pp. 826–829.

May, R.M., Nowak, M.A. (1993) The spatial dilemmas of evolution. *International Journal of Bifurcations and Chaos*, p. 3, pp. 35–78.

Meadows, D., Richardson, J. , Bruckmann, G. (1982) *Groping in the dark*, New York: Wiley.

Meehl, P.E. (1978) Theoretical risks and tabular asterisks: Sir Karl, Sir Ronald, and the slow progress of soft psychology. *Journal of Consulting and Clinical Psychology*, 46, pp. 806–834.

Messick, D.M., and Liebrand, W.B.G. (1995) Individual heuristics and dynamics of cooperation in large groups. *Psychological Review*, 102, pp. 131–145.

Moon, F.C. (1987) *Chaotic Vibrations.* New York: J. Wiley and Sons.

Moscovici, S. (1963) Attitudes and opinions. *Annual Review of Psychology*, pp. 231–260.

Moscovici, S., and Zavalloni, M. (1969) The group as a polarizer of attitudes. *Journal of Personality and Social Psychology*, 12, pp. 125-135.

Myers, D., and Lamm, H. (1976) The group polarization phenomena. *Psychological Bulletin*, 83, pp. 602–627.

Myrdal G. (1958) *Rich lands and poor: the road to world prosperity.* New York: Harper.

Noelle-Neumann, E. (1984) *The spiral of silence: Public opinion – our social skin.* Chicago: University of Chicago Press.

Newtson, D. (1994) The perception and coupling of behavior waves. R. Vallacher and A. Nowak (eds.), Dynamical systems in social psychology. San Diego, CA: Academic Press.

Nowak, A., and Latané, B. (1994) Simulating the emergence of social order from individual behavior. N. Gilbert and J.E. Doran (eds.), *Simulating societies: The computer simulations of social processes.* London: University College London Press, pp. 63–84

Nowak, A., Latané, B., and Lewenstein, M. (1994) Social Dilemmas exist in space. U. Schulz, W. Alberts and U. Mueller (eds.), *Social dilemmas and corporation.* Heidelberg: Springer.

Nowak, A., and Lewenstein, M. (1994) Dynamical systems: A tool for social psychology? R. Vallacher and A. Nowak (eds.), *Dynamical systems in social psychology.* San Diego, CA: Academic Press.

Nowak, A., Lewenstein, M., and Szamrej, J. (1993) Bąble modelem przemian społecznych (Social transitions occur through bubbles). *Świat Nauki*, 12, 28, pp. 16–25.

Nowak, A., Lewenstein, M., and Szamrej, J. (in press) Bubbles – a model of social transitions. *Scientific American.*

Nowak, A., Szamrej, J., and Latané, B. (1990) From private attitude to public opinion: a dynamic theory of social impact. *Psychological Review*, 97, pp. 362–376.

Nowak, A., Zienkowski, L., and Urbaniak, K. Clustering processes in economic transition. *Research Bulletin RECESS*, 3, 4, pp. 43–61.

Ostrom, T.M., Skowronski, J.J., and Nowak, A. (1994) The cognitive foundations of attitudes: It's a wonderful construct. P. G. Devine, D. L. Hamilton and T. M. Ostrom (eds.), *Social cognition: Impact on social psychology*. New York: Springer Verlag

Othmer, H.G. (ed.) (1986) *Non-linear Oscillations in Biology and Chemistry*. Lecture Notes in Mathematics, 66, Berlin: Springer.

Peitgen, H.-O., and Richter, P.H. (1986) *The Beauty of Fractals*. Berlin: Springer.

Pines, D. (1987) *Emerging synthesis in science*, Redwood City, Addison Wesley.

Rasband, N.S. (1990) *Chaotic Dynamics of Non-linear Systems*. New York: J. Wiley and Sons.

Ruelle, D. (1989) *Elements of Differentiable Dynamics and Bifurcation Theory*. New York: Academic Press.

Ruelle, D., and Takens, F. (1971) On the nature of turbulence. *Communications in Mathematical Physics*, 20, pp. 167–192.

Schuster, H.G. (1984) *Deterministic Chaos*. Weinheim: Physik Verlag.

Shaw, M.E. (1976) *Group Dynamics* (2nd ed.). New York: McGraw-Hill.

Shaw, M.E., and Constanzo, P.R. (1982) *Theories of Social Psychology*. New York: McGraw-Hill.

Schelling, T.C. (1969) Models of segregation. *American Economical Review*, 59, pp. 488–493.

Schelling, T.C. (1971). Dynamic models of segregation. *Journal of Mathematical Sociology*, 1, pp. 143–186.

Séror, A.C. (1994) Simulation of complex organizational processes: a review of methods and their epistomological foundations. N. Gilbert and J.E. Doran (eds.), *Simulating societies: The computer simulations of social processes*. London: University College London Press, pp. 19–40.

Sigmund, K. (1993) *Games of life: explorations in ecology, evolution and behaviour*, Oxford, Oxford University Press.

Skarda, C.A., and Freeman, W.J. (1987) How brains make chaos in order to make sense of the world. *Behavioral and Brain Sciences*, 10, pp. 161–195.

Stein, D.L. (1988) *Lectures in the sciences of complexity*. Redwood City, Addison Wesley.

Sperber, D. (1990) The epidemiology of beliefs. Fraser, C., Gaskel, G. (eds.) *The social psychological study of widespread beliefs*. Oxford: Clareden.

Szamrej, J., Nowak, A., and Latané, B. (1992) *Self-organizing attitudinal structures in society: visual display of dynamic social processes*. Poster presented at XXV-th International Congress of Psychology, Brussels, Belgium.

Tesser, A., and Achee, J. (1994) Aggression, love, conformity and other social psychological catastrophys. Vallacher, R., and Nowak, A. (eds.) *Dynamical systems in social psychology*. San Diego, CA: Academic Press

Thibaut, J.W., and Kelley, H.H. (1959) *The social psychology of groups*. New York: Wiley.

Vallacher, R.R., Nowak, A., and Kaufman, J. (1994) Intrinsic dynamics of social judgment. *Journal of Personality and Social Psychology*, 67, pp. 20–34.

Vallacher, R.R., and Nowak, A. (eds.) (1994) *Dynamical systems in social psychology*. San Diego, CA: Academic Press.

Vallacher, R., and Nowak, A. (in press) The emergence of dynamical social psychology. *Psychological Inquiry*, in press.

Weidlich, W. (1991) Physics and social science — the approach of synergetics. *Physics Reports, 204*, pp. 1–163.

Weidlich, W., and Haag, G. (1983) *Concepts and Models of Quantitative Sociology*. Berlin: Springer.

Weisbuch, G. (1992) *Complex systems dynamics*, Redwood City: Addison Wesley.

Whicker M.L.and Sigelman, L. (1991) *Computer simulation applications: an introduction.*, Newbury Park: Sage.

Wolfram, S. (ed.) (1986) *Theory and applications of cellular automata*. Singapore: World

Scientific.

Zaslavsky, G.M., and Sagdeev, R.Z. (1988) *Vvedenye v nyelinyeynuyu fizyku*. Moscow: Nauka.

DYNAMIC SOCIAL IMPACT

Robust Predictions from Simple Theory

BIBB LATANÉ
Department of Psychology
Florida Atlantic University
Boca Raton FL, USA

1. Introduction

In this chapter, I discuss several conceptions of simulation as a tool for doing social science and describe a specific approach, dynamic social impact theory, for understanding the self organization of society. I then discuss some ways in which simulation has and has not been useful in advancing this theory and compare it with leading examples of sucessful social science simulations based on approaches from statistical physics, microeconomics, and distributed artificial intelligence. I do this, not as a philosopher of science, a physicist, or a computer scientist expert in the logic or techniques of computer simulation. Instead I am a working social psychologist, writing about computer simulation from a user's perspective. My kind of social psychology is not just a subfield of psychology but an interdisciplinary focus on the individual human being as both the nexus of cultural, social, historical, economic, political, and biological influences, and the agent for societal change. People are the eyes and the brains and the arms of society, which, I believe, must be seen as a complex, self-organizing system whose nature can be explored with the help of computer simulation.

1.1. COMPUTER SIMULATION AS A SCIENTIFIC TOOL.

Physicists sometimes think of science as a stool with three legs — theory, simulation, and experimentation — each helping us interpret and understand the others. If we agree with Ruelle (1991) that science is simply taking a bit of mathematics and trying to paste it to the real world, theory can be seen as the realization of mathematics in terms of symbolic representations, simulation as the realization of mathematics in the circuits of a computer

R. Hegselmann et al. (eds.),
Modelling and Simulation in the Social Sciences from the Philosophy of Science Point of View, 287–310.
© *1996 Kluwer Academic Publishers. Printed in the Netherlands.*

chip, and experiment as the realization of mathematics in physical objects and processes. Thus, you can think of our physical universe as a giant device for the expression and evolution of a particular subset of all possible mathematical rules. If theory is regarded as a simplified map of that universe, computer simulation and experimentation can be seen as different ways of evaluating and correcting theory, with overlapping domains of usefulness depending on such things as the scale of the phenomena. Just as theory can be tested by simulation or by experimentation, simulation can be used to instantiate theoretical propositions or to describe experimental data. In fact, each of the legs — theory, simulation, and experiment — can be seen as intermediate between the others, and together they constitute a structure which would not stand without all three.

1.2. COMPUTER SIMULATION AS A LANGUAGE FOR EXPRESSING THEORY.

A somewhat different perspective was provided by my former colleague Thomas Ostrom (1988). Like many others, Ostrom saw science as the creative interplay between theory and data: you develop some theory, you collect some data, the data suggesting ways to modify the theory, the theory suggesting new data to collect. Ostrom proposed that just as there are different sources of data — everyday knowledge, systematic observation, and experiment — so too there are different ways of expressing theory, and he distinguished three:

1. Much of social science consists of *verbal statements.* Propositions are advanced, phenomena are described, processes are explained by stringing together words and sentences which refer to objects and events in the real world. These statements link together more or less coherently and constitute the theory. Derivations from the theory consist of more or less logical deductions or expectations which can be compared with data. Verbally stated theories may be better at dealing with uncertainty, fuzziness and ambiguity than mathematical models, and they are certainly necessary for the easy communication of theoretical ideas.

2. In some other sciences, especially economics and physics, theoretical statements are also expressed in *mathematical equations* such as Einstein's famous $e = mc^2$. The big advantage of a mathematical statement is that although it loses something for most of us in ease of intuitive understanding, it can be manipulated in certain well-specified ways to generate necessary consequences, which again can be tested against data. By expressing theoretical ideas in mathematical terms, we gain access to an armamentarium of analytic techniques for producing derivations.

3. Following this line of thought, computer simulation is simply a way of stating theory in the form of *computer programs*. The program consists of a series of symbols that can be taken to represent entities and processes in the real world. Running the program corresponds to deriving logical deductions from a verbally stated set of premises or solving a series of mathematical equations. Computer programs are especially adept at specifying complex conditionals and interactions exactly.

Ostrom's idea naturalizes computer simulation, suggesting that we should not regard it as anything magical but simply as another way of stating theory. Of course, it would probably not be a good idea to write your theoretical ideas *only* in the form of computer programs (just as it may not be wise to express them only in the form of equations). Not only will many people fail to understand them in that form, another advantage of using several languages is lost. Just as new insights sometimes come from trying to find new ways of expressing an old idea, it is often helpful to state something in a different form. Ostrom's view may locate simulation too much on the theory rather than the data side of the scientific road to suit everyone's taste, but there is no need to choose between these two or any other views of computer simulation. Different conceptions may suit your purposes at different times, and, as long as a simulation is well specified, everyone is free to draw their own conclusions from it.

1.3. COMPUTER SIMULATION AS AN EASY ALTERNATIVE TO THINKING.

Some social scientists disdain computer simulation as a quick and dirty substitute for creative thought or empirical effort, believing that simulators simply whip off some code based on ad hoc assumpt   nd let the computer do the work. "That's too easy", they say. "You   ldn't get a publication out of that!" Unfortunately, I have found comput simulation to be much harder than experimentation. Simulations ar   remely susceptible to error, and errors are extremely hard to detect. For example, the SITSIM program (Nowak and Latané 1994) I describe below has been independently checked by four highly skilled programmers, recompiled for several machines, and extensively tested by students in a course on computer simulation. Yet we still discover occasional bugs. Fortunately, most are in the sections of code involving the user interface, rather than the simulation itself, and none call into question the results I report here. I think we now have a quite reliable program, but I am pleased that its results are also qualitatively consistent with three other programs written in Pascal, three in FORTRAN, and one each in C++ and Objective C by five programmers and run on computers ranging from a Russian mini-computer, through several different types of PC, to a supercomputer. Although these programs differ somewhat in details and assumptions and thus their results may differ in a quantitative way, they are consistent in showing the same phenomena under the appropriate conditions. Rather than being a panacea, simulation is an opportunity for hard work and especially deep thinking.

1.4. COMPUTER SIMULATION AS A MACHINE FOR MAKING DERIVATIONS.

My own use of simulation, similar to Ostrom's, is as a derivation machine, a way to discover the consequences of theory. A major problem in science is we often fail to draw valid inferences from our theoretical ideas (Harris 1976). This problem is especially severe when it comes to phenomena that involve different levels of organization, such as when we try to infer the behavior of groups, institutions, or nations from the attributes and motives of the individuals that constitute them. The character of an organization is not simply an aggregation of the characteristics of its members, and the actions of a country may bear no necessary relationship to the personalities of its citizens. Verbal theories about the group-level consequences of individual behavior are particularly unhelpful when it comes to predicting what groups of people will do.

If you know something about such demographics as the age, sex, and race of a pool of potential jurors, you might be able to predict how each will be initially inclined to vote in the case of O.J. Simpson. If you have further information about such things as their experience with family violence, belief in science, and interest in sports, you might be able to predict how they will react to the testimony. By observing them further, you might even be able to estimate which will be leaders and which will be passive recipients of influence from the others. If you're an attorney for either the prosecution or the defense, such individual-level information might help you decide which jurors to challenge. Ultimately, however, you care not which individuals vote guilty, but rather what the final distribution of verdicts will be and whether the jury will hang.

Unfortunately, it is often difficult to predict what will happen as a result of social interaction and mutual influence even in such a small group as a jury. We need to know not only what the first-order influences will be, but what will evolve as individuals react to one another over time. The dynamics of larger groups and societies are undoubtedly even more complex. The problem is compounded by the fact that such social systems often are inherently nonlinear, because individuals display non-incremental, disproportionate reactions to particular pieces of information. This makes it difficult to apply many techniques of analysis developed for simple linear systems. In such cases, since the phenomena are complex beyond the current human capacity for logical or mathematical analysis, computer simulation may be the only alternative for generating predictions.

A computer simulation can be designed to grind out derivations from theoretical input. As with a sausage machine, the output can taste much better than the meat it starts with, but it still will be spoiled if it starts with rotten meat. It is not easy to make a computer simulation, and like scientific theories in general, most turn out to tell us little, or even worse, to tell us wrong. If all goes well, however, a simulation may turn out results that confirm what we already know about society. Even better, it may lead to theoretical "discoveries", predicted phenomena that we might not otherwise have thought to look for.

It is often claimed that computer programs provide a built-in check for gaps and inconsistencies in your thinking because your computer simply won't run a deficient program. Unfortunately, your computer's reaction to incompletely specified conditionals or contradictory commands is often simply to execute whatever comes first, so it is not wise to count on the computer for checking the coherence and completeness of your theory. A more important advantage is that by systematically varying them, you can determine which features of your theory are necessary to your derivations (sensitivity analysis) and which make little difference (determination of robustness).

2. Simulating Dynamic Social Impact

Social impact, borrowing from Gordon Allport's definition of social psychology, is "any of the great variety of changes in physiological states and subjective feelings, motives and emotions, cognitions and beliefs, values and behavior, that occur in an individual, human or animal, as a result of the real, implied, or imagined presence or actions of other individuals" (Latané 1981, p. 343). Many specific social processes lead to influence. For example, people may accommodate their speech patterns to express solidarity with their listeners (Giles 1984), they may respond to direct or implied bribes or threats (Tedeschi 1983), or they may persuade each other with rational arguments (Eagly and Chaiken 1993). These processes differ in such important respects as whether it is the talker or the talkee who is most influenced or even whether they require verbal communication at all, what attributes are influenced, what constitutes strength, and what is the time scale and permanence of change.

However, following a field theoretic conception of social impact (Latané 1981), all such social processes seem to be governed by a single principle: Social impact is a multiplicative function of the strength, immediacy, and number of people influencing an individual: $\hat{\imath} = SIN$, where $\hat{\imath}$ is the total impact experienced by an individual, S is the average strength (attractiveness, status, persuasiveness, etc.) of the other members of the group, I is their average immediacy (closeness in space or time, visual as well as vocal contact, etc.) and N is simply the number of individuals who are the source of impact.

This principle has been shown to hold in a variety of social situations, including determining what people are interested in, how nervous they will be before an audience, whether they will intervene in an emergency, work hard in a group, or tip in a restaurant, and, in particular, how much they will respond to majority and minority influence (Latané and Wolf 1981; Wolf and Latané 1983). The domain of the theory is any socially influencable attribute of an individual, including the development of norms and conventions such as sex roles, life style behaviors such as wearing ties, smoking, and exercising, emotions and emotionally tinged behaviors such as rudeness and hostility, moods such as consumer confidence and depression, and values such as the development of a strong work ethic, as well, of course, as political and social attitudes. The problem comes when we ask "What if we have a population of people, each influenced by and influencing each other?" In other words, how can this set of static principles be transformed into a theory of dynamic social impact?

2.1. DYNAMIC SOCIAL IMPACT.

Imagine 400 people, differing with respect to a wide variety of individual attributes but living in an unusually regular social environment. Figure 1 is an aerial view showing their positions on three different attributes before social influence: their accent (how they pronounce "arm"), their drink preferences (do they like ale) and their sexual practices (whether they use condoms). Each happy face[1] is a person. Their attributes are represented as simple, bipolar choices with the open faces representing the 60 % majority view, and the dark faces the 40% minority on each issue. The placement of the faces represents their location in social space, which is important because they should be influenced most by their neighbors, less by those who live further away. Initial attitudes are random with respect both to their distribution in space and across issues. Finally, although it cannot be seen in the figures, the people differ in how influential they are; some being more educated, articulate, attractive, and/or powerful than others.

What happens as these people influence each other? Presumably, any given individual will be swayed both by his or her own taste and experi-

---

[1] For technical reasons, the "happy faces" had to be replaced by open circles, and the "dark faces" by black circles. (Editors' note)

Figure 1. Initial values of 400 people differing in strength with respect to three socially influenceable attributes. Each face represents a person whose location in space is fixed. Each will be influenced by the others in proportion to the relative strength, immediacy, and number of people holding each position.

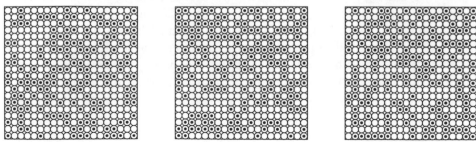

ence and by the influence of other, nearby people. However, the specific nature of social influence may differ for each attribute-sometimes involving persuasion, sometimes accommodation, sometimes audience-tailored self-presentation, and a given individual may be "stronger" with respect to some processes than others. Further, it is difficult to predict what will happen as each person in turn affects the others in his or her environment. For example, consider the drink preference of the fourth person in the third-to-last row-a majority member who happens to be surrounded by people with an opposite preference and would seem relatively likely to switch. The neighbors, however, are each themselves vulnerable to influence and may themselves change, leaving the target individual again isolated and susceptible to pressure. If we can make the simplifying assumption that people faced with disagreement change incrementally in proportion to the degree of difference in opinion, the problem can be solved analytically (Abelson 1964), but if the change rule is nonlinear, the problem becomes much more complex and can presently be solved only by computer simulation.

For the last five years, with the help of a number of colleagues (including most notably Andrzej Nowak), I have been studying the group-level consequences of different rules of influence in simple social systems. We approach the question by breaking it into two components. First, what will be the effect on any given individual of being exposed to the views of those around them at a given moment in time? Second, what will be the emergent, cumulative result of the iterative, interactive, recursive operation of these individual processes on individuals and on the groups as a whole? Basically, we use SITSIM, a Pascal program described more fully by Nowak and Latané (1994), to calculate, for each person in turn, whether they will change their opinion as a result of the influence of those around them.

The pressure on a person to change opinion is calculated as a multiplicative function of the strength, immediacy, and number of people who disagree. The pressure to stay is calculated as a similar function of the strength, immediacy, and number of persons, including the self, who agree. A person will change if and only if the pressure to change, plus a random term representing personal tastes or experiences, outweighs the pressure to stay (note that this is a nonlinear change rule). Changes are computed separately for each person in turn, until the system stabilizes with respect to group parameters (individuals of course may continue to change). Changes are computed separately for each attribute or issue; the only constant is the starting distribution of strength. Finally, the computer looks for signs of emergent, group-level organization arising from these purely individual processes.

Figure 2.   Attributes after 30 rounds of influence with high noise (dynamic equilibria). The minority position with respect to each attribute initially becomes reduced in size, but spatial clustering protects it from further erosion. Clustering also results in the emergence of correlation among attributes.

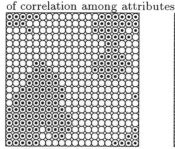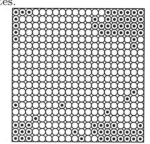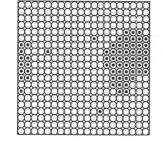

2.2. A SELF-ORGANIZING SYSTEM.

Please note four important results in Figure 2, which shows predicted opinions after discussion: (1) The system achieved stable diversity. The minority was able to survive, contrary to the belief that social influence inexorably leads to uniformity (eg. Abelson 1964; Woelfel and Fink 1980). Individuals are still changing, but the size of the minority has become more or less constant. (2) Nevertheless, the minority is reduced, from 40% to an average of 23% (as shown on the top right of each array). (3) Attitudes have become spatially clustered, not through individuals changing their location, but simply through attitude change processes. It is this clustering that protects minorities from extirpation, since only those people on the edges of clusters are exposed to the social pressure of contrary viewpoints (Latané in press a). Latané *et al.* (1994) developed a clustering index, shown on the top left of each array, based on the probability that neighbors share a common view as compared to individuals spaced further apart, normalized to equal 0 if the system is randomly ordered and 1 if the system attains its maximum possible order.

Finally, although it cannot be seen in Figure 2, (4) attitudes have become correlated (Latané in press b). Although pre-discussion attitudes were randomly assigned, the social influence process leads to the development of relationships among them, as shown by the emergence of correlations between attributes. The degree of correlation is small (the average correlation under the conditions above is only .14) but it is pervasive (over two-thirds of relationships are statistically reliable at the .05 level). This primitive form of ideology emerges from nothing-there is no content to these attitudes and no reason for them to be correlated other than that they reflect social processes occurring in a social matrix outside the individual minds.

2.3. SIMULATION, THEORY, AND EXPERIMENT.

It is often assumed that the true test of a scientific model, whether it is a verbal or mathematical theory or a simulation, is whether it can predict empirical outcomes. Yet, I have found relatively little occasion so far to compare the results of these simulations of dynamic social impact directly to data from the real world. Instead, in addition to making unexpected theoretical discoveries such as the predicted emergence of clustering and correlation and the ability of social systems to maintain stable diversity in the face of strong pressures to uniformity, I have found simulation especially useful for three other non-empirical purposes:

1. Determining whether the theory is robust with respect to variations in stochastic and other theoretically uninteresting variations (models that are not robust in this way are useless and probably wrong).
2. Identifying which elements of the theory are critical to the resulting dynamics (by narrowing the focus of the model, we can separate the theoretical wheat from the chaff, avoiding controversy about irrelevant details and achieving a maximally simple or parsimonious model).
3. Extending the theory by adding complexity and recursion.

A forthcoming symposium in the *Journal of Communication* (Latané in press c) relates dynamic social impact theory to recent developments in social networks, cognitive psychology, social representations, and evolutionary biology, suggesting specific ways in which the theory can be extended in scope. Here, I describe briefly some tests of robustness and sensitivity.

3. Determining the Necessary and Sufficient Conditions for Polarization and Clustering

Latané, and Nowak (in press.) explored the necessary and sufficient conditions for the emergence of clustering and incomplete polarization by conducting hundreds of thousands of simulations factorially combining conditions and procedures. Although they could not test every possible combination of factors which might affect polarization and clustering (indeed, a single iteration of a design including just the 19 SITSIM variables, with two to five values each, would require 200 billion simulations and some 4 million years on a 386 PC to complete), they were able to draw some strong conclusions.

3.1. ROBUSTNESS.

First, these phenomena are robust under a wide variety of assumptions, parameters, and procedures for computer simulation systematically chosen to sample their possible range: they occur for two quite different versions of the basic formula, four different assumptions concerning the coupling and reassignment of strength parameters, three different methods of modeling parallel processes, in groups with or without borders, in groups ranging in size from nine to sixteen hundred people, and with initial minorities ranging from 10% to 50%. Although many of these factors have quantitative effects, they do not control the basic qualitative outcomes. I doubt whether other variables will be found to limit the generality of the qualitative results to any great extent.

Most remarkably, these phenomena, rather than being fragile results of an overdetermined system, are actually enhanced by looseness and variation in the determination of individual behavior. In the simple system I have described, attitudes are a joint function of social influence and individual experience represented as a random term added to the change formula. Surprisingly, increasing the size of the random component actually increases the degree of self-organization. Compared to zero-noise systems, randomness increases clustering, increases both the mean and variability of polarization, and increases the degree of correlation among attitudes. Even at extremely high degrees of randomness, the system retains some organization with respect both to spatial clustering and cross-issue correlation. Why? Because, randomness makes local equilibria unstable, allowing a higher degree of global organization to emerge (Latané, in press).

3.2. CRITICAL ELEMENTS.

Second, it seems that at least three factors are essential for the maintenance of organized diversity: (1) People must be located in some sort of social space such that they have more influence on neighbors than on strangers, allowing clustering to occur. (2) Individual attitude change processes must have some degree of nonlinearity associated with them, rather than being simply incremental responses to social pressures, preventing the system from compromising itself into uniformity. (3) There must be individual differences in strength among the people in the population, such that "strong" individuals can anchor the borders of minority clusters, preventing them from eroding into the majority sea. Eliminating these features leads to large qualitative changes in the expected outcomes of social influence.

Thus we can understand the results: Briefly, clustering is a necessary consequence of the fact that people are most influenced by their neighbors in social space, polarization results from the minority being necessarily more exposed, nonlinear change prevents people from simply compromising, and clusters protect minority members in their interiors from exposure to the majority position while strong personalities anchor the borders, allowing the maintenance of stable diversity. Finally, correlation results from the fact that social influence reduces the degree of independence in the system, increasing the variation in correlation coefficients, and thus their average absolute value.

These results, of course, are not empirical findings. Rather they should be regarded as theoretical discoveries — necessary consequences of a simple, relatively non-controversial theory of individual social influence applied iteratively to a complex social system. Although the underlying theory is boringly simple, it leads to unexpected, counter-intuitive predictions which should have important consequences for society. The discovery that stable diversity and increasing correlation are characteristic of this class of dynamic nonlinear social systems has considerable generality — any system which shares the basic characteristics outlined above should behave similarly.

4. Comparing Simulations to Data

Although comparison with data may not be a necessary component of a program of research on social simulation, I believe simulation should also spend some time on the data side of the scientific road, or it is in danger not only of becoming too theoretical and abstract but of simply being wrong. In particular, we should try to find out whether human social systems possess the characteristics and behave the way the theory says. In my own research program, we are pursuing empirical investigations of four types:

1. *Checking the theory of the agent.* A major effort has been devoted to finding out whether the individual-level assumptions made by the simulation model are reasonable. Guided by the critical features as determined from the simulation results, my colleagues and I have focused our investigations on two issues. First, to what extent is influence determined by distance? We have found strong evidence that distance does matter when it comes to social impact, even in populations varying greatly in access to advanced communication and transportation

technology (Latané *et al.* in press). Second, under what conditions is change nonlinear? Surprisingly, according to a new catastrophe model, this seems most true for important rather than unimportant issues (Harton and Latané 1995; Latané and Nowak 1994; Liu and Latané in press). These new results, together with the extensive research which went into the original development of social impact theory (Latané 1981), reassure us that the individual assumptions of the model provide a good representation of how individuals actually affect each other in society.

2. *Determining whether predicted phenomena actually emerge.* A key question is whether the results of the simulation are "correct" in the sense that they accurately capture the nature of social organization in the outside world. Certainly they show an intriguing similarity to some familiar phenomena. For instance, polarization seems to characterize many elections, which seem "too close to call" just weeks in advance, but turn out not to be close at all, as voter preferences "crystallize" in the final days (Crespi 1988; Noelle-Neumann 1984). Clustering, too, is pervasive — almost too obvious to notice. Consider, for example, the distribution of religious attitudes shown in most world atlases. Religious clusters, of course, represent social processes taking place over a time scale of centuries, and the forms of influence include violent as well as friendly persuasion. Perhaps the most familiar empirical example of clustering clearly caused by short-term social influence comes from Festinger, Schachter, and Back (1950). This seminal study documented the emergence of spatially clustered attitudes toward a tenant's council in a student housing project for WWII veterans at MIT, even though all residents faced highly similar life circumstances and housing units were randomly assigned. Thus the predictions of dynamic social impact theory seem qualitatively correct.

3. *Qualitative modelling of different types of issue and geometries of social space.* Unfortunately, since society exists on such a large scale, it would be expensive and logistically difficult to test the more specific predictions of the model on this scale. Instead, we have conducted experiments on actual groups of 12-24 people communicating by e-mail about both trivial and important social and political questions, including judgments on jury cases, using communication geometries with such names as Torus, Ribbon, and Family (Jackson and Latané 1995; Latané and Bourgeois in press; Latané and L'Herrou in press). Studies have now been done on hundreds of people and dozens of issues, and they show strong evidence for the emergence of polarization, clustering, correlation, and stable diversity. Further, they suggest that the phenomena will actually be enhanced in the presumably "clumpi-

er" social spaces of the outside world.

4. *Quantitative predictions of specific groups and individuals.* A final level of testing is to see whether the model can predict the post-discussion opinions of individual members of actual groups of people from knowing their initial positions on a variety of different topics. This would be a remarkable feat, if it worked. Wim Liebrand, Martin Bourgeois, and I are trying to see how far we can carry such precise prediction.

5. Comparing Dynamic Social Impact with Other Approaches to Simulation

To conclude, I would like to consider some other approaches to social simulation. The simulations I shall describe come from different traditions — statistical physics, microeconomics, and computer science — but, like the dynamic social impact model from social psychology, they are all reductionist or "bottom-up" in that they conceive group-level order as emerging out of individual interactions and try to derive the macroscopic development of social systems from the rules governing the behavior of their individual elements. The system is more than the sum of the parts only because of their interaction; order results, not from external or superimposed influence, but from within. In other words, the systems are self-organizing.

5.1. STATISTICAL PHYSICS: TOP-DOWN ANALYSIS THROUGH SYNERGETICS.

Developed around the turn of the century, statistical physics explains how the magnetic spin of individual molecules of iron or the movement of individual particles of a fluid are jointly determined by the physical forces exerted by the neighboring particles and by some stochastic individual fluctuations. Like the Ising model of ferromagnets, the main rule of dynamics is for each element to adopt the state of its neighbors. The system "loses" information as it evolves, in that the number of independent parameters needed to describe it is reduced. This dynamics is a "relaxation", in that as time progresses, each element no longer maintains a difference from its neighbors. As a relaxation, it converges on a stable point attractor and does not display more complex or chaotic dynamics.

In the presence of randomness (noise), such models belong to the field of statistical physics, and the tools of statistical physics can be applied to analyze the dynamics of the model's behavior. Although formulated on the micro level, such models can be analyzed on the macro level in the form of mean field equations. Such equations are much more complex than the simple logistic equation, but if they can be solved, they allow the analytic derivation of the distribution of individual positions.

The eminent physicist Hermann Haken (1988) has developed "syner-getics", an approach to nonlinear dynamics that describes how systems organize themselves through feedback as the individual elements together produce a mean field which in turn determines their behavior. Fast-moving individual processes become determined or "entrained" by slower, global ones, reducing the degrees of freedom in a system, as they each become de-pendent on the state of the system as a whole. For example, the contraction of different muscles in a given individual become coordinated in the beating of a heart, breathing, and other physiological systems. This entrainment al-lows the action of individual elements to culminate in such complex feats of coordination as temperature regulation and bodily movement.

In like fashion, the behavior of individuals in a society may become en-trained by slow moving social trends, as each individual reacts to the social "field" created by all the other members of the society. Physicist Wolf-gang Weidlich (1991, see also Helbing 1995) and political scientist Klaus G. Troitzsch (this volume) have shown how to develop mathematical models based on the master equation of statistical physics to analyse such cases. Conceiving social trends as inputs to as well as outputs of individual at-titudes, Weidlich and Troitzsch use stochastical simulations of a nonlinear Langevin equation to predict the probability density function of attitudes. In this case, individuals are not seen as interacting directly with each oth-er, but rather as reacting to the global state of the system, which in turn represents the net effect of many such individual changes. A coupling pa-rameter, κ, is used to represent the degree to which individuals want to concur with the global majority or mean social field. As in other top-down approaches, individual behavior is assumed to be dependent on or deter-mined by the macroscopic (aggregate/average) dynamics of the system as a whole.

One problem with this model is specifying to what population members of a group are trying to couple, since they may be led in very different di-rections depending on whether they attempt to adopt the views of other residents of their city, their province, their country, or the world. A sec-ond problem comes in imagining how individuals could come to know and thus couple themselves to whichever global majority they select. "Morphic resonance", a form of ESP proposed by Sheldrake (1989) provides an ex-planation, but it is unlikely to be accepted by the primary audience for this class of models. A more plausible possibility is that public opinion polls and the media provide a mechanism for social entrainment, serving as a (possibly biased) mirror reflecting the master trends of the society back to individual citizens. In this regard, the model is consistent with the appar-ent perspective of many political scientists, who act as if public opinion were determined primarily by the aggregate response of self-interested individuals interacting only with their TV sets. Thus, top-down theories such as Troitsch's and Weidlich's may be best suited for modern societies and big issues where the mass media can serve as the physical embodiment of a global feedback mechanism.

The Troitzsch models have been implemented in a well-designed and sophisticated package called MIMOSE, which is particularly useful for teaching. Because of their origin in very large physical systems, they can be solved for very large populations of people and these solutions can be compared to data from sample surveys. The major cost of these advantages is the simplifying assumption that people are homogeneously equivalent. In order to apply the analytic techniques, spatial location and other individual differences among the members of the population must be ignored (although some can be brought in by aggregating different subpopulations). As we know from dynamic social impact theory, social influence under these conditions should always result in unanimity, as spatial clusters cannot protect minority opinions from elimination. When applied to attitudes, the Troitzsch model indeed does predict that, for reasonably high levels of k, social systems always will converge on unanimity at one or another extreme of the initial distribution. This predicted outcome, ultimate universal agreement, has so far been elusive in most modern societies.

The dynamic social impact model is also similar to a ferromagnet, but in a random magnetic field. Individual differences can be seen as analogous to small random magnetic fields affecting individual elements. The existence of self-influence makes our model "non-Hamiltonian" and in this respect it is basically different from ferromagnets. These features make it difficult to apply the analytic techniques of statistical physics (but see Lewenstein *et al.* 1992). These features also make the dynamics different from a ferromagnet (or from the Troitzsch models) in that they develop stable diversity and do not converge on unanimity.

5.2. MICROECONOMICS: CELLULAR AUTOMATA AND THE PRISONER'S DILEMMA.

Recently, scientists from a variety of disciplines have become interested in the application of cellular automaton models (von Neumann 1966) to dynamical systems. In a cellular automaton, spatially distributed entities are each affected by their nearest neighbors according to a well-specified set of rules to produce complex patterns in the system as a whole. In the classic tradition of social automat, entities reside on a two-dimensional grid, each with four (North, South, East, and West) or eight (plus Northeast, Southeast, Northwest, and Southwest) neighbors. cellular automata usually have no memory, that is, your situation at time $t + 1$ is determined only by the state of your neighbors at time t. In typical cellular automata, elements are affected only by their nearest neighbors, do not influence themselves, and do not differ with respect to strength. They can be considered "multi-agent" in that the elements of the system respond only to the other individuals in their immediate local environment and not to the global properties of the system as a whole (indeed, the elements do not have any information about the state of the global system). However, the agents typically obey very simple rules.

cellular automata are often conceived with arbitrary rules or rules chosen after the fact. In John Conway's famous Game of Life, for example, a cell is "born" only if surrounded by exactly three living neighbors and "lives" only with two or three neighbors, otherwise being too lonely or too crowded. This arbitrary rule, which leads to the creation of wonderful shapes and patterns such as gliders and battleships, was obviously chosen for its consequences, rather than a priori. Although there are a great variety of approaches in the tradition of research on cellular automata, these studies often attempt to systematically explore all possible rules in the search for interesting dynamics, and, to keep the task manageable, are thereby led to limit themselves to restricted sets of possibility. Wolfram (1986), for example, systematically examines the outcomes for all 256 possible rules for two-state cells communicating only with their two nearest neighbors in a one-dimensional space. With a larger or less regular space or interactions at longer distances, or more attributes of each cell, the number of possible rules quickly becomes astronomical. For example, even the simplest of 2D automata have literally billions of possible rules. Therefore, some method is needed for choosing among them.

The rules of the dynamic social impact model were chosen to accord with hypotheses about the psychology of human beings rather than with more abstract mathematical imperatives, to represent a plausible description of what people do, rather than because their application leads to a plausible outcome. Another interesting set of rules has its roots in Game theory, which, like cellular automata themselves, also originated with von Neumann (Neumann and Morgenstern 1944). Game theory provides prescriptions for rational action in a variety of social situations, although it is debatable how often people follow such rules. In perhaps the earliest application of game theory to cellular automata, Axelrod (1984)) demonstrated that with a strategy of tit-for-tat, cooperation can survive and even prosper in a two-dimensional world where people repeatedly play a prisoner's dilemma game with their neighbors in social space. The system does this by forming clusters of cooperating individuals who reward each other sufficiently over the long term to compensate for the immediate cost of cooperation.

Hegselmann (this volume) simulates an N-person prisoner's dilemma in which each person can take 1 unit of reward for herself or cause x units to be distributed among all her neighbors in proportion to their distance. Like the 2-person case, such a dilemma is defined by two characteristics for all $x > N$: at any given time it always pays more to take than to give, but a neighborhood of givers will always be better off than a neighborhood of takers. Because your actions may have consequences for what your neighbors will do, it becomes difficult to decide what action is rationally best for you — taking pays, but giving allows you to set a desirable example for others. Hegselmann describes some consequences of one possible rule of bounded rationality — imitating the most successful person or strategy in your neighborhood. This social comparison heuristic should allow people to achieve a satisfactory outcome without having to engage in complex negotiation or calculation. Hegselmann has conducted many simulations varying the spatial and temporal horizon as well as a number of versions of this decision rule.

Hegselmann reports that all these experiments evolved clusters, islands of cooperation in a sea of defectors. Although there was an initial breakdown of cooperation as sparsely distributed cooperators converted to defection, any initial clusters of cooperators survived and even grew. By comparing Hegselmann's Figs. 7 and 9 (page 221), we see that the scale of clustering is increased with a greater spatial or temporal horizon. Sometimes, when influence is proportional to $1/d$ (which results in more net influence from people at a distance), the islands of cooperation become so large that they resemble an aerial view of the Netherlands, with large areas of fertile land drained by narrow canals of defection (Fig. 8, page 221).

Liebrand and Messick (this volume) have studied the consequences of bounded rationality with a tit-for-tat rule and two social comparison heuristics based on the idea that people are motivated not by the direct utility of their choices, but whether they earn more points than their neighbors. They also restrict their analysis to short-range influence (nearest neighbors in the immediately preceding time step) among members of a completely homogeneous population. With such a competitive rule, of course, it is hard for cooperative clusters to form, for occasional defection may be the only way to get a satisfactory number of wins. Perhaps for one or another of these reasons, they found substantial clustering only with tit-for-tat rules.

The Hegselmann and Liebrand results suggest the desirability of extending the dynamic social impact model to account for how externally imposed payoff structures can shape social influence. In the case of dilemma games, influence is likely to be asymmetric. Cooperators are like herds of sheep — happy to be similar to their neighbors because they all contribute resources to one another. Defectors, however, are like wolves, happier the more sheep are in their neighborhood and the fewer other predators to compete with.

Cooperators win by being similar to their neighbors, defectors by being different from theirs. This research tradition might benefit from introducing more heterogeneity into the descriptions of the cells, allowing them to develop something of a personality.

Cellular automaton models can be made quite flexible, even allowing for movement in space. Schelling's (1971) checkerboard model of neighborhood "tipping" shows that individuals moving to the nearest place where they will not be in the minority soon find themselves living unwittingly in a completely segregated society. Hegselmann (1994) has extended such migration models to predict emergent clusters of people willing to provide social support for one another. In society, people do indeed move, but usually not very often nor very far. Although such movement is not necessary to the development of clustering, it should increase its scale.

Formally, the dynamic social impact model may also be described as a cellular automaton with a weighted stochastic majority rule, in that people are influenced primarily by their local environment. It differs from typical cellular automata in that the state of each cell depends on *all* the other cells, the degree of dependence decaying with distance, individual differences in strength make different cells react differently to the same external situation, and self-supportiveness and individual experience allow individuals to resist influence from their social environment. Simulations of dynamic social impact under conditions characteristic of a standard cellular automation always show an immediate formation of clusters, but these clusters tend to be small in scale and gradually erode at their edges until the system unifies. This result shows that the present model generally leads to different predictions than simple cellular automata, and again reaffirms the importance of being able to represent in a simulation not only space, but also non-homogeneous populations of people.

5.3. COMPUTER SCIENCE: DISTRIBUTED ARTIFICIAL INTELLIGENCE (DAI).

In a fascinating attempt to simulate the world of 15–30,000 years ago, Doran *et al.* (1994) developed a manual for how to organize a Paleolithic mastodon hunt. In their program, written in Prolog, "agents" structured as dynamic production systems with rules connecting sensory input to action move about a changing landscape, harvesting food and interacting with one another. Agents have "beliefs" about other agents and their social relations, communicate "plans" about which resources to target to one another, and follow "rules" designed to help them choose and coordinate their activities. Rules cover such matters as the optimal size and organization of hunting parties, how to arrange a rendezvous while avoiding accidental interference from other parties, who to select as partners and how far to trust them, and how to divide the spoils — a total of 54 different rules altogether. Although the model can in principle be tested by comparing the predicted dispersion of mastodon kills with the spatial distribution of weapons and bones found at archeological sites, most of us will undoubtedlyt evaluate it with respect to face validity — does it provide a recognizable account of what we imagine to be a primitive society?

When I first heard about this model I was stunned by its impressive ability to model a large numer of intelligent agents interacting in a life-like landscale. I was also astonished by the large number of rules written into the system. Why so many?

I believe the answer may reflect a mind set related to its disciplinary background. The Doran *et al.* model originated in the artifical intelligence tradition of computer science, which is rooted in engineering. Engineers are trained to make things work. A typical engineering task might be to design a lunar lander able to move about on the surface of the moon, gather and analyze samples, and communicate the resulting information back to Earth without being incapacitated by cold, airlessness, radiation, or weightlessness. Failure in any respect will waste billions. The solution is often to design many redundant subsystems, covering every aspect of performance. Although complexity is not an end in itself, it often is the best way of maximizing the chance that a system will work under a variety of demanding conditions.

The artificial intelligence tradition seems to have developed a similar emphasis on performance. In the popular mind at least, the prototypical challenge for artificial intelligence is the Turing test — can you write a computer program that will be indistinguishable from a human being? As in building a working robot, you need satisfactory performance with respect to each of a number of complex criteria — vocabulary, syntax, factual knowledge, humor, politeness — and success would seem to require writing comparably complex programs. Likewise, many people believe that one cannot capture all the rich complexity of social life in just one or two simple principles or that a pretzel-shaped universe requires pretzel-shaped theories. From this perspective, it is easy to see why one would be tempted to keep adding subsystems when trying to incorporate the key processes of society in a computer program, resulting, of course, in a rich set of rules.

Scientists are trained to understand phenomena. To succeed, you need not explain everything at once, but simply to explain something. The prototypical challenge in the scientific community is to formulate or confirm an interesting hypothesis. The question is not usually whether a theory fails to cover some of the many aspects of reality, but whether it succeeds in covering at least one. From a scientific perspective it would seem that the more rules you include, the greater the chance that at least one will be wrong (the chance for error probably grows exponentially, since the effects of each rule may interact with all the others). With a more parsimonious set of rules, there might be less chance for error.

The goal of a scientific simulation is to simplify and abstract. You look for the core, central features of the system and you leave out all the extraneous elements. The problem is to decide what can be left out and what must stay. Sensitivity analysis can be used to tell. If you write a simulation program that is rich with detail and includes a lot of rules, you may be able to duplicate many aspects of the social system that you are studying. This strategy will allow you to guard against critics who point to imagined complexities in social life to refute your model (it also makes it harder for people to understand and thus rebut your model). A problem with including many rules, however, is that they make the simulation overdetermined. With enough parameters and variables, one can reproduce anything. My own preference is obviously for the simplest plausible rule sets.

Unfortunately, models often leave out one or both of two features which our simulations show to be absolutely critical to the most interesting dynamics — namely spatial location and individual differences. Representing spatial location makes it possible to understand how clusters can emerge from both social influence and migration, and how these clusters enable individuals to perceive themselves in a local majority with respect to their practices, beliefs, and values, even when they are in the minority with regard to some larger definition of the population. Individual differences provide mechanisms for the maintenance of stable diversity, as the edges of minority clusters may be anchored by strong individuals. Regrettably, these features make it impossible to easily adapt analytic techniques from statistical physics, but with the increasing availability of computational power, this limitation is not debilitating.

Science, of course, is not the only use of simulation. In one of the first applications of simulation to social phenomena, Pool *et al.* (1961) attempted to model the outcomes of community referenda on whether to fluoridate the water supply. Their simulation ultimately included 473 rules, refuting any thought that first steps are necessarily simple ones. This project reminds us that simply *understanding* the mechanisms of social life may not be the only motive for simulating social processes. In fact, the day may be all too close when simulations are used to predict the future time course of cultural trends. As with current weather forecasting systems, one can imagine an extensive array of spatially distributed social measurement stations each feeding information into a central computer as it generates daily forecasts of whether the attitudinal climate will be sunny or cold, what ideological storms next will sweep across the intellectual landscape, or where the next outbreak of nonconformity will arise. Such forecasting models will probably need to be extremely complicated and may include many thousands of complex rules relating demographic and economic factors to specific beliefs and practices as well as detailed information about their clustering and correlation.

Those who are less than enthusiastic about such a vision may take comfort in the fact that the science and measurement of public opinion still lags far behind our understanding of the physical systems that create weather. Technology is quickly providing the computing resources to make such prediction possible, however, while new developments in genetic algorithms open the possibility of simulation programs that write themselves, ever adapting to feedback and evolving to make more and more precise predictions. Unfortunately, although such an oracle might be extremely useful in predicting what will happen, it would contribute nothing to understanding why!

6. Conclusion

It is my belief that multi-agent systems are undoubtedly the way of the future in computer simulation, and they will certainly become ever more complex. Some of this complexity is highly desirable. In particular, I believe simulations should include the capacity to model large numbers of spatially distributed individuals differing in a variety of ways, as is now done by dynamic social impact theory. In addition, they need more satisfactory ways of interpreting and representing social space (Latané *et al.* in press) as well as the ability to incorporate individual motives based on externally imposed payoff structures, allow individuals to perceive the emergent spatial clusters and correlated attributes of their social world, and on the basis of these perceptions form social representations, stereotypes, and social identities. These perceptions and identities, in turn, should lead to the development of more complex forms of social space which rechannel the exertion of social influence, leading ever to higher levels of social self-organization such as coalitions, alliances, hierarchies, organizations, and institutions. These features, of course, should have to prove their theoretical worth through rigourous sensitivity testing, being repaired, replaced, or removed if they make little difference in the outcomes of simulation or lead to trivial or incorrect predictions. Not every theoretical innovation will lead to robust predictions; through computer simulation we can simplify the rules. To grind good theory, you should continuously maintain and sharpen your derivation machine.

In this chapter, I have discussed a number of computer models of society from a user's perspective. These models each reflect their origins in their underlying assumptions — of aggregate homogeneous particles in the case of statistical physics, of rational decision rules in the case of game theoretic cellular automata, of richly specified spatially distributed agents in the case of artificial intelligence. I believe these models show great promise, especially when taken together, and that they have much to learn from each other. Statistical physics provides a general force field conception plus analytic techniques useful under limited circumstance. cellular automata introduce the notion of space, but in too limited a fashion and with too limited a repertoire of individual variations. In its conception of agents as distributed computing entities, DAI makes us aware of the need to understand the internal processes of individual agents, but is in some danger of proliferating ad hoc prescriptive principles. It is up to pychology to contribute a workable theory of the individual human actors whose responses to one another culminate in our complex social world.

7. Acknowledgement

Preparation of this chapter was supported by grants from the National Science Foundation and the Netherlands Organization for Scientific Research and was completed during a stay at the Netherlands Institute for Advanced Study in Wassenaar. I thank Rainer Hegselmann, Dirk Helbing, and Wim Liebrand for helpful comments on an earlier version. Address comments to Bibb Latané, Department of Psychology, Florida Atlantic University, Boca Raton, FL 33431, or by e-mail to latane@socpsy.sci.fau.edu.

References

Abelson, R.P. (1964) Mathematical models of the distributions of attitudes under controversy. In N. Fredericksen and H. Gullicksen (eds.), *Contributions to mathematical psychology*, New York: Holt, Rinehart and Winston.

Axelrod, R. (1984) *The evolution of cooperation*. New York: Basic Books.

Crespi, I. (1988) *Pre-election polling*. New York: Russell Sage Foundation.

Doran, J.E., Palmer, M., Gilbert, N., and Mellars, P. (1994) The EOS project: Modelling Upper Paleolithic social change. In N. Gilbert and J. Doran (eds.), *Simulating society: The computer simulation of social phenomena*. London: UCL Press.

Eagly, A.H., and Chaiken, S. (1993) *The psychology of attitudes*. Fort Worth, TX: Harcourt Brace Jovanovich.

Festinger, L., Schachter, S., and Back, K. (1950) *Social pressures in informal groups*. Stanford: Stanford University Press.

Giles, H. (1984) *The dynamics of speech accommodation*. International Journal Of the Sociology of Language, 46, whole issue.

Haken, H. (1988) *Information and self-organization: A macroscopic approach to complex systems*. Berlin: Springer.

Harris, (1976) The uncertain connection between verbal theories and research hypotheses in social psychology. *Journal of Experimental Social Psychology*, 12, pp. 210–219.

Harton, H.C., and Latané, B. (1995) *Thought- and information-induced polarization: The mediating role of involvement in making attitudes extreme*. Paper presented at annual meetings of the American Association of Public Opinion Research, Ft. Lauderdale.

Hegselmann, R. (1994) Zur Selbstorganisation von Solidarnetzwerken unter Ungleichen.

In K. Homann (Ed.), *Wirtschaftsethische Perspektiven I: Theorie, Ordnungsfragen, Internationale Institutionen.* Berlin: Duncker and Humblot.

Hegselmann, R. (this volume) Cellular automata in the social sciences: Perspectives, restrictions, and artifacts.

Helbing, D. (1995) *Quantitative sociodynamics: Stochastic methods and models of social interaction processes.* Dordrecht: Kluwer.

Jackson, C., and Latané, B. (1995) *Dynamic social impact on trial: The emergence of subgroups in electronic juries.* Paper presented at the annual meetings of the American Psychological Society, New York.

Latané, B. (1981). The psychology of social impact. *American Psychologist*, 36, pp. 343–356.

Latané, B. (in press a) Strength from weakness: The fate of opinion minorities in spatially distributed groups. In E. Witte and J.H. Davis (eds.), *Understanding group behavior: Consensual action by small groups.* Hillsdale, NJ: LEA.

Latané, B. (in press b) The emergence of clustering and correlation from social interaction. In R. Hegselmann and H.O. Peitgen (eds.), *Ordnung und Chaos in Natur und Gesellschaft.* Vienna: Hoelder-Pichler-Tempski.

Latané, B. (in press c) Dynamic social impact: The creation of culture by communication. *Journal of Communication.*

Latané, B., and Bourgeois, M. (in press) Empirical evidence for dynamic social impact: The emergence of subcultures in electronic groups. *Journal of Communication.*

Latané, B., and L'Herrou, T. (in press) Spatial clustering in the Conformity Game: Dynamic social impact in electronic groups. *Journal of Personality and Social Psychology.*

Latané, B., and Liu, J.H. (in press) The intersubjective geometry of social space. *Journal of Communication.*

Latané, B., Liu, J.H., Nowak, A., Bonevento, M., and Zheng, L. (in press) Distance matters: Physical space and social impact. *Personality and Social Psychology Bulletin.*

Latané, B., and Nowak, A. (1994) Attitudes as catastrophes: From dimensions to categories with increasing involvement. In R. Vallacher and A. Nowak (eds.), *Dynamical systems in social psychology,* New York: Academic Press.

Latané, B., and Nowak, A. (in press) Self-organizing social systems: Necessary and sufficient conditions for the emergence of clustering and polarization in groups. In G. Barnett and F. Boster (Eds.). *Progress in communication sciences: Persuasion.* Norwood, NJ: Ablex.

Latané, B., Nowak, A., and Liu, J.H. (1994) The measurement of emergent social phenomena: Dynamism, polarization, and clustering as order parameters of social systems. *Behavioral Science*, 39, pp. 1–24.

Latané, B., and Wolf, S. (1981) The social impact of majorities and minorities. *Psychological Review*, 88, pp. 438–453.

Lewenstein, M., Nowak, A., and Latané, B. (1992) Statistical mechanics of social impact. *Physical Review A*, 45, pp. 703–716.

Liebrand, W.B.G., and Messick, D.M. (this volume). Computer simulations of sustainable cooperation in social dilemmas.

Liu, J.H., and Latané, B. (in press). The catastrophic link between the importance and extremity of political attitudes. *Political Behavior.*

Noelle-Neumann, E. (1984) *The spiral of silence: Public opinion — our social skin.* Chicago: University of Chicago Press.

Nowak, A., and Latané, B. (1994) Simulating the emergence of social order from individual behavior. In N. Gilbert and J. Doran (eds.), *Simulating societies: The computer simulation of social processes,* London: University College Press.

Ostrom, T. (1988) Computer simulation: The third symbol system. *Journal of Experimental Social Psychology*, 24, pp. 381–392.

Pool, I. deS. and Abelson, R.P. (1961) The simulmatics project. *Public Opinion Quarterly*, 25, pp. 167–183.

Ruelle, D. (1991) *Chance and chaos*. Princeton, NJ: Princeton University Press.

Schelling, T.C. (1971) Dynamic models of segregation. *Journal of Mathematical Sociology.* 1, pp. 143–186.

Sheldrake, R. (1989) *The presence of the past: Morphic resonance and the habits of nature.* London: Fontana.

Tedeschi, J. (1983) Social influence theory and aggression. In R.G. Geen and E.E. Donnerstein (eds.), *Aggression: Theoretical and empirical reviews.* vol. 1. New York: Academic Press, pp. 135–162.

Troitzsch, K.G. (this volume) Simulation, and structuralism.

von Neumann, J. (1966). *Theory of self-reproducing automata.* Urbana: University of Illinois Press.

von Neumann, J., and Morgenstern, O. (1944) *Theory of games and economic behavior.* Princeton: Princeton University Press.

Weidlich, W. (1991) Physics and social science — The approach of synergetics. *Physics Reports*, 204, pp. 1–163.

Woelfel, J. and Fink, E.L. (1980) *The measurement of communication processes: Galileo theory and method.* New York: Academic Press.

Wolf, S., and Latané, B. (1983) Majority and Minority influence on restaurant preferences. *Journal of Personality and Social Psychology* 45, pp. 282–292.

Wolfram, S. (1986) *Theory and applications of cellular automata.* Singapore: World Scientific.

Author Index

Subject Index

THEORY AND DECISION LIBRARY

SERIES A: PHILOSOPHY AND METHODOLOGY OF THE SOCIAL SCIENCES
Editors: W. Leinfellner (*Vienna*) and G. Eberlein (*Munich*)

23. R. Hegselmann, U. Mueller and K.G. Troitzsch (eds.): *Modelling and Simulation in the Social Sciences from the Philosophy of Science Point of View.* 1996
ISBN 0-7923-4125-2

KLUWER ACADEMIC PUBLISHERS – DORDRECHT / BOSTON / LONDON